Representations and Characters of Groups

Now in its second edition, this text provides a modern introduction to the representation theory of finite groups. The authors have revised the popular first edition and added a considerable amount of new material. The theory is developed in terms of modules, since this is appropriate for more advanced work, but considerable emphasis is placed upon constructing characters. The character tables of many groups are given, including all groups of order less than 32, and all simple groups of order less than 1000.

Among the applications covered are Burnside's $p^a q^b$ theorem, the use of character theory in studying subgroup structure and permutation groups, and a description of how to use representation theory to investigate molecular vibration.

Each chapter is accompanied by a variety of exercises, and full solutions to all the exercises are provided at the end of the book. This will be ideal as a text for a course in representation theory, and in view of the applications of the subject, will be of interest to mathematicians, chemists and physicists alike.

REPRESENTATIONS AND CHARACTERS OF GROUPS

GORDON JAMES AND MARTIN LIEBECK

Department of Mathematics,
Imperial College, London

Second Edition

CAMBRIDGE
UNIVERSITY PRESS

CAMBRIDGE
UNIVERSITY PRESS

University Printing House, Cambridge CB2 8BS, United Kingdom

Cambridge University Press is part of the University of Cambridge.

It furthers the University's mission by disseminating knowledge in the pursuit of education, learning and research at the highest international levels of excellence.

www.cambridge.org
Information on this title: www.cambridge.org/9780521003926

First published 1993
Reprinted 1995
Seven printing 1997
Second edition 2004
Fourth printing 2008

A catalogue record for this publication is available from the British Library

ISBN 978-0-521-81205-4 Hardback
ISBN 978-0-521-00392-6 Paperback

Contents

Preface

We have attempted in this book to provide a leisurely introduction to the representation theory of groups. But why should this subject interest you?

Representation theory is concerned with the ways of writing a group as a group of matrices. Not only is the theory beautiful in its own right, but it also provides one of the keys to a proper understanding of finite groups. For example, it is often vital to have a concrete description of a particular group; this is achieved by finding a representation of the group as a group of matrices. Moreover, by studying the different representations of the group, it is possible to prove results which lie outside the framework of representation theory. One simple example: all groups of order p^2 (where p is a prime number) are abelian; this can be shown quickly using only group theory, but it is also a consequence of basic results about representations. More generally, all groups of order $p^a q^b$ (p and q primes) are soluble; this again is a statement purely about groups, but the best proof, due to Burnside, is an outstanding example of the use of representation theory. In fact, the range of applications of the theory extends far beyond the boundaries of pure mathematics, and includes theoretical physics and chemistry – we describe one such application in the last chapter.

The book is suitable for students who have taken first undergraduate courses involving group theory and linear algebra. We have included two preliminary chapters which cover the necessary background material. The basic theory of representations is developed in Chapters 3–23, and our methods concentrate upon the use of modules; although this accords with the more modern style of algebra, in several instances our proofs differ from those found in other textbooks. The main results are elegant and surprising, but at first sight they sometimes have an air of mystery about them; we have chosen the approach which we believe to be the most transparent.

We also emphasize the practical aspects of the subject, and the text is illustrated with a wealth of examples. A feature of the book is the wide variety of groups which we investigate in detail. By the end of Chapter 28, we have presented the character tables of all groups of order less than 32, of all p-groups of order at most p^4, and of all the simple groups of order less than 1000.

Every chapter is accompanied by a set of Exercises, and the solutions to all of these are provided at the end of the book.

We would like to thank Dr Hans Liebeck for his careful reading of our manuscript and the many helpful suggestions which he made.

Preface to Second Edition

In this second edition, we have included two new chapters; one (Chapter 28) deals with the character tables of an infinite series of groups, and the other (Chapter 29) covers aspects of the representation theory of permutation groups. We have also added a considerable amount of new material to Chapters 20, 23 and 30, and made minor amendments elsewhere.

1
Groups and homomorphisms

This book is devoted to the study of an aspect of group theory, so we begin with a résumé of facts about groups, most of which you should know already. In addition, we introduce several examples, such as dihedral groups and symmetric groups, which we shall use extensively to illustrate the later theory. An elementary course on abstract algebra would normally cover all the material in the chapter, and any book on basic group theory will supply you with further details. One or two results which we shall use only infrequently are demoted to the exercises at the end of the chapter – you can refer to the solutions if necessary.

Groups

A *group* consists of a set G, together with a rule for combining any two elements g, h of G to form another element of G, written gh; this rule must satisfy the following axioms:

(1) for all g, h, k in G,

$$(gh)k = g(hk);$$

(2) there exists an element e in G such that for all g in G,

$$eg = ge = g;$$

(3) for all g in G, there exists an element g^{-1} in G such that

$$gg^{-1} = g^{-1}g = e.$$

We refer to the rule for combining elements of G as the *product operation* on G.

Axiom (1) states that the product operation is *associative*; the element e in axiom (2) is an *identity* element of G; and g^{-1} is an *inverse* of g in axiom (3).

It is elementary to see that G has just one identity element, and that every g in G has just one inverse. Usually we write 1, rather than e, for the identity element of G.

The product of an element g with itself, gg, is written g^2; similarly $g^3 = g^2 g$, $g^{-2} = (g^{-1})^2$, and so on. Also, $g^0 = 1$.

If the number of elements in G is finite, then we call G a *finite group*; the number of elements in G is called the *order* of G, and is written $|G|$.

1.1 Examples

(1) Let n be a positive integer, and denote by $\mathbb{C}$ the set of all complex numbers. The set of nth roots of unity in $\mathbb{C}$, with the usual multiplication of complex numbers, is a group of order n. It is written as C_n and is called the *cyclic* group of order n. If $a = e^{2\pi i/n}$, then

$$C_n = \{1, a, a^2, \ldots, a^{n-1}\},$$

and $a^n = 1$.

(2) The set $\mathbb{Z}$ of all integers, under addition, is a group.

(3) Let n be an integer with $n \geqslant 3$, and consider the rotation and reflection symmetries of a regular n-sided polygon.

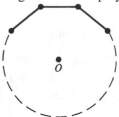

There are n rotation symmetries: these are $\rho_0, \rho_1, \ldots, \rho_{n-1}$ where ρ_k is the (clockwise) rotation about the centre O through an angle $2\pi k/n$. There are also n reflection symmetries: these are reflections in the n lines passing through O and a corner or the mid-point of a side of the polygon.

These $2n$ rotations and reflections form a group under the product operation of *composition* (that is, for two symmetries f and g, the product fg means 'first do f, then do g'). This group is called the *dihedral group* of order $2n$, and is written D_{2n}.

Let A be a corner of the polygon. Write b for the reflection in the line through O and A, and write a for the rotation ρ_1. Then the n rotations are

$$1, a, a^2, \ldots, a^{n-1}$$

(where 1 denotes the identity, which leaves the polygon fixed); and the n reflections are

$$b, ab, a^2b, \ldots, a^{n-1}b.$$

Thus all elements of D_{2n} are products of powers of a and b – that is, D_{2n} is *generated* by a and b.

Check that

$$a^n = 1, \ b^2 = 1 \text{ and } b^{-1}ab = a^{-1}.$$

These relations determine the product of any two elements of the group. For example, we have $ba^j = a^{-j}b$ (using the relation $ba = a^{-1}b$), and hence

$$(a^ib)(a^jb) = a^iba^jb = a^ia^{-j}bb = a^{i-j}.$$

We summarize all this in the presentation

$$D_{2n} = \langle a, b: a^n = 1, b^2 = 1, b^{-1}ab = a^{-1} \rangle.$$

(4) For n a positive integer, the set of all permutations of $\{1, 2, \ldots, n\}$, under the product operation of composition, is a group. It is called the *symmetric group* of degree n, and is written S_n. The order of S_n is $n!$.

(5) Let F be either $\mathbb{R}$ (the set of real numbers) or $\mathbb{C}$ (the set of complex numbers). The set of all invertible $n \times n$ matrices with entries in F, under matrix multiplication, forms a group. This group is called the *general linear group* of degree n over F, and is denoted by $GL(n, F)$. It is an infinite group. The identity of $GL(n, F)$ is of course the identity matrix, which we denote by I_n or just I.

A group G is said to be *abelian* if $gh = hg$ for all g and h in G. While C_n and $\mathbb{Z}$ are abelian, most of the other examples given above are non-abelian groups.

Subgroups

Let G be a group. A subset H of G is said to be a *subgroup* if H is itself a group under the product operation inherited from G. We use the notation $H \leq G$ to indicate that H is a subgroup of G.

It is easy to see that a subset H of a group G is a subgroup if and only if the following two conditions hold:

(1) $1 \in H$, and
(2) if h, $k \in H$ then $hk^{-1} \in H$.

1.2 Examples

(1) For every group G, both $\{1\}$ and G are subgroups of G.

(2) Let G be a group and $g \in G$. The subset

$$\langle g \rangle = \{g^n : n \in \mathbb{Z}\}$$

is a subgroup of G, called the *cyclic subgroup generated by g*. If $g^n = 1$ for some $n \geqslant 1$, then $\langle g \rangle$ is finite. In this case, let r be the least positive integer such that $g^r = 1$; then r is equal to the number of elements in $\langle g \rangle$ – indeed,

$$\langle g \rangle = \{1, g, g^2, \ldots, g^{r-1}\}.$$

We call r the *order* of the element g.

If $G = \langle g \rangle$ for some $g \in G$ then we call G a *cyclic* group. The groups C_n and $\mathbb{Z}$ in Examples 1.1 are cyclic.

(3) Let G be a group and let a, $b \in G$. Define H to be the subset of G consisting of all elements which are products of powers of a and b – that is, all elements of the form

$$a^{i_1} b^{j_1} a^{i_2} b^{j_2} \ldots a^{i_n} b^{j_n}$$

for some n, where $i_k, j_k \in \mathbb{Z}$ for $1 \leqslant k \leqslant n$. Then H is a subgroup of G; we call H the subgroup *generated* by a and b, and write

$$H = \langle a, b \rangle.$$

Given any finite set S of elements of G, we can similarly define $\langle S \rangle$, the subgroup of G generated by S.

This construction gives a powerful method of finding new groups as subgroups of given groups, such as general linear or symmetric groups. We illustrate the construction in the next example, and again in Example 1.5 below.

(4) Let $G = \mathrm{GL}(2, \mathbb{C})$, the group of invertible 2×2 matrices with entries in $\mathbb{C}$, and let

$$A = \begin{pmatrix} i & 0 \\ 0 & -i \end{pmatrix}, B = \begin{pmatrix} 0 & 1 \\ -1 & 0 \end{pmatrix}.$$

Put $H = \langle A, B \rangle$, the subgroup of G generated by A and B. Check that

$$A^4 = I, \ A^2 = B^2, \ B^{-1}AB = A^{-1}.$$

Using the third relation, we see that every element of H has the form $A^i B^j$ for some integers i, j; and using the first two relations, we can take $0 \leqslant i \leqslant 3$ and $0 \leqslant j \leqslant 1$. Hence H has at most eight elements. Since the matrices

$$A^i B^j \quad (0 \leqslant i \leqslant 3, 0 \leqslant j \leqslant 1)$$

are all distinct, in fact $|H| = 8$.

The group H is called the *quaternion group* of order 8, and is written Q_8. The above three relations determine the product of any two elements of Q_8, so we have the presentation

$$Q_8 = \langle A, B \colon A^4 = I, \ A^2 = B^2, \ B^{-1}AB = A^{-1} \rangle.$$

(5) A *transposition* in the symmetric group S_n is a permutation which interchanges two of the numbers $1, 2, \ldots, n$ and fixes the other $n - 2$ numbers. Every permutation g in S_n can be expressed as a product of transpositions. It can be shown that either all such expressions for g have an even number of transpositions, or they all have an odd number of transpositions; we call g an *even* or an *odd* permutation, accordingly. The subset

$$A_n = \{g \in S_n \colon \ g \text{ is an even permutation}\}$$

is a subgroup of S_n, called the *alternating group* of degree n.

Direct products

We describe a construction which produces a new group from given ones.

Let G and H be groups, and consider

$$G \times H = \{(g, h) \colon g \in G \text{ and } h \in H\}.$$

Define a product operation on $G \times H$ by

$$(g, h)(g', h') = (gg', hh')$$

for all $g, g' \in G$ and all $h, h' \in H$. With this product operation, $G \times H$ is a group, called the *direct product* of G and H.

More generally, if $G_1, \ldots, G_r$ are groups, then the direct product $G_1 \times \ldots \times G_r$ is

$$\{(g_1, \ldots, g_r): g_i \in G_i \text{ for } 1 \leqslant i \leqslant r\},$$

with product operation defined by

$$(g_1, \ldots, g_r)(g_1', \ldots, g_r') = (g_1 g_1', \ldots, g_r g_r').$$

If all the groups G_i are finite, then $G_1 \times \ldots \times G_r$ is also finite, of order $|G_1| \ldots |G_r|$.

1.3 Example
The group $C_2 \times \ldots \times C_2$ (r factors) has order 2^r and all its non-identity elements have order 2.

Functions

A *function* from one set G to another set H is a rule which assigns a unique element of H to each element of G. In this book, we generally apply functions on the *right* – that is, the image of g under a function ϑ is written as $g\vartheta$, not as ϑg. We often indicate that ϑ is a function from G to H by the notation $\vartheta \colon G \to H$. By an expression $\vartheta \colon g \to h$, where $g \in G$ and $h \in H$, we mean that $h = g\vartheta$.

A function $\vartheta \colon G \to H$ is *invertible* if there is a function $\phi \colon H \to G$ such that for all $g \in G$, $h \in H$,

$$(g\vartheta)\phi = g \text{ and } (h\phi)\vartheta = h.$$

Then ϕ is called the *inverse* of ϑ, and is written as ϑ^{-1}. A function ϑ from G to H is invertible if and only if it is both *injective* (that is, $g_1\vartheta = g_2\vartheta$ for $g_1, g_2 \in G$ implies that $g_1 = g_2$) and *surjective* (that is, for every $h \in H$ there exists $g \in G$ such that $g\vartheta = h$). An invertible function is also called a *bijection*.

Homomorphisms

Given groups G and H, those functions from G to H which 'preserve the group structure' – the so-called homomorphisms – are of particular importance.

If G and H are groups, then a *homomorphism* from G to H is a function $\vartheta \colon G \to H$ which satisfies

$$(g_1 g_2)\vartheta = (g_1\vartheta)(g_2\vartheta) \quad \text{for all } g_1, g_2 \in G.$$

An invertible homomorphism is called an *isomorphism*. If there is an isomorphism ϑ from G to H, then G and H are said to be *isomorphic*, and we write $G \cong H$; also, ϑ^{-1} is an isomorphism from H to G, so $H \cong G$.

The following example displays a technique which can often be used to prove that certain functions are homomorphisms.

1.4 Example

Let $G = D_{2n} = \langle a, b : a^n = b^2 = 1, \ b^{-1}ab = a^{-1} \rangle$, and write the $2n$ elements of G in the form $a^i b^j$ with $0 \leqslant i \leqslant n - 1$, $0 \leqslant j \leqslant 1$. Let H be any group, and suppose that H contains elements x and y which satisfy

$$x^n = y^2 = 1, \ y^{-1}xy = x^{-1}.$$

We shall prove that the function $\vartheta : G \to H$ defined by

$$\vartheta : a^i b^j \to x^i y^j \quad (0 \leqslant i \leqslant n - 1, 0 \leqslant j \leqslant 1)$$

is a homomorphism.

Suppose that $0 \leqslant r \leqslant n - 1$, $0 \leqslant s \leqslant 1$, $0 \leqslant t \leqslant n - 1$, $0 \leqslant u \leqslant 1$. Then

$$a^r b^s a^t b^u = a^i b^j$$

for some i, j with $0 \leqslant i \leqslant n - 1$, $0 \leqslant j \leqslant 1$. Moreover, i and j are determined by repeatedly using the relations

$$a^n = b^2 = 1, \ b^{-1}ab = a^{-1}.$$

Since we have $x^n = y^2 = 1$, $y^{-1}xy = x^{-1}$, we can also deduce that

$$x^r y^s x^t y^u = x^i y^j.$$

Therefore,

$$(a^r b^s a^t b^u)\vartheta = (a^i b^j)\vartheta = x^i y^j = x^r y^s x^t y^u$$

$$= (a^r b^s)\vartheta \cdot (a^t b^u)\vartheta,$$

and so ϑ is a homomorphism.

We now demonstrate the technique of Example 1.4 in action.

1.5 Example

Let $G = S_5$ and let x, y be the following permutations in G:

$$x = (1\ 2\ 3\ 4\ 5), \ y = (2\ 5)(3\ 4).$$

(Here we adopt the usual cycle notation – thus, $(1\ 2\ 3\ 4\ 5)$ denotes the permutation $1 \to 2 \to 3 \to 4 \to 5 \to 1$, and so on.) Check that

$$x^5 = y^2 = 1, \ y^{-1}xy = x^{-1}.$$

Let H be the subgroup $\langle x, y \rangle$ of G. Using the above relations, we see that

$$H = \{x^i y^j \colon 0 \leqslant i \leqslant 4, 0 \leqslant j \leqslant 1\},$$

a group of order 10.

Now recall that

$$D_{10} = \langle a, b \colon a^5 = b^2 = 1, b^{-1}ab = a^{-1} \rangle.$$

By Example 1.4, the function $\vartheta \colon D_{10} \to H$ defined by

$$\vartheta \colon a^i b^j \to x^i y^j \quad (0 \leqslant i \leqslant 4, 0 \leqslant j \leqslant 1)$$

is a homomorphism. Since ϑ is invertible, it is an isomorphism. Thus, $H = \langle x, y \rangle \cong D_{10}$.

Cosets

Let G be a group and let H be a subgroup of G. For x in G, the subset

$$Hx = \{hx \colon h \in H\}$$

of G is called a *right coset* of H in G. The distinct right cosets of H in G form a partition of G (that is, every element of G is in precisely one of the cosets).

Suppose now that G is finite, and let $Hx_1, \ldots, Hx_r$ be all the distinct right cosets of H in G. For all i, the function

$$h \to hx_i \quad (h \in H)$$

is a bijection from H to Hx_i, and so $|Hx_i| = |H|$. Since

$$G = Hx_1 \cup \ldots \cup Hx_r, \ \text{and}$$
$$Hx_i \cap Hx_j \ \text{is empty if } i \neq j,$$

we deduce that

$$|G| = r|H|.$$

In particular, we have

1.6 Lagrange's Theorem
If G is a finite group and H is a subgroup of G, then $|H|$ divides $|G|$.

The number r of distinct right cosets of H in G is called the *index* of H in G, and is written as $|G: H|$. Thus

$$|G: H| = |G|/|H|$$

when G is finite.

Normal subgroups

A subgroup N of a group G is said to be a *normal* subgroup of G if $g^{-1}Ng = N$ for all $g \in G$ (where $g^{-1}Ng = \{g^{-1}ng: n \in N\}$); we write $N \lhd G$ to indicate that N is a normal subgroup of G.

Suppose that $N \lhd G$ and let G/N be the set of right cosets of N in G. The importance of the condition $g^{-1}Ng = N$ (for all $g \in G$) is that it can be used to show that for all $g, h \in G$, we have

$$\{xy: x \in Ng \text{ and } y \in Nh\} = Ngh.$$

Hence we can define a product operation on G/N by

$$(Ng)(Nh) = Ngh \quad \text{for all } g, h \in G.$$

This makes G/N into a group, called the *factor group* of G by N.

1.7 Examples
(1) For every group G, the sub-groups $\{1\}$ and G are normal subgroups of G.
(2) For $n \geq 1$, we have $A_n \lhd S_n$. If $n \geq 2$ then there are just two right cosets of A_n in S_n, namely

$$A_n = \{g \in S_n: g \text{ even}\}, \quad \text{and}$$

$$A_n(1\ 2) = \{g \in S_n: g \text{ odd}\}.$$

Thus $|S_n:A_n| = 2$, and so $S_n/A_n \cong C_2$.
(3) Let $G = D_8 = \langle a, b: a^4 = b^2 = 1, \quad b^{-1}ab = a^{-1} \rangle$ and let $N = \langle a^2 \rangle = \{1, a^2\}$. Then $N \lhd G$ and

$$G/N = \{N, Na, Nb, Nab\}.$$

Since $(Na)^2 = (Nb)^2 = (Nab)^2 = N$, we see that $G/N \cong C_2 \times C_2$.
The subgroup $\langle a \rangle$ is also normal in G, but the subgroup $H = \langle b \rangle$ is not normal in G, since $b \in H$ while $a^{-1}ba = a^2b \notin H$.

Simple groups

A group G is said to be *simple* if $G \neq \{1\}$ and the only normal subgroups of G are $\{1\}$ and G. For example, the cyclic group C_p, with p a prime number, is simple. We shall give examples of non-abelian simple groups in later chapters – the smallest one is A_5.

If G is a finite group which is not simple, then G has a normal subgroup N such that both N and G/N have smaller order than G; and in a sense, G is 'built' out of these two smaller groups. Continuing this process with the smaller groups, we eventually see that G is 'built' out of a collection of simple groups. (This is analogous to the fact that every positive integer is built out of its prime factors.) Thus, simple groups are fundamental to the study of finite groups.

Kernels and images

To conclude the chapter, we relate normal subgroups and factor groups to homomorphisms. Let G and H be groups and suppose that $\vartheta: G \to H$ is a homomorphism. We define the *kernel* of ϑ by

$$(1.8) \qquad \operatorname{Ker} \vartheta = \{ g \in G \colon g\vartheta = 1 \}.$$

Then $\operatorname{Ker} \vartheta$ is a normal subgroup of G. Also, the *image* of ϑ is

$$(1.9) \qquad \operatorname{Im} \vartheta = \{ g\vartheta \colon g \in G \},$$

and $\operatorname{Im} \vartheta$ is a subgroup of H.

The following result describes the way in which the kernel and image of ϑ are related.

1.10 Theorem
Suppose that G and H are groups and let $\vartheta: G \to H$ be a homomorphism. Then

$$G/\operatorname{Ker} \vartheta \cong \operatorname{Im} \vartheta.$$

An isomorphism is given by the function

$$Kg \to g\vartheta \quad (g \in G)$$

where $K = \operatorname{Ker} \vartheta$.

1.11 Example

The function $\vartheta: S_n \rightarrow C_2$ given by

$$\vartheta: g \rightarrow \begin{cases} 1, & \text{if } g \text{ is an even permutation,} \\ -1, & \text{if } g \text{ is an odd permutation,} \end{cases}$$

is a homomorphism. We have $\text{Ker}\,\vartheta = A_n$, and for $n \geqslant 2$, $\text{Im}\,\vartheta = C_2$. We know from Example 1.7(2) that $S_n/A_n \cong C_2$, illustrating Theorem 1.10.

Summary of Chapter 1

1. Examples of groups are

$C_n = \langle a: a^n = 1 \rangle$,
$D_{2n} = \langle a, b: a^n = b^2 = 1, b^{-1}ab = a^{-1} \rangle$,
$Q_8 = \langle a, b: a^4 = 1, a^2 = b^2, b^{-1}ab = a^{-1} \rangle$,
$S_n =$ the symmetric group of degree n,
$A_n =$ the alternating group of degree n,
$GL(n, \mathbb{C}) =$ the group of invertible $n \times n$ matrices over $\mathbb{C}$,
$G_1 \times \ldots \times G_r$, the direct product of the groups $G_1, \ldots, G_r$.

2. A normal subgroup N of G is a subgroup such that $g^{-1}Ng = N$ for all g in G. The factor group G/N consists of the right cosets Ng ($g \in G$), with multiplication

$$(Ng)(Nh) = Ngh.$$

3. A homomorphism $\vartheta: G \rightarrow H$ is a function such that

$$(g_1 g_2)\vartheta = (g_1\vartheta)(g_2\vartheta)$$

for all g_1, g_2 in G. The kernel, $\text{Ker}\,\vartheta$, is a normal subgroup of G, and the image, $\text{Im}\,\vartheta$, is a subgroup of H. The factor group $G/\text{Ker}\,\vartheta$ is isomorphic to $\text{Im}\,\vartheta$.

Exercises for Chapter 1

1. Show that if G is an abelian group which is simple, then G is cyclic of prime order.

2. Suppose that G and H are groups, with G simple, and that $\vartheta: G \rightarrow H$ is a surjective homomorphism. Show that either ϑ is an isomorphism or $H = \{1\}$.

3. Suppose that G is a subgroup of S_n, and that G is not contained in A_n. Prove that $G \cap A_n$ is a normal subgroup of G, and $G/(G \cap A_n) \cong C_2$.

4. Let

$$G = D_8 = \langle a, b: a^4 = b^2 = 1, b^{-1}ab = a^{-1} \rangle, \text{ and}$$

$$H = Q_8 = \langle c, d: c^4 = 1, c^2 = d^2, d^{-1}cd = c^{-1} \rangle.$$

   (a) Let x, y be the permutations in S_4 which are given by

$$x = (1\ 2),\ y = (3\ 4),$$

   and let K be the subgroup $\langle x, y \rangle$ of S_4. Show that both the functions $\phi: G \to K$ and $\psi: H \to K$, defined by

$$\phi: a^r b^s \to x^r y^s,$$

$$\psi: c^r d^s \to x^r y^s \quad (0 \le r \le 3, 0 \le s \le 1),$$

   are homomorphisms. Find $\operatorname{Ker} \phi$ and $\operatorname{Ker} \psi$.

   (b) Let X, Y be the 2×2 matrices which are given by

$$X = \begin{pmatrix} 0 & i \\ i & 0 \end{pmatrix}, Y = \begin{pmatrix} 0 & -1 \\ 1 & 0 \end{pmatrix},$$

   and let L be the subgroup $\langle X, Y \rangle$ of $GL(2, \mathbb{C})$. Show that just one of the functions $\lambda: G \to L$ and $\mu: H \to L$, defined by

$$\lambda: a^r b^s \to X^r Y^s,$$

$$\mu: c^r d^s \to X^r Y^s \quad (0 \le r \le 3, 0 \le s \le 1),$$

   is a homomorphism. Prove that this homomorphism is an isomorphism.

5. Prove that $D_{4m} \cong D_{2m} \times C_2$ if m is odd.

6. (a) Show that every subgroup of a cyclic group is cyclic.

   (b) Let G be a finite cyclic group, and let n be a positive integer which divides $|G|$. Prove that

$$\{g \in G: g^n = 1\}$$

   is a cylic subgroup of G of order n.

   (c) If G is a finite cyclic group and x, y are elements of G with the same order, show that x is a power of y.

7. Show that the set of non-zero complex numbers, under the usual multiplication, is a group. Prove that every finite subgroup of this group is cyclic.

8. Show that every group of even order contains an element of order 2.

9. Find elements A and B of GL$(2, \mathbb{C})$ such that A has order 8, B has order 4, and

$$B^2 = A^4 \text{ and } B^{-1}AB = A^{-1}.$$

Show that the group $\langle A, B \rangle$ has order 16.
(Hint: compare Q_8 in Example 1.2(4).)

10. Suppose that H is a subgroup of G with $|G:H| = 2$. Prove that $H \triangleleft G$.

2

Vector spaces and linear transformations

An attractive feature of representation theory is that it combines two strands of mainstream mathematics, namely group theory and linear algebra. For reference purposes, we gather the results from linear algebra concerning vector spaces, linear transformations and matrices which we shall use later. Most of the material will be familiar to you if you have taken a first course on linear algebra, so we omit the proofs. An exception occurs in the last section, where we deal with projections; here, we explain in detail how the results work, in case you have not come across projections before.

Vector spaces

Let F be either $\mathbb{R}$ (the set of real numbers) or $\mathbb{C}$ (the set of complex numbers). A *vector space* over F is a set V, together with a rule for adding any two elements u, v of V to form an element $u + v$ of V, and a rule for multiplying any element v of V by any element λ of F to form an element λv of V. (The latter rule is called *scalar multiplication*.) Moreover, these rules must satisfy:

(2.1) (a) V is an abelian group under addition;
      (b) for all u, v in V and all λ, μ in F,
         (1) $\lambda(u + v) = \lambda u + \lambda v$,
         (2) $(\lambda + \mu)v = \lambda v + \mu v$,
         (3) $(\lambda\mu)v = \lambda(\mu v)$,
         (4) $1v = v$.

The elements of V are called *vectors*, and those of F are called *scalars*. We write 0 for the identity element of the abelian group V under addition.

14

2.2 Examples

(1) Let $\mathbb{R}^2$ denote the set of all ordered pairs (x, y) where x and y are real numbers. Define addition and scalar multiplication on $\mathbb{R}^2$ by

$$(x, y) + (x', y') = (x + x', y + y'),$$
$$\lambda(x, y) = (\lambda x, \lambda y).$$

Then $\mathbb{R}^2$ is a vector space over $\mathbb{R}$.

(2) More generally, for each positive integer n, we consider *row vectors*

$$(x_1, x_2, \ldots, x_n)$$

where $x_1, x_2, \ldots, x_n$ belong to F. We denote the set of all such row vectors by F^n, and define addition and scalar multiplication on F^n by

$$(x_1, \ldots, x_n) + (x_1', \ldots, x_n') = (x_1 + x_1', \ldots, x_n + x_n'),$$
$$\lambda(x_1, \ldots, x_n) = (\lambda x_1, \ldots, \lambda x_n).$$

Then F^n is a vector space over F.

Bases of vector spaces

Let $v_1, \ldots, v_n$ be vectors in a vector space V over F. A vector v in V is a *linear combination* of $v_1, \ldots, v_n$ if

$$v = \lambda_1 v_1 + \ldots + \lambda_n v_n$$

for some $\lambda_1, \ldots, \lambda_n$ in F. The vectors $v_1, \ldots, v_n$ are said to *span* V if every vector in V is a linear combination of $v_1, \ldots, v_n$.

We say that $v_1, \ldots, v_n$ are *linearly dependent* if

$$\lambda_1 v_1 + \ldots + \lambda_n v_n = 0$$

for some $\lambda_1, \ldots, \lambda_n$ in F, not all of which are zero; otherwise, $v_1, \ldots, v_n$ are *linearly independent*.

The vectors $v_1, \ldots, v_n$ form a *basis* of V if they span V and are linearly independent.

Throughout this book, we shall consider only vector spaces V which are finite-dimensional – this means that V has a basis consisting of finitely many vectors, as above. It turns out that any two bases of V have the same number of vectors. The number of vectors in a basis of V is called the *dimension* of V and is written as dim V. If $V = \{0\}$ then dim $V = 0$. The vector space V is *n-dimensional* if dim $V = n$.

2.3 Example
Let $V = F^n$. Then

$$(1, 0, 0, \ldots, 0), (0, 1, 0, \ldots, 0), \ldots, (0, 0, 0, \ldots, 1)$$

is a basis of V, so dim $V = n$. Another basis is

$$(1, 0, 0, \ldots, 0), (1, 1, 0, \ldots, 0), \ldots, (1, 1, 1, \ldots, 1).$$

Given a basis $v_1, \ldots, v_n$ of a vector space V, each vector v in V can be written in a *unique* way as

$$v = \lambda_1 v_1 + \ldots + \lambda_n v_n,$$

with $\lambda_1, \ldots, \lambda_n$ in F. The vector v therefore determines the scalars $\lambda_1, \ldots, \lambda_n$. Except in the case where $V = \{0\}$, there are many bases of V. Indeed, the next result says that any linearly independent vectors can be extended to a basis.

(2.4)   If $v_1, \ldots, v_k$ are linearly independent vectors in V, then there exist $v_{k+1}, \ldots, v_n$ in V such that $v_1, \ldots, v_n$ form a basis of V.

Subspaces

A *subspace* of a vector space V over F is a subset of V which is itself a vector space under the addition and scalar multiplication inherited from V. For a subset U of V to be a subspace, it is necessary and sufficient that all the following conditions hold:

(2.5)   (1) $0 \in U$;
          (2) if $u, v \in U$ then $u + v \in U$;
          (3) if $\lambda \in F$ and $u \in U$ then $\lambda u \in U$.

2.6 Examples
(1) $\{0\}$ and V are subspaces of V.

(2) Let $u_1, \ldots, u_r$ be vectors in V. We define $\mathrm{sp}(u_1, \ldots, u_r)$ to be the set of all linear combinations of $u_1, \ldots, u_r$; that is,

$$\mathrm{sp}(u_1, \ldots, u_r) = \{\lambda_1 u_1 + \ldots + \lambda_r u_r \colon \lambda_1, \ldots, \lambda_r \in F\}.$$

By (2.5), $\mathrm{sp}(u_1, \ldots, u_r)$ is a subspace of V, and it is called the subspace *spanned* by $u_1, \ldots, u_r$.

Notice that the following fact is a consequence of (2.4).

(2.7)   *Suppose that U is a subspace of the vector space V. Then* $\dim U \leqslant \dim V$. *Also,* $\dim U = \dim V$ *if and only if $U = V$.*

Direct sums of subspaces

If $U_1, \ldots, U_r$ are subspaces of a vector space V, then the sum $U_1 + \ldots + U_r$ is defined by

$$U_1 + \ldots + U_r = \{u_1 + \ldots + u_r \colon u_i \in U_i \text{ for } 1 \leqslant i \leqslant r\}.$$

By (2.5), $U_1 + \ldots + U_r$ is a subspace of V.

We say that the sum $U_1 + \ldots + U_r$ is a *direct sum* if every element of the sum can be written in a *unique* way as $u_1 + \ldots + u_r$ with $u_i \in U_i$ for $1 \leqslant i \leqslant r$. If the sum is direct, then we write it as

$$U_1 \oplus \ldots \oplus U_r.$$

2.8 Examples

(1) Suppose that $v_1, \ldots, v_n$ is a basis of V, and for $1 \leqslant i \leqslant n$, let U_i be the subspace spanned by v_i. Then

$$V = U_1 \oplus \ldots \oplus U_n.$$

(2) Let U be a subspace of V and let $v_1, \ldots, v_k$ be a basis of U. Extend $v_1, \ldots, v_k$ to a basis $v_1, \ldots, v_n$ of V (see (2.4)), and let $W = \mathrm{sp}\,(v_{k+1}, \ldots, v_n)$. Then

$$V = U \oplus W.$$

From this construction it follows that there are infinitely many subspaces W with $V = U \oplus W$, unless U is $\{0\}$ or V.

The next result is frequently useful when dealing with the direct sum of two subspaces. You should consult the solutions to Exercises 2.3 and 2.4 if you have difficulty with the proof.

(2.9)   *Suppose that $V = U + W$, that $u_1, \ldots, u_r$ is a basis of U and that $w_1, \ldots, w_s$ is a basis of W. Then the following three conditions are equivalent:*

    (1) $V = U \oplus W$,

    (2) $u_1, \ldots, u_r, w_1, \ldots, w_s$ *is a basis of V,*

    (3) $U \cap W = \{0\}$.

Our next result, involving the direct sum of several subspaces, can be deduced immediately from the definition of a direct sum.

(2.10)　　*Suppose that U, W, $U_1, \ldots, U_a$, $W_1, \ldots, W_b$ are subspaces of a vector space V. If $V = U \oplus W$ and also*

$$U = U_1 \oplus \ldots \oplus U_a, \text{ and}$$

$$W = W_1 \oplus \ldots \oplus W_b,$$

　then
$$V = U_1 \oplus \ldots \oplus U_a \oplus W_1 \oplus \ldots \oplus W_b.$$

We now introduce a construction for vector spaces which is analogous to the construction of direct products for groups.

Let $U_1, \ldots, U_r$ be vector spaces over F, and let

$$V = \{(u_1, \ldots, u_r): u_i \in U_i \text{ for } 1 \leqslant i \leqslant r\}.$$

Define addition and scalar multiplication on V as follows: for all u_i, u_i' in U_i ($1 \leqslant i \leqslant r$) and all λ in F, let

$$(u_1, \ldots, u_r) + (u_1', \ldots, u_r') = (u_1 + u_1', \ldots, u_r + u_r'),$$

$$\lambda(u_1, \ldots, u_r) = (\lambda u_1, \ldots, \lambda u_r).$$

With these definitions, V is a vector space over F. If, for $1 \leqslant i \leqslant r$, we put
$$U_i' = \{(0, \ldots, u_i, \ldots, 0): u_i \in U_i\}$$

(where the u_i is in the ith position), then it is immediate that

$$V = U_1' \oplus \ldots \oplus U_r'.$$

We call V the *external direct sum* of $U_1, \ldots, U_r$, and, abusing notation slightly, we write

$$V = U_1 \oplus \ldots \oplus U_r.$$

Linear transformations

Let V and W be vector spaces over F. A *linear transformation* from V to W is a function $\vartheta: V \to W$ which satisfies

$$(u + v)\vartheta = u\vartheta + v\vartheta \quad \text{for all } u, v \in V, \text{ and}$$

$$(\lambda v)\vartheta = \lambda(v\vartheta) \qquad \text{for all } \lambda \in F \text{ and } v \in V.$$

Just as a group homomorphism preserves the group multiplication, so a linear transformation preserves addition and scalar multiplication.

Notice that if $\vartheta\colon V \to W$ is a linear transformation and $v_1, \ldots, v_n$ is a basis of V, then for $\lambda_1, \ldots, \lambda_n$ in F we have

$$(\lambda_1 v_1 + \ldots + \lambda_n v_n)\vartheta = \lambda_1(v_1\vartheta) + \ldots + \lambda_n(v_n\vartheta).$$

Thus, ϑ is determined by its action on a basis. Furthermore, given any basis $v_1, \ldots, v_n$ of V and any n vectors $w_1, \ldots, w_n$ in W, there is a unique linear transformation $\phi\colon V \to W$ such that $v_i\phi = w_i$ for all i; the linear transformation ϕ is given by

$$(\lambda_1 v_1 + \ldots + \lambda_n v_n)\phi = \lambda_1 w_1 + \ldots + \lambda_n w_n.$$

We sometimes construct a linear transformation $\phi\colon V \to W$ in this way, by specifying the values of ϕ on a basis of V, and then saying 'extend the action of ϕ to be linear'.

Kernels and images

Suppose that $\vartheta\colon V \to W$ is a linear transformation. The kernel of ϑ (written $\mathrm{Ker}\,\vartheta$) and the image of ϑ (written $\mathrm{Im}\,\vartheta$) are defined as follows:

(2.11) $$\mathrm{Ker}\,\vartheta = \{v \in V\colon v\vartheta = 0\},$$

$$\mathrm{Im}\,\vartheta = \{v\vartheta\colon v \in V\}.$$

Using (2.5), it is easy to check that $\mathrm{Ker}\,\vartheta$ is a subspace of V and $\mathrm{Im}\,\vartheta$ is a subspace of W. Their dimensions are connected by the following equation, which is known as the Rank–Nullity Theorem:

(2.12) $$\dim V = \dim(\mathrm{Ker}\,\vartheta) + \dim(\mathrm{Im}\,\vartheta).$$

2.13 Examples
(1) If $\vartheta\colon V \to W$ is defined by $v\vartheta = 0$ for all $v \in V$, then ϑ is a linear transformation, and

$$\mathrm{Ker}\,\vartheta = V, \quad \mathrm{Im}\,\vartheta = \{0\}.$$

(2) If $\vartheta\colon V \to V$ is defined by $v\vartheta = 3v$ for all $v \in V$, then ϑ is a linear transformation, and

$$\mathrm{Ker}\,\vartheta = \{0\}, \quad \mathrm{Im}\,\vartheta = V.$$

(3) If $\vartheta: \mathbb{R}^3 \rightarrow \mathbb{R}^2$ is given by

$$(x, y, z)\vartheta = (x + 2y + z, -y + 3z)$$

for all $x, y, z \in \mathbb{R}$, then ϑ is a linear transformation; we have

$$\mathrm{Ker}\,\vartheta = \mathrm{sp}\,((7, -3, -1)), \quad \mathrm{Im}\,\vartheta = \mathbb{R}^2,$$

so $\dim\,(\mathrm{Ker}\,\vartheta) = 1$ and $\dim\,(\mathrm{Im}\,\vartheta) = 2$.

Invertible linear transformations

Again, let V and W be vector spaces over F. A linear transformation ϑ from V to W is injective if and only if $\mathrm{Ker}\,\vartheta = \{0\}$, and hence ϑ is invertible precisely when ϑ is surjective and $\mathrm{Ker}\,\vartheta = \{0\}$. It turns out that the inverse of an invertible linear transformation is also a linear transformation (see Exercise 2.1).

If there exists an invertible linear transformation from V to W, then V and W are said to be *isomorphic* vector spaces. By applying (2.12), we see that isomorphic vector spaces have the same dimension. By also taking (2.7) into account, we obtain the next result (see Exercise 2.2).

(2.14) *Let ϑ be a linear transformation from V to itself. Then the following three conditions are equivalent:*

(1) ϑ *is invertible*;
(2) $\mathrm{Ker}\,\vartheta = \{0\}$;
(3) $\mathrm{Im}\,\vartheta = V.$

Endomorphisms

A linear transformation from a vector space V to itself is called an *endomorphism* of V.

Suppose that ϑ and ϕ are endomorphisms of V and $\lambda \in F$. We define the functions $\vartheta + \phi$, $\vartheta\phi$ and $\lambda\vartheta$ from V to V by

(2.15) $$v(\vartheta + \phi) = v\vartheta + v\phi,$$

$$v(\vartheta\phi) = (v\vartheta)\phi,$$

$$v(\lambda\vartheta) = \lambda(v\vartheta),$$

for all $v \in V$. Then $\vartheta + \phi$, $\vartheta\phi$ and $\lambda\vartheta$ are endomorphisms of V. We write ϑ^2 for $\vartheta\vartheta$.

2.16 Examples

(1) The identity function 1_V defined by

$$1_V: v \rightarrow v \quad \text{for all } v \in V$$

is an endomorphism of V. If ϑ is an endomorphism of V, then so is $\vartheta - \lambda 1_V$, for all $\lambda \in F$. Note that

$$\text{Ker}\,(\vartheta - \lambda 1_V) = \{v \in V: v\vartheta = \lambda v\}.$$

(2) Let $V = \mathbb{R}^2$, and let ϑ, ϕ be the functions from V to V defined by

$$(x, y)\vartheta = (x + y, x - 2y),$$

$$(x, y)\phi = (x - 2y, -2x + 4y).$$

Then ϑ and ϕ are endomorphisms of V, and $\vartheta + \phi$, $\vartheta\phi$, 3ϑ and ϑ^2 are given by

$$(x, y)(\vartheta + \phi) = (2x - y, -x + 2y),$$

$$(x, y)(\vartheta\phi) = (-x + 5y, 2x - 10y),$$

$$(x, y)(3\vartheta) = (3x + 3y, 3x - 6y),$$

$$(x, y)\vartheta^2 = (2x - y, -x + 5y).$$

Matrices

Let V be a vector space over F, and let ϑ be an endomorphism of V. Suppose that $v_1, \ldots, v_n$ is a basis of V and call it $\mathcal{B}$. Then there are scalars a_{ij} in F $(1 \leq i \leq n, 1 \leq j \leq n)$ such that for all i,

$$v_i\vartheta = a_{i1}v_1 + \ldots + a_{in}v_n.$$

2.17 Definition

The $n \times n$ matrix (a_{ij}) is called the matrix of ϑ *relative to the basis* $\mathcal{B}$, and is denoted by $[\vartheta]_{\mathcal{B}}$.

2.18 Examples

(1) If $\vartheta = 1_V$ (so that $v\vartheta = v$ for all $v \in V$), then $[\vartheta]_{\mathcal{B}} = I_n$ for all bases $\mathcal{B}$ of V, where I_n denotes the $n \times n$ *identity matrix*.

(2) Let $V = \mathbb{R}^2$ and let ϑ be the endomorphism $(x, y) \rightarrow (x + y, x - 2y)$ of V. If $\mathcal{B}$ is the basis $(1, 0)$, $(0, 1)$ of V and $\mathcal{B}'$ is the basis

(1, 0), (1, 1) of V, then

$$[\vartheta]_{\mathscr{B}} = \begin{pmatrix} 1 & 1 \\ 1 & -2 \end{pmatrix}, \quad [\vartheta]_{\mathscr{B}'} = \begin{pmatrix} 0 & 1 \\ 3 & -1 \end{pmatrix}.$$

If we wish to indicate that the entries in a matrix A come from F, then we describe A as a *matrix over F*.

Given two $m \times n$ matrices $A = (a_{ij})$ and $B = (b_{ij})$ over F, their sum $A + B$ is the $m \times n$ matrix over F whose ij-entry is $a_{ij} + b_{ij}$ for all i, j; and for $\lambda \in F$, the matrix λA is the $m \times n$ matrix over F obtained from A by multiplying all the entries by λ.

As you know, the product of two matrices is defined in a less transparent way. Given an $m \times n$ matrix $A = (a_{ij})$ and an $n \times p$ matrix $B = (b_{ij})$, their product AB is the $m \times p$ matrix whose ij-entry is

$$\sum_{k=1}^{n} a_{ik} b_{kj}.$$

2.19 Example

Let

$$A = \begin{pmatrix} -1 & 2 \\ 3 & 1 \end{pmatrix}, B = \begin{pmatrix} 0 & -4 \\ 2 & -1 \end{pmatrix}.$$

Then

$$A + B = \begin{pmatrix} -1 & -2 \\ 5 & 0 \end{pmatrix}, 3A = \begin{pmatrix} -3 & 6 \\ 9 & 3 \end{pmatrix}, AB = \begin{pmatrix} 4 & 2 \\ 2 & -13 \end{pmatrix},$$

$$BA = \begin{pmatrix} -12 & -4 \\ -5 & 3 \end{pmatrix}.$$

The matrix of the sum or product of two endomorphisms (relative to some basis) is related to the matrices of the individual endomorphisms in the way you would expect:

(2.20) *Suppose that $\mathscr{B}$ is a basis of the vector space V, and that ϑ and ϕ are endomorphisms of V. Then*

$$[\vartheta + \phi]_{\mathscr{B}} = [\vartheta]_{\mathscr{B}} + [\phi]_{\mathscr{B}}, \quad and$$
$$[\vartheta\phi]_{\mathscr{B}} = [\vartheta]_{\mathscr{B}}[\phi]_{\mathscr{B}}.$$

Also, for all scalars λ,

$$[\lambda \vartheta]_{\mathscr{B}} = \lambda [\vartheta]_{\mathscr{B}}.$$

We showed you in (2.17) how to get a matrix from an endomorphism of a vector space V, given a basis of V. It is easy enough to reverse this process and use a matrix to define an endomorphism. We concentrate on a particular way of doing this. Suppose that A is an $n \times n$ matrix over F, and let $V = F^n$, the vector space of row vectors $(x_1, \ldots, x_n)$ with each x_i in F. Then for all v in V, the matrix product vA also lies in V. The following remark is easily justified.

(2.21) *If A is an $n \times n$ matrix over F, then the function*

$$v \rightarrow vA \quad (v \in F^n)$$

is an endomorphism of F^n.

2.22 Example
Let

$$A = \begin{pmatrix} 1 & -1 \\ 3 & 2 \end{pmatrix}.$$

Then A gives us an endomorphism ϑ of F^2, where

$$(x, y)\vartheta = (x, y) \begin{pmatrix} 1 & -1 \\ 3 & 2 \end{pmatrix} = (x + 3y, -x + 2y).$$

Invertible matrices

An $n \times n$ matrix A is said to be *invertible* if there exists an $n \times n$ matrix B with $AB = BA = I_n$. Such a matrix B, if it exists, is unique; it is called the *inverse* of A and is written as A^{-1}. Write $\det A$ for the determinant of A. Then a necessary and sufficient condition for A to be invertible is that $\det A \neq 0$.

The connection between invertible endomorphisms and invertible matrices is straightforward, and follows from (2.20): given a basis $\mathscr{B}$ of V, an endomorphism ϑ of V is invertible if and only if the matrix $[\vartheta]_{\mathscr{B}}$ is invertible.

Invertible matrices turn up when we relate two bases of a vector space. An invertible matrix converts one basis into another, and this same matrix is used to describe the way in which the matrix of an endomorphism depends upon the basis. The precise meaning of these remarks is revealed in the definition (2.23) and the result (2.24) below.

2.23 Definition

Let $v_1, \ldots, v_n$ be a basis $\mathscr{B}$ of the vector space V, and let $v'_1, \ldots, v'_n$ be a basis $\mathscr{B}'$ of V. Then for $1 \leqslant i \leqslant n$,

$$v'_i = t_{i1}v_1 + \ldots + t_{in}v_n$$

for certain scalars t_{ij}. The $n \times n$ matrix $T = (t_{ij})$ is invertible, and is called the *change of basis matrix* from $\mathscr{B}$ to $\mathscr{B}'$.

The inverse of T is the change of basis matrix from $\mathscr{B}'$ to $\mathscr{B}$.

(2.24) *If $\mathscr{B}$ and $\mathscr{B}'$ are bases of V and ϑ is an endomorphism of V, then*

$$[\vartheta]_{\mathscr{B}} = T^{-1}[\vartheta]_{\mathscr{B}'}T,$$

where T is the change of basis matrix from $\mathscr{B}$ to $\mathscr{B}'$.

2.25 Example

Suppose that $V = \mathbb{R}^2$. Let $\mathscr{B}$ be the basis $(1, 0)$, $(0, 1)$ and $\mathscr{B}'$ the basis $(1, 0)$, $(1, 1)$ of V. Then

$$T = \begin{pmatrix} 1 & 0 \\ 1 & 1 \end{pmatrix}, \quad T^{-1} = \begin{pmatrix} 1 & 0 \\ -1 & 1 \end{pmatrix}.$$

If ϑ is the endomorphism

$$\vartheta: (x, y) \rightarrow (x + y, x - 2y)$$

of V, as in Example 2.18(2), then

$$[\vartheta]_{\mathscr{B}} = \begin{pmatrix} 1 & 1 \\ 1 & -2 \end{pmatrix} = \begin{pmatrix} 1 & 0 \\ -1 & 1 \end{pmatrix} \begin{pmatrix} 0 & 1 \\ 3 & -1 \end{pmatrix} \begin{pmatrix} 1 & 0 \\ 1 & 1 \end{pmatrix} = T^{-1}[\vartheta]_{\mathscr{B}'}T.$$

Eigenvalues

Let V be an n-dimensional vector space over F, and suppose that ϑ is an endomorphism of V. The scalar λ is said to be an *eigenvalue* of ϑ if $v\vartheta = \lambda v$ for some non-zero vector v in V. Such a vector v is called an *eigenvector* of ϑ.

Now λ is an eigenvalue of ϑ if and only if $\mathrm{Ker}\,(\vartheta - \lambda 1_V) \neq \{0\}$, which occurs if and only if $\vartheta - \lambda 1_V$ is not invertible. Therefore, if $\mathcal{B}$ is a basis of V, then the eigenvalues of ϑ are those scalars λ in F which satisfy the equation

$$\det\left([\vartheta]_{\mathcal{B}} - \lambda I_n\right) = 0.$$

Solving this equation involves finding the roots of a polynomial of degree n. Since every non-constant polynomial with coefficients in $\mathbb{C}$ has a root in $\mathbb{C}$, we deduce the following result.

(2.26) *Let V be a non-zero vector space over $\mathbb{C}$, and let ϑ be an endomorphism of V. Then ϑ has an eigenvalue.*

2.27 Examples
(1) Let $V = \mathbb{C}^2$ and let ϑ be the endomorphism of V which is given by

$$(x, y)\vartheta = (-y, x).$$

If $\mathcal{B}$ is the basis $(1, 0)$, $(0, 1)$ of V, then

$$[\vartheta]_{\mathcal{B}} = \begin{pmatrix} 0 & 1 \\ -1 & 0 \end{pmatrix}.$$

We have $\det\left([\vartheta]_{\mathcal{B}} - \lambda I_2\right) = \lambda^2 + 1$, so i and $-\mathrm{i}$ are the eigenvalues of ϑ. Corresponding eigenvectors are $(1, -\mathrm{i})$ and $(1, \mathrm{i})$. Note that if $\mathcal{B}'$ is the basis $(1, -\mathrm{i})$, $(1, \mathrm{i})$ of V, then

$$[\vartheta]_{\mathcal{B}'} = \begin{pmatrix} \mathrm{i} & 0 \\ 0 & -\mathrm{i} \end{pmatrix}.$$

(2) Let $V = \mathbb{R}^2$ and let ϑ again be the endomorphism which is given by

$$(x, y)\vartheta = (-y, x).$$

This time, V is a vector space over $\mathbb{R}$, and ϑ has no eigenvalues in $\mathbb{R}$. Thus we depend upon F being $\mathbb{C}$ in result (2.26).

For an $n \times n$ matrix A over F, the element λ of F is said to be an eigenvalue of A if $vA = \lambda v$ for some non-zero row vector v in F^n. The eigenvalues of A are those elements λ of F which satisfy

$$\det\left(A - \lambda I_n\right) = 0.$$

2.28 Example
We say that an $n \times n$ matrix $A = (a_{ij})$ is *diagonal* if $a_{ij} = 0$ for all i and j with $i \neq j$. We often display such a matrix in the form

$$A = \begin{pmatrix} \lambda_1 & & 0 \\ & \ddots & \\ 0 & & \lambda_n \end{pmatrix}$$

which indicates, in addition, that $a_{ii} = \lambda_i$ for $1 \leq i \leq n$. For this diagonal matrix A, the eigenvalues are $\lambda_1, \ldots, \lambda_n$.

Projections

If a vector space V is a direct sum of two subspaces U and W, then we can construct a special endomorphism of V which depends upon the expression $V = U \oplus W$:

2.29 Proposition
Suppose that $V = U \oplus W$. Define $\pi: V \to V$ by

$$(u + w)\pi = u \quad \text{for all } u \in U, w \in W.$$

Then π is an endomorphism of V. Further,

$$\operatorname{Im} \pi = U, \ \operatorname{Ker} \pi = W \ \text{and} \ \pi^2 = \pi.$$

Proof Since every vector in V has a *unique* expression in the form $u + w$ with $u \in U$, $w \in W$, it follows that π is a function on V.

Let v and v' belong to V. Then $v = u + w$ and $v' = u' + w'$ for some u, u' in U and w, w' in W. We have

$$(v + v')\pi = (u + u' + w + w')\pi = u + u'$$

$$= (u + w)\pi + (u' + w')\pi$$

$$= v\pi + v'\pi.$$

Also, for λ in F,

$$(\lambda v)\pi = (\lambda u + \lambda w)\pi = \lambda u = \lambda(v\pi).$$

Therefore, π is an endomorphism of V.

Clearly $\operatorname{Im} \pi \subseteq U$; and since $u\pi = u$ for all u in U, we have $\operatorname{Im} = U$. Also,

$$(u + w)\pi = 0 \Leftrightarrow u = 0 \Leftrightarrow u + w \in W,$$

and so $\operatorname{Ker} \pi = W$.

Finally,

$$(u + w)\pi^2 = u\pi = u = (u + w)\pi,$$

and so $\pi^2 = \pi$.   ■

2.30 Definition

An endomorphism π of a vector space V which satisfies $\pi^2 = \pi$ is called a *projection* of V.

2.31 Example

The endomorphism

$$(x, y) \rightarrow (2x + 2y, -x - y)$$

of $\mathbb{R}^2$ is a projection.

We now show that every projection can be constructed using a direct sum, as in Proposition 2.29.

2.32 Proposition

Suppose that π is a projection of a vector space V. Then

$$V = \operatorname{Im} \pi \oplus \operatorname{Ker} \pi.$$

Proof If $v \in V$ then $v = v\pi + (v - v\pi)$. Clearly the first term $v\pi$ belongs to $\operatorname{Im} \pi$; and the second term $v - v\pi$ lies in $\operatorname{Ker} \pi$, since

$$(v - v\pi)\pi = v\pi - v\pi^2 = v\pi - v\pi = 0.$$

This establishes that $V = \operatorname{Im} \pi + \operatorname{Ker} \pi$.

Now suppose that v lies in $\operatorname{Im} \pi \cap \operatorname{Ker} \pi$. As $v \in \operatorname{Im} \pi$, we have $v = u\pi$ for some $u \in V$. Therefore

$$v\pi = u\pi^2 = u\pi = v.$$

Since $v \in \operatorname{Ker} \pi$, it follows that $v = v\pi = 0$. Thus

$$\operatorname{Im} \pi \cap \operatorname{Ker} \pi = \{0\},$$

and (2.9) now shows that $V = \operatorname{Im} \pi \oplus \operatorname{Ker} \pi$.   ■

2.33 Example

If $\pi: (x, y) \rightarrow (2x + 2y, -x - y)$ is the projection of $\mathbb{R}^2$ which appears in Example 2.31, then

$$\text{Im}\,\pi = \{(2x, -x): x \in \mathbb{R}\}, \text{Ker}\,\pi = \{(x, -x): x \in \mathbb{R}\}.$$

Summary of Chapter 2

1. All our vector spaces are finite-dimensional over F, where $F = \mathbb{C}$ or $\mathbb{R}$. For example, F^n is the set of row vectors $(x_1, \ldots, x_n)$ with each x_i in F, and $\dim F^n = n$.

2. $V = U_1 \oplus \ldots \oplus U_r$ if each U_i is a subspace of V, and every element v of V has a unique expression of the form $v = u_1 + \ldots + u_r$ $(u_i \in U_i)$.

   Also, $V = U \oplus W$ if and only if $V = U + W$ and $U \cap W = \{0\}$.

3. A linear transformation $\vartheta: V \rightarrow W$ satisfies

$$(u + v)\vartheta = u\vartheta + v\vartheta \text{ and } (\lambda v)\vartheta = \lambda(v\vartheta)$$

   for all u, v in V and all λ in F. $\text{Ker}\,\vartheta$ is a subspace of V and $\text{Im}\,\vartheta$ is a subspace of W, and

$$\dim V = \dim(\text{Ker}\,\vartheta) + \dim(\text{Im}\,\vartheta).$$

4. A linear transformation $\vartheta: V \rightarrow W$ is invertible if and only if $\text{Ker}\,\vartheta = \{0\}$ and $\text{Im}\,\vartheta = W$.

5. Given a basis $\mathscr{B}$ of the n-dimensional vector space V, there is a correspondence between the endomorphisms ϑ of V and the $n \times n$ matrices $[\vartheta]_{\mathscr{B}}$ over F.

   Given two bases $\mathscr{B}$ and $\mathscr{B}'$ of V, and an endomorphism ϑ of V, there exists an invertible matrix T such that

$$[\vartheta]_{\mathscr{B}} = T^{-1}[\vartheta]_{\mathscr{B}'} T.$$

6. Eigenvalues λ of an endomorphism ϑ satisfy $v\vartheta = \lambda v$ for some non-zero v in V.

7. A projection is an endomorphism π of V which satisfies $\pi^2 = \pi$.

Exercises for Chapter 2

1. Show that if V and W are vector spaces and $\vartheta: V \rightarrow W$ is an invertible linear transformation then ϑ^{-1} is a linear transformation.

2. Suppose that ϑ is an endomorphism of the vector space V. Show that the following are equivalent:

(1) ϑ is invertible;
(2) $\operatorname{Ker}\vartheta = \{0\}$;
(3) $\operatorname{Im}\vartheta = V$.

3. Let U and W be subspaces of the vector space V. Prove that $V = U \oplus W$ if and only if $V = U + W$ and $U \cap W = \{0\}$.

4. Let U and W be subspaces of the vector space V. Suppose that $u_1, \ldots, u_r$ is a basis of U and $w_1, \ldots, w_s$ is a basis of W. Show that $V = U \oplus W$ if and only if $u_1, \ldots, u_r, w_1, \ldots, w_s$ is a basis of V.

5. (a) Let U_1, U_2 and U_3 be subspaces of a vector space V, with $V = U_1 + U_2 + U_3$. Show that

$$V = U_1 \oplus U_2 \oplus U_3 \Leftrightarrow$$
$$U_1 \cap (U_2 + U_3) = U_2 \cap (U_1 + U_3) = U_3 \cap (U_1 + U_2) = \{0\}.$$

(b) Give an example of a vector space V with three subspaces U_1, U_2 and U_3 such that $V = U_1 + U_2 + U_3$ and

$$U_1 \cap U_2 = U_1 \cap U_3 = U_2 \cap U_3 = \{0\},$$

but $V \neq U_1 \oplus U_2 \oplus U_3$.

6. Suppose that $U_1, \ldots, U_r$ are subspaces of the vector space V, and that $V = U_1 \oplus \ldots \oplus U_r$. Prove that

$$\dim V = \dim U_1 + \ldots + \dim U_r.$$

7. Give an example of a vector space V with endomorphisms ϑ and ϕ such that $V = \operatorname{Im}\vartheta \oplus \operatorname{Ker}\vartheta$, but $V \neq \operatorname{Im}\phi \oplus \operatorname{Ker}\phi$.

8. Let V be a vector space and let ϑ be an endomorphism of V. Show that ϑ is a projection if and only if there is a basis $\mathcal{B}$ of V such that $[\vartheta]_{\mathcal{B}}$ is diagonal, with all diagonal entries equal to 1 or 0.

9. Suppose that ϑ is an endomorphism of the vector space V and $\vartheta^2 = 1_V$. Show that $V = U \oplus W$, where

$$U = \{v \in V : v\vartheta = v\}, \quad W = \{v \in V : v\vartheta = -v\}.$$

Deduce that V has a basis $\mathcal{B}$ such that $[\vartheta]_{\mathcal{B}}$ is diagonal, with all diagonal entries equal to $+1$ or -1.

3

Group representations

A representation of a group G gives us a way of visualizing G as a group of matrices. To be precise, a representation is a homomorphism from G into a group of invertible matrices. We set out this idea in more detail, and give some examples of representations. We also introduce the concept of equivalence of representations, and consider the kernel of a representation.

Representations

Let G be a group and let F be $\mathbb{R}$ or $\mathbb{C}$. Recall from the first chapter that $\mathrm{GL}(n, F)$ denotes the group of invertible $n \times n$ matrices with entries in F.

3.1 Definition

A *representation* of G over F is a homomorphism ρ from G to $\mathrm{GL}(n, F)$, for some n. The *degree* of ρ is the integer n.

Thus if ρ is a function from G to $\mathrm{GL}(n, F)$, then ρ is a representation if and only if

$$(gh)\rho = (g\rho)(h\rho) \quad \text{for all } g, h \in G.$$

Since a representation is a homomorphism, it follows that for every representation $\rho: G \to \mathrm{GL}(n, F)$, we have

$$1\rho = I_n, \text{ and}$$

$$g^{-1}\rho = (g\rho)^{-1} \quad \text{for all } g \in G,$$

where I_n denotes the $n \times n$ identity matrix.

3.2 Examples

(1) Let G be the dihedral group $D_8 = \langle a, b: a^4 = b^2 = 1, \ b^{-1}ab = a^{-1} \rangle$. Define the matrices A and B by

$$A = \begin{pmatrix} 0 & 1 \\ -1 & 0 \end{pmatrix}, \ B = \begin{pmatrix} 1 & 0 \\ 0 & -1 \end{pmatrix}$$

and check that

$$A^4 = B^2 = I, \ B^{-1}AB = A^{-1}.$$

It follows (see Example 1.4) that the function $\rho: G \to \mathrm{GL}\,(2, F)$ which is given by

$$\rho: a^i b^j \to A^i B^j \quad (0 \leqslant i \leqslant 3, 0 \leqslant j \leqslant 1)$$

is a representation of D_8 over F. The degree of ρ is 2.

The matrices $g\rho$ for g in D_8 are given in the following table:

g	1	a	a^2	a^3
$g\rho$	$\begin{pmatrix} 1 & 0 \\ 0 & 1 \end{pmatrix}$	$\begin{pmatrix} 0 & 1 \\ -1 & 0 \end{pmatrix}$	$\begin{pmatrix} -1 & 0 \\ 0 & -1 \end{pmatrix}$	$\begin{pmatrix} 0 & -1 \\ 1 & 0 \end{pmatrix}$

g	b	ab	a^2b	a^3b
$g\rho$	$\begin{pmatrix} 1 & 0 \\ 0 & -1 \end{pmatrix}$	$\begin{pmatrix} 0 & -1 \\ -1 & 0 \end{pmatrix}$	$\begin{pmatrix} -1 & 0 \\ 0 & 1 \end{pmatrix}$	$\begin{pmatrix} 0 & 1 \\ 1 & 0 \end{pmatrix}$

(2) Let G be any group. Define $\rho: G \to \mathrm{GL}\,(n, F)$ by

$$g\rho = I_n \quad \text{for all } g \in G,$$

where I_n is the $n \times n$ identity matrix, as usual. Then

$$(gh)\rho = I_n = I_n I_n = (g\rho)(h\rho)$$

for all $g, h \in G$, so ρ is a representation of G. This shows that every group has representations of arbitrarily large degree.

Equivalent representations

We now look at a way of converting a given representation into another one.

Let $\rho: G \rightarrow \mathrm{GL}\,(n, F)$ be a representation, and let T be an invertible $n \times n$ matrix over F. Note that for all $n \times n$ matrices A and B, we have

$$(T^{-1}AT)(T^{-1}BT) = T^{-1}(AB)T.$$

We can use this observation to produce a new representation σ from ρ; we simply define

$$g\sigma = T^{-1}(g\rho)T \quad \text{for all } g \in G.$$

Then for all $g, h \in G$,

$$
\begin{aligned}
(gh)\sigma &= T^{-1}((gh)\rho)T \\
&= T^{-1}((g\rho)(h\rho))T \\
&= T^{-1}(g\rho)T \cdot T^{-1}(h\rho)T \\
&= (g\sigma)(h\sigma),
\end{aligned}
$$

and so σ is, indeed, a representation.

3.3 Definition
Let $\rho: G \rightarrow \mathrm{GL}\,(m, F)$ and $\sigma: G \rightarrow \mathrm{GL}\,(n, F)$ be representations of G over F. We say that ρ is *equivalent* to σ if $n = m$ and there exists an invertible $n \times n$ matrix T such that for all $g \in G$,

$$g\sigma = T^{-1}(g\rho)T.$$

Note that for all representations ρ, σ and τ of G over F, we have (see Exercise 3.4):

(1) ρ is equivalent to ρ;
(2) if ρ is equivalent to σ then σ is equivalent to ρ;
(3) if ρ is equivalent to σ and σ is equivalent to τ, then ρ is equivalent to τ.

In other words, equivalence of representations is an equivalence relation.

3.4 Examples
(1) Let $G = D_8 = \langle a, b: a^4 = b^2 = 1, b^{-1}ab = a^{-1} \rangle$, and consider the representation ρ of G which appears in Example 3.2(1). Thus $a\rho = A$

and $b\rho = B$, where

$$A = \begin{pmatrix} 0 & 1 \\ -1 & 0 \end{pmatrix}, \quad B = \begin{pmatrix} 1 & 0 \\ 0 & -1 \end{pmatrix}.$$

Assume that $F = \mathbb{C}$, and define

$$T = \frac{1}{\sqrt{2}} \begin{pmatrix} 1 & 1 \\ i & -i \end{pmatrix}.$$

Then

$$T^{-1} = \frac{1}{\sqrt{2}} \begin{pmatrix} 1 & -i \\ 1 & i \end{pmatrix}.$$

In fact, T has been constructed so that $T^{-1}AT$ is diagonal; we have

$$T^{-1}AT = \begin{pmatrix} i & 0 \\ 0 & -i \end{pmatrix}, \quad T^{-1}BT = \begin{pmatrix} 0 & 1 \\ 1 & 0 \end{pmatrix},$$

and so we obtain a representation σ of D_8 for which

$$a\sigma = \begin{pmatrix} i & 0 \\ 0 & -i \end{pmatrix}, \quad b\sigma = \begin{pmatrix} 0 & 1 \\ 1 & 0 \end{pmatrix}.$$

The representations ρ and σ are equivalent.

(2) Let $G = C_2 = \langle a: a^2 = 1 \rangle$ and let

$$A = \begin{pmatrix} -5 & 12 \\ -2 & 5 \end{pmatrix}.$$

Check that $A^2 = I$. Hence $\rho: 1 \to I, a \to A$ is a representation of G. If

$$T = \begin{pmatrix} 2 & -3 \\ 1 & -1 \end{pmatrix},$$

then

$$T^{-1}AT = \begin{pmatrix} 1 & 0 \\ 0 & -1 \end{pmatrix},$$

and so we obtain a representation σ of G for which

$$1\sigma = \begin{pmatrix} 1 & 0 \\ 0 & 1 \end{pmatrix}, \quad a\sigma = \begin{pmatrix} 1 & 0 \\ 0 & -1 \end{pmatrix},$$

and σ is equivalent to ρ.

There are two easily recognized situations where the only representation which is equivalent to ρ is ρ itself; these are when the degree of ρ is 1, and when $g\rho = I_n$ for all g in G. However, there are usually lots of representations which are equivalent to ρ.

Kernels of representations

We conclude the chapter with a discussion of the kernel of a representation $\rho: G \to GL(n, F)$. In agreement with Definition 1.8, this consists of the group elements g in G for which $g\rho$ is the identity matrix. Thus

$$\text{Ker}\,\rho = \{g \in G: g\rho = I_n\}.$$

Note that $\text{Ker}\,\rho$ is a normal subgroup of G.

It can happen that the kernel of a representation is the whole of G, as is shown by the following definition.

3.5 Definition
The representation $\rho: G \to GL(1, F)$ which is defined by

$$g\rho = (1) \quad \text{for all } g \in G,$$

is called the *trivial* representation of G.

To put the definition another way, the trivial representation of G is the representation where every group element is sent to the 1×1 identity matrix.

Of particular interest are those representations whose kernel is just the identity subgroup.

3.6 Definition
A representation $\rho: G \to GL(n, F)$ is said to be *faithful* if $\text{Ker}\,\rho = \{1\}$; that is, if the identity element of G is the only element g for which $g\rho = I_n$.

3.7 Proposition
A representation ρ of a finite group G is faithful if and only if $\text{Im}\,\rho$ is isomorphic to G.

Proof We know that $\operatorname{Ker}\rho \triangleleft G$ and by Theorem 1.10, the factor group $G/\operatorname{Ker}\rho$ is isomorphic to $\operatorname{Im}\rho$. Therefore, if $\operatorname{Ker}\rho = \{1\}$ then $G \cong \operatorname{Im}\rho$. Conversely, if $G \cong \operatorname{Im}\rho$, then these two groups have the same (finite) order, and so $|\operatorname{Ker}\rho| = 1$; that is, ρ is faithful. ∎

3.8 Examples
(1) The representation ρ of D_8 given by

$$(a^i b^j)\rho = \begin{pmatrix} 0 & 1 \\ -1 & 0 \end{pmatrix}^i \begin{pmatrix} 1 & 0 \\ 0 & -1 \end{pmatrix}^j$$

as in Example 3.2(1) is faithful, since the identity is the only element g which satisfies $g\rho = I$. The group generated by the matrices

$$\begin{pmatrix} 0 & 1 \\ -1 & 0 \end{pmatrix} \text{ and } \begin{pmatrix} 1 & 0 \\ 0 & -1 \end{pmatrix}$$

is therefore isomorphic to D_8.

(2) Since $T^{-1}AT = I_n$ if and only if $A = I_n$, it follows that all representations which are equivalent to a faithful representation are faithful.

(3) The trivial representation of a group G if faithful if and only if $G = \{1\}$.

In Chapter 6 we shall show that every finite group has a faithful representation.

The basic problem of representation theory is to discover and understand representations of finite groups.

Summary of Chapter 3

1. A representation of a group G is a homomorphism from G into $GL(n, F)$, for some n.

2. Representations ρ and σ of G are equivalent if and only if there exists an invertible matrix T such that for all $g \in G$,

$$g\sigma = T^{-1}(g\rho)T.$$

3. A representation is faithful if it is injective.

Exercises for Chapter 3

1. Let G be the cyclic group of order m, say $G = \langle a: a^m = 1 \rangle$. Suppose that $A \in \mathrm{GL}\,(n, \mathbb{C})$, and define $\rho: G \to \mathrm{GL}\,(n, \mathbb{C})$ by

$$\rho: a^r \to A^r \quad (0 \leqslant r \leqslant m - 1).$$

Show that ρ is a representation of G over $\mathbb{C}$ if and only if $A^m = I$.

2. Let

$$A = \begin{pmatrix} 1 & 0 \\ 0 & 1 \end{pmatrix}, \ B = \begin{pmatrix} 1 & 0 \\ 0 & e^{2\pi i/3} \end{pmatrix}, \ C = \begin{pmatrix} 0 & 1 \\ -1 & -1 \end{pmatrix}$$

and let $G = \langle a: a^3 = 1 \rangle \cong C_3$. Show that each of the functions $\rho_j: G \to \mathrm{GL}\,(2, \mathbb{C})$ $(1 \leqslant j \leqslant 3)$, defined by

$$\rho_1: a^r \to A^r,$$

$$\rho_2: a^r \to B^r,$$

$$\rho_3: a^r \to C^r \quad (0 \leqslant r \leqslant 2),$$

is a representation of G over $\mathbb{C}$. Which of these representations are faithful?

3. Suppose that $G = D_{2n} = \langle a, b: a^n = b^2 = 1, \ b^{-1}ab = a^{-1} \rangle$, and $F = \mathbb{R}$ or $\mathbb{C}$. Show that there is a representation $\rho: G \to \mathrm{GL}\,(1, F)$ such that $a\rho = (1)$ and $b\rho = (-1)$.

4. Suppose that ρ, σ and τ are representations of G over F. Prove:

   (1) ρ is equivalent to ρ;
   (2) if ρ is equivalent to σ, then σ is equivalent to ρ;
   (3) if ρ is equivalent to σ, and σ is equivalent to τ, then ρ is equivalent to τ.

5. Let $G = D_{12} = \langle a, b: a^6 = b^2 = 1, \ b^{-1}ab = a^{-1} \rangle$. Define the matrices A, B, C, D over $\mathbb{C}$ by

$$A = \begin{pmatrix} e^{i\pi/3} & 0 \\ 0 & e^{-i\pi/3} \end{pmatrix}, \ B = \begin{pmatrix} 0 & 1 \\ 1 & 0 \end{pmatrix},$$

$$C = \begin{pmatrix} 1/2 & \sqrt{3}/2 \\ -\sqrt{3}/2 & 1/2 \end{pmatrix}, \ D = \begin{pmatrix} 1 & 0 \\ 0 & -1 \end{pmatrix}.$$

Prove that each of the functions $\rho_k: G \to \mathrm{GL}\,(2, \mathbb{C})$ $(k = 1, 2, 3, 4)$, given by

$\rho_1: a^r b^s \rightarrow A^r B^s,$

$\rho_2: a^r b^s \rightarrow A^{3r}(-B)^s,$

$\rho_3: a^r b^s \rightarrow (-A)^r B^s,$

$\rho_4: a^r b^s \rightarrow C^r D^s \quad (0 \leqslant r \leqslant 5, 0 \leqslant s \leqslant 1),$

is a representation of G. Which of these representations are faithful? Which are equivalent?

6. Give an example of a faithful representation of D_8 of degree 3.

7. Suppose that ρ is a representation of G of degree 1. Prove that $G/\operatorname{Ker} \rho$ is abelian.

8. Let ρ be a representation of the group G. Suppose that g and h are elements of G such that $(g\rho)(h\rho) = (h\rho)(g\rho)$. Does it follow that $gh = hg$?

4

FG-modules

We now introduce the concept of an *FG*-module, and show that there is a close connection between *FG*-modules and representations of *G* over *F*. Much of the material in the remainder of the book will be presented in terms of *FG*-modules, as there are several advantages to this approach to representation theory.

FG-modules

Let *G* be a group and let *F* be $\mathbb{R}$ or $\mathbb{C}$.

Suppose that $\rho: G \to \mathrm{GL}\,(n,\,F)$ is a representation of *G*. Write $V = F^n$, the vector space of all row vectors $(\lambda_1, \ldots, \lambda_n)$ with $\lambda_i \in F$. For all $v \in V$ and $g \in G$, the matrix product

$$v(g\rho),$$

of the row vector *v* with the $n \times n$ matrix $g\rho$, is a row vector in *V* (since the product of a $1 \times n$ matrix with an $n \times n$ matrix is again a $1 \times n$ matrix).

We now list some basic properties of the multiplication $v(g\rho)$. First, the fact that ρ is a homomorphism shows that

$$v((gh)\rho) = v(g\rho)(h\rho)$$

for all $v \in V$ and all $g, h \in G$. Next, since 1ρ is the identity matrix, we have

$$v(1\rho) = v$$

for all $v \in V$. Finally, the properties of matrix multiplication give

$$(\lambda v)(g\rho) = \lambda(v(g\rho)),$$

$$(u + v)(g\rho) = u(g\rho) + v(g\rho)$$

for all $u, v \in V, \lambda \in F$ and $g \in G$.

4.1 Example

Let $G = D_8 = \langle a, b: a^4 = b^2 = 1, b^{-1}ab = a^{-1}\rangle$, and let $\rho: G \rightarrow$ GL $(2, F)$ be the representation of G over F given in Example 3.2(1). Thus

$$a\rho = \begin{pmatrix} 0 & 1 \\ -1 & 0 \end{pmatrix}, b\rho = \begin{pmatrix} 1 & 0 \\ 0 & -1 \end{pmatrix}.$$

If $v = (\lambda_1, \lambda_2) \in F^2$ then, for example,

$$v(a\rho) = (-\lambda_2, \lambda_1),$$

$$v(b\rho) = (\lambda_1, -\lambda_2),$$

$$v(a^3\rho) = (\lambda_2, -\lambda_1).$$

Motivated by the above observations on the product $v(g\rho)$, we now define an *FG*-module.

4.2 Definition

Let V be a vector space over F and let G be a group. Then V is an *FG-module* if a multiplication vg ($v \in V, g \in G$) is defined, satisfying the following conditions for all $u, v \in V, \lambda \in F$ and $g, h \in G$:

(1) $vg \in V$;
(2) $v(gh) = (vg)h$;
(3) $v1 = v$;
(4) $(\lambda v)g = \lambda(vg)$;
(5) $(u + v)g = ug + vg$.

We use the letters F and G in the name '*FG*-module' to indicate that V is a vector space over F and that G is the group from which we are taking the elements g to form the products vg ($v \in V$).

Note that conditions (1), (4) and (5) in the definition ensure that for all $g \in G$, the function

$$v \rightarrow vg \quad (v \in V)$$

is an endomorphism of V.

4.3 Definition
Let V be an FG-module, and let $\mathcal{B}$ be a basis of V. For each $g \in G$, let

$$[g]_{\mathcal{B}}$$

denote the matrix of the endomorphism $v \to vg$ of V, relative to the basis $\mathcal{B}$.

The connection between FG-modules and representations of G over F is revealed in the following basic result.

4.4 Theorem
(1) *If $\rho: G \to \mathrm{GL}(n, F)$ is a representation of G over F, and $V = F^n$, then V becomes an FG-module if we define the multiplication vg by*

$$vg = v(g\rho) \quad (v \in V, g \in G).$$

Moreover, there is a basis $\mathcal{B}$ of V such that

$$g\rho = [g]_{\mathcal{B}} \quad \text{for all } g \in G.$$

(2) *Assume that V is an FG-module and let $\mathcal{B}$ be a basis of V. Then the function*

$$g \to [g]_{\mathcal{B}} \quad (g \in G)$$

is a representation of G over F.

Proof (1) We have already observed that for all $u, v \in F^n$, $\lambda \in F$ and $g, h \in G$, we have

$$v(g\rho) \in F^n,$$
$$v((gh)\rho) = (v(g\rho))(h\rho),$$
$$v(1\rho) = v,$$
$$(\lambda v)(g\rho) = \lambda(v(g\rho)),$$
$$(u + v)(g\rho) = u(g\rho) + v(g\rho).$$

Therefore, F^n becomes an FG-module if we define

$$vg = v(g\rho) \quad \text{for all } v \in F^n, g \in G.$$

Moreover, if we let $\mathcal{B}$ be the basis

$$(1, 0, 0, \ldots, 0), (0, 1, 0, \ldots, 0), \ldots, (0, 0, 0, \ldots, 1)$$

of F^n, then $g\rho = [g]_{\mathcal{B}}$ for all $g \in G$.

(2) Let V be an FG-module with basis $\mathscr{B}$. Since $v(gh) = (vg)h$ for all $g,\ h \in G$ and all v in the basis $\mathscr{B}$ of V, it follows that

$$[gh]_{\mathscr{B}} = [g]_{\mathscr{B}}[h]_{\mathscr{B}}.$$

In particular,

$$[1]_{\mathscr{B}} = [g]_{\mathscr{B}}[g^{-1}]_{\mathscr{B}}$$

for all $g \in G$. Now $v1 = v$ for all $v \in V$, so $[1]_{\mathscr{B}}$ is the identity matrix. Therefore each matrix $[g]_{\mathscr{B}}$ is invertible (with inverse $[g^{-1}]_{\mathscr{B}}$).

We have proved that the function $g \to [g]_{\mathscr{B}}$ is a homomorphism from G to $\mathrm{GL}(n, F)$ (where $n = \dim V$), and hence is a representation of G over F. ∎

Our next example illustrates part (1) of Theorem 4.4.

4.5 Examples
(1) Let $G = D_8 = \langle a, b \colon a^4 = b^2 = 1,\ b^{-1}ab = a^{-1}\rangle$ and let ρ be the representation of G over F given in Example 3.2(1), so

$$a\rho = \begin{pmatrix} 0 & 1 \\ -1 & 0 \end{pmatrix},\quad b\rho = \begin{pmatrix} 1 & 0 \\ 0 & -1 \end{pmatrix}.$$

Write $V = F^2$. By Theorem 4.4(1), V becomes an FG-module if we define

$$vg = v(g\rho) \quad (v \in V, g \in G).$$

For instance,

$$(1, 0)a = (1, 0)\begin{pmatrix} 0 & 1 \\ -1 & 0 \end{pmatrix} = (0, 1).$$

If v_1, v_2 is the basis $(1, 0), (0, 1)$ of V, then we have

$$v_1 a = v_2, \quad v_1 b = v_1,$$
$$v_2 a = -v_1, \quad v_2 b = -v_2.$$

If $\mathscr{B}$ denotes the basis v_1, v_2, then the representation

$$g \to [g]_{\mathscr{B}} \quad (g \in G)$$

is just the representation ρ (see Theorem 4.4(1) again).
(2) Let $G = Q_8 = \langle a, b \colon a^4 = 1,\ a^2 = b^2,\ b^{-1}ab = a^{-1}\rangle$. In Example

1.2(4) we defined Q_8 to be the subgroup of $GL(2, \mathbb{C})$ generated by

$$A = \begin{pmatrix} i & 0 \\ 0 & -i \end{pmatrix} \text{ and } B = \begin{pmatrix} 0 & 1 \\ -1 & 0 \end{pmatrix},$$

so we already have a representation of G over $\mathbb{C}$. To illustrate Theorem 4.4(1) we must this time take $F = \mathbb{C}$. We then obtain a $\mathbb{C}G$-module with basis v_1, v_2 such that

$$v_1 a = iv_1, \quad v_1 b = v_2,$$
$$v_2 a = -iv_2, \quad v_2 b = -v_1.$$

Notice that in the above examples, the vectors $v_1 a$, $v_2 a$, $v_1 b$ and $v_2 b$ determine vg for all $v \in V$ and $g \in G$. For instance, in Example 4.5(1),

$$(v_1 + 2v_2)ab = v_1 ab + 2v_2 ab$$

$$= v_2 b - 2v_1 b$$

$$= -v_2 - 2v_1.$$

A similar remark holds for all FG-modules V: if $v_1, \ldots, v_n$ is a basis of V and $g_1, \ldots, g_r$ generate G, then the vectors $v_i g_j$ $(1 \le i \le n, 1 \le j \le r)$ determine vg for all $v \in V$ and $g \in G$.

Shortly, we shall show you various ways of constructing FG-modules directly, without using a representation. To do this, we turn a vector space V over F into an FG-module by specifying the action of group elements on a basis $v_1, \ldots, v_n$ of V and then extending the action to be linear on the whole of V; that is, we first define $v_i g$ for each i and each g in G, and then define

$$(\lambda_1 v_1 + \ldots + \lambda_n v_n)g \quad (\lambda_i \in F)$$

to be

$$\lambda_1(v_1 g) + \ldots + \lambda_n(v_n g).$$

As you might expect, there are restrictions on how we may define the vectors $v_i g$. The next result will often be used to show that our chosen multiplication turns V into an FG-module.

4.6 Proposition
Assume that $v_1, \ldots, v_n$ is a basis of a vector space V over F. Suppose that we have a multiplication vg for all v in V and g in G which

satisfies the following conditions for all i with $1 \leqslant i \leqslant n$, for all $g, h \in G$, and for all $\lambda_1, \ldots, \lambda_n \in F$:

(1) $v_i g \in V$;
(2) $v_i(gh) = (v_i g)h$;
(3) $v_i 1 = v_i$;
(4) $(\lambda_1 v_1 + \ldots + \lambda_n v_n)g = \lambda_1(v_1 g) + \ldots + \lambda_n(v_n g)$.

Then V is an FG-module.

Proof It is clear from (3) and (4) that $v1 = v$ for all $v \in V$.

Conditions (1) and (4) ensure that for all g in G, the function $v \to vg$ ($v \in V$) is an endomorphism of V. That is,

$$vg \in V,$$

$$(\lambda v)g = \lambda(vg),$$

$$(u + v)g = ug + vg,$$

for all $u, v \in V$, $\lambda \in F$ and $g \in G$. Hence

(4.7) $\qquad (\lambda_1 u_1 + \ldots + \lambda_n u_n)h = \lambda_1(u_1 h) + \ldots + \lambda_n(u_n h)$

for all $\lambda_1, \ldots, \lambda_n \in F$, all $u_1, \ldots, u_n \in V$ and all $h \in G$.

Now let $v \in V$ and $g, h \in G$. Then $v = \lambda_1 v_1 + \ldots + \lambda_n v_n$ for some $\lambda_1, \ldots, \lambda_n \in F$, and

$$v(gh) = \lambda_1(v_1(gh)) + \ldots + \lambda_n(v_n(gh)) \quad \text{by condition (4)}$$

$$= \lambda_1((v_1 g)h) + \ldots + \lambda_n((v_n g)h) \quad \text{by condition (2)}$$

$$= (\lambda_1(v_1 g) + \ldots + \lambda_n(v_n g))h \quad \text{by (4.7)}$$

$$= (vg)h \quad \text{by condition (4).}$$

We have now checked all the axioms which are required for V to be an *FG*-module. ∎

Our next definitions translate the concepts of the trivial representation and a faithful representation into module terms.

4.8 Definitions
(1) The *trivial FG-module* is the 1-dimensional vector space V over F with

$$vg = v \quad \text{for all } v \in V, g \in G.$$

(2) An *FG*-module V is *faithful* if the identity element of G is the only element g for which

$$vg = v \quad \text{for all } v \in V.$$

For instance, the FD_8-module which appears in Example 4.5(1) is faithful.

Our next aim is to use Proposition 4.6 to construct faithful *FG*-modules for all subgroups of symmetric groups.

Permutation modules

Let G be a subgroup of S_n, so that G is a group of permutations of $\{1, \ldots, n\}$. Let V be an n-dimensional vector space over F, with basis $v_1, \ldots, v_n$. For each i with $1 \leqslant i \leqslant n$ and each permutation g in G, define

$$v_i g = v_{ig}.$$

Then $v_i g \in V$ and $v_i 1 = v_i$. Also, for g, h in G,

$$v_i(gh) = v_{i(gh)} = v_{(ig)h} = (v_i g)h.$$

We now extend the action of each g linearly to the whole of V; that is, for all $\lambda_1, \ldots, \lambda_n$ in F and g in G, we define

$$(\lambda_1 v_1 + \ldots + \lambda_n v_n)g = \lambda_1(v_1 g) + \ldots + \lambda_n(v_n g).$$

Then V is an *FG*-module, by Proposition 4.6.

4.9 Example
Let $G = S_4$ and let $\mathscr{B}$ denote the basis v_1, v_2, v_3, v_4 of V. If $g = (1\ 2)$, then

$$v_1 g = v_2, \ v_2 g = v_1, \ v_3 g = v_3, \ v_4 g = v_4.$$

And if $h = (1\ 3\ 4)$, then

$$v_1 h = v_3, \ v_2 h = v_2, \ v_3 h = v_4, \ v_4 h = v_1.$$

We have

$$[g]_{\mathscr{B}} = \begin{pmatrix} 0 & 1 & 0 & 0 \\ 1 & 0 & 0 & 0 \\ 0 & 0 & 1 & 0 \\ 0 & 0 & 0 & 1 \end{pmatrix}, \ [h]_{\mathscr{B}} = \begin{pmatrix} 0 & 0 & 1 & 0 \\ 0 & 1 & 0 & 0 \\ 0 & 0 & 0 & 1 \\ 1 & 0 & 0 & 0 \end{pmatrix}.$$

4.10 Definition

Let G be a subgroup of S_n. The FG-module V with basis $v_1, \ldots, v_n$ such that

$$v_i g = v_{ig} \quad \text{for all } i, \text{ and all } g \in G,$$

is called the *permutation module* for G over F. We call $v_1, \ldots, v_n$ the *natural basis* of V.

Note that if we write $\mathscr{B}$ for the basis $v_1, \ldots, v_n$ of the permutation module, then for all g in G, the matrix $[g]_{\mathscr{B}}$ has precisely one non-zero entry in each row and column, and this entry is 1. Such a matrix is called a *permutation matrix*.

Since the only element of G which fixes every v_i is the identity, we see that the permutation module is a faithful FG-module. If you are aware of the fact that every group G of order n is isomorphic to a subgroup of S_n, then you should be able to see that G has a faithful FG-module of dimension n. We shall go into this in more detail in Chapter 6.

4.11 Example

Let $G = C_3 = \langle a \colon a^3 = 1 \rangle$. Then G is isomorphic to the cyclic subgroup of S_3 which is generated by the permutation $(1\ 2\ 3)$. This alerts us to the fact that if V is a 3-dimensional vector space over F, with basis v_1, v_2, v_3, then we may make V into an FG-module in which

$$v_1 1 = v_1,\ v_2 1 = v_2,\ v_3 1 = v_3,$$

$$v_1 a = v_2,\ v_2 a = v_3,\ v_3 a = v_1,$$

$$v_1 a^2 = v_3,\ v_2 a^2 = v_1,\ v_3 a^2 = v_2.$$

Of course, we define vg, for v an arbitrary vector in V and $g = 1$, a or a^2, by

$$(\lambda_1 v_1 + \lambda_2 v_2 + \lambda_3 v_3)g = \lambda_1(v_1 g) + \lambda_2(v_2 g) + \lambda_3(v_3 g)$$

for all $\lambda_1, \lambda_2, \lambda_3 \in F$. Proposition 4.6 can be used to verify that V is an FG-module, but we have been motivated by the definition of permutation modules in our construction.

FG-modules and equivalent representations

We conclude the chapter with a discussion of the relationship between FG-modules and equivalent representations of G over F. An FG-

module gives us many representations, all of the form

$$g \to [g]_{\mathscr{B}} \quad (g \in G)$$

for some basis $\mathscr{B}$ of V. The next result shows that all these representations are equivalent to each other (see Definition 3.3); and moreover, any two equivalent representations of G arise from some FG-module in this way.

4.12 Theorem
Suppose that V is an FG-module with basis $\mathscr{B}$, and let ρ be the representation of G over F defined by

$$\rho \colon g \to [g]_{\mathscr{B}} \quad (g \in G).$$

   (1) If $\mathscr{B}'$ is a basis of V, then the representation

$$\phi \colon g \to [g]_{\mathscr{B}'} \quad (g \in G)$$

of G is equivalent to ρ.

   (2) If σ is a representation of G which is equivalent to ρ, then there is a basis $\mathscr{B}''$ of V such that

$$\sigma \colon g \to [g]_{\mathscr{B}''} \quad (g \in G).$$

Proof (1) Let T be the change of basis matrix from $\mathscr{B}$ to $\mathscr{B}'$ (see Definition 2.23). Then by (2.24), for all $g \in G$, we have

$$[g]_{\mathscr{B}} = T^{-1}[g]_{\mathscr{B}'}T.$$

Therefore ϕ is equivalent to ρ.

   (2) Suppose that ρ and σ are equivalent representations of G. Then for some invertible matrix T, we have

$$g\rho = T^{-1}(g\sigma)T \quad \text{for all } g \in G.$$

Let $\mathscr{B}''$ be the basis of V such that the change of basis matrix from $\mathscr{B}$ to $\mathscr{B}''$ is T. Then for all $g \in G$,

$$[g]_{\mathscr{B}} = T^{-1}[g]_{\mathscr{B}''}T,$$

and so $g\sigma = [g]_{\mathscr{B}''}$.   ∎

4.13 Example
Again let $G = C_3 = \langle a \colon a^3 = 1 \rangle$. There is a representation ρ of G which is given by

$$1\rho = \begin{pmatrix} 1 & 0 \\ 0 & 1 \end{pmatrix}, \; a\rho = \begin{pmatrix} 0 & 1 \\ -1 & -1 \end{pmatrix}, \; a^2\rho = \begin{pmatrix} -1 & -1 \\ 1 & 0 \end{pmatrix}.$$

(To see this, simply note that $(a\rho)^2 = a^2\rho$ and $(a\rho)^3 = I$; see Exercise 3.2.)

If V is a 2-dimensional vector space over $\mathbb{C}$, with basis v_1, v_2 (which we call $\mathscr{B}$), then we can turn V into a $\mathbb{C}G$-module as in Theorem 4.4(1) by defining

$$v_1 1 = v_1, \quad v_1 a = v_2, \quad v_1 a^2 = -v_1 - v_2,$$

$$v_2 1 = v_2, \quad v_2 a = -v_1 - v_2, \quad v_2 a^2 = v_1.$$

We then have

$$[1]_{\mathscr{B}} = \begin{pmatrix} 1 & 0 \\ 0 & 1 \end{pmatrix}, [a]_{\mathscr{B}} = \begin{pmatrix} 0 & 1 \\ -1 & -1 \end{pmatrix}, [a^2]_{\mathscr{B}} = \begin{pmatrix} -1 & -1 \\ 1 & 0 \end{pmatrix}.$$

Now let $u_1 = v_1$ and $u_2 = v_1 + v_2$. Then u_1, u_2 is another basis of V, which we call $\mathscr{B}'$. Since

$$u_1 1 = u_1, \quad u_1 a = -u_1 + u_2, \quad u_1 a^2 = -u_2,$$

$$u_2 1 = u_2, \quad u_2 a = -u_1, \quad u_2 a^2 = u_1 - u_2,$$

we obtain the representation $\phi: g \to [g]_{\mathscr{B}'}$ where

$$[1]_{\mathscr{B}'} = \begin{pmatrix} 1 & 0 \\ 0 & 1 \end{pmatrix}, [a]_{\mathscr{B}'} = \begin{pmatrix} -1 & 1 \\ -1 & 0 \end{pmatrix}, [a^2]_{\mathscr{B}'} = \begin{pmatrix} 0 & -1 \\ 1 & -1 \end{pmatrix}.$$

Note that if

$$T = \begin{pmatrix} 1 & 0 \\ 1 & 1 \end{pmatrix}$$

then for all g in G, we have

$$[g]_{\mathscr{B}} = T^{-1}[g]_{\mathscr{B}'}T,$$

and so ρ and ϕ are equivalent, in agreement with Theorem 4.12(1).

Summary of Chapter 4

1. An *FG*-module is a vector space over F, together with a multiplication by elements of G on the right. The multiplication satisfies properties (1)–(5) of Definition 4.2.

2. There is a correspondence between representations of G over F and FG-modules, as follows.

    (a)   Suppose that $\rho: G \to \mathrm{GL}\,(n, F)$ is a representation of G. Then F^n is an FG-module, if we define

$$vg = v(g\rho) \quad (v \in F^n, g \in G).$$

    (b)   If V is an FG-module, with basis $\mathscr{B}$, then $\rho: g \to [g]_{\mathscr{B}}$ is a representation of G over F.

3. If G is a subgroup of S_n, then the permutation FG-module has basis $v_1, \ldots, v_n$, and $v_i g = v_{ig}$ for all i with $1 \leqslant i \leqslant n$, and all g in G.

Exercises for Chapter 4

1. Suppose that $G = S_3$, and that $V = \mathrm{sp}\,(v_1, v_2, v_3)$ is the permutation module for G over $\mathbb{C}$, as in Definition 4.10. Let $\mathscr{B}_1$ be the basis v_1, v_2, v_3 of V and let $\mathscr{B}_2$ be the basis $v_1 + v_2 + v_3$, $v_1 - v_2$, $v_1 - v_3$. Calculate the 3×3 matrices $[g]_{\mathscr{B}_1}$ and $[g]_{\mathscr{B}_2}$ for all g in S_3. What do you notice about the matices $[g]_{\mathscr{B}_2}$?

2. Let $G = S_n$ and let V be a vector space over F. Show that V becomes an FG-module if we define, for all v in V,

$$vg = \begin{cases} v, & \text{if } g \text{ is an even permutation,} \\ -v, & \text{if } g \text{ is an odd permutation.} \end{cases}$$

3. Let $Q_8 = \langle a, b: a^4 = 1, \ b^2 = a^2, \ b^{-1}ab = a^{-1} \rangle$, the quaternion group of order 8. Show that there is an $\mathbb{R}Q_8$-module V of dimension 4 with basis v_1, v_2, v_3, v_4 such that

$$v_1 a = v_2, \quad v_2 a = -v_1, \quad v_3 a = -v_4, \quad v_4 a = v_3, \text{ and}$$

$$v_1 b = v_3, \quad v_2 b = v_4, \quad v_3 b = -v_1, \quad v_4 b = -v_2.$$

4. Let A be an $n \times n$ matrix and let B be a matrix obtained from A by permuting the rows. Show that there is an $n \times n$ permutation matrix P such that $B = PA$. Find a similar result for a matrix obtained from A by permuting the columns.

5
FG-submodules and reducibility

We begin the study of *FG*-modules by introducing the basic building blocks of the theory – the irreducible *FG*-modules. First we require the notion of an *FG*-submodule of an *FG*-module.

Throughout, *G* is a group and *F* is $\mathbb{R}$ or $\mathbb{C}$.

FG-submodules

5.1 Definition
Let *V* be an *FG*-module. A subset *W* of *V* is said to be an *FG*-submodule of *V* if *W* is a subspace and $wg \in W$ for all $w \in W$ and all $g \in G$.

Thus an *FG*-submodule of *V* is a subspace which is also an *FG*-module.

5.2 Examples
(1) For every *FG*-module *V*, the zero subspace $\{0\}$, and *V* itself, are *FG*-submodules of *V*.

(2) Let $G = C_3 = \langle a: a^3 = 1 \rangle$, and let *V* be the 3-dimensional *FG*-module defined in Example 4.11. Thus, *V* has basis v_1, v_2, v_3, and

$$v_1 1 = v_1, \ v_2 1 = v_2, \ v_3 1 = v_3,$$

$$v_1 a = v_2, \ v_2 a = v_3, \ v_3 a = v_1,$$

$$v_1 a^2 = v_3, \ v_2 a^2 = v_1, \ v_3 a^2 = v_2.$$

Put $w = v_1 + v_2 + v_3$, and let $W = \text{sp}(w)$, the 1-dimensional subspace spanned by *w*. Since

$$w1 = wa = wa^2 = w,$$

W is an FG-submodule of V. However, $\mathrm{sp}\,(v_1 + v_2)$ is not an FG-submodule, since

$$(v_1 + v_2)a = v_2 + v_3 \notin \mathrm{sp}\,(v_1 + v_2).$$

Irreducible FG-modules

5.3 Definition
An FG-module V is said to be *irreducible* if it is non-zero and it has no FG-submodules apart from $\{0\}$ and V.

If V has an FG-submodule W with W not equal to $\{0\}$ or V, then V is *reducible*.

Similarly, a representation $\rho\colon G \to \mathrm{GL}\,(n, F)$ is *irreducible* if the corresponding FG-module F^n given by

$$vg = v(g\rho) \quad (v \in F^n, \, g \in G)$$

(see Theorem 4.4(1)) is irreducible; and ρ is *reducible* if F^n is reducible.

Suppose that V is a reducible FG-module, so that there is an FG-submodule W with $0 < \dim W < \dim V$. Take a basis $\mathcal{B}_1$ of W and extend it to a basis $\mathcal{B}$ of V. Then for all g in G, the matrix $[g]_{\mathcal{B}}$ has the form

(5.4)
$$\left(\begin{array}{c|c} X_g & 0 \\ \hline Y_g & Z_g \end{array} \right)$$

for some matrices X_g, Y_g and Z_g, where X_g is $k \times k$ ($k = \dim W$).

A representation of degree n is reducible if and only if it is equivalent to a representation of the form (5.4), where X_g is $k \times k$ and $0 < k < n$.

Notice that in (5.4), the functions $g \to X_g$ and $g \to Z_g$ are representations of G: to see this, let $g, \, h \in G$ and multiply the matrices $[g]_{\mathcal{B}}$ and $[h]_{\mathcal{B}}$ given by (5.4). Notice also that if V is reducible then $\dim V \geq 2$.

5.5 Examples
(1) Let $G = C_3 = \langle a\colon a^3 = 1 \rangle$ and let V be the 3-dimensional FG-module with basis v_1, v_2, v_3 such that

$$v_1 a = v_2, \; v_2 a = v_3, \; v_3 a = v_1,$$

as in Example 4.11. We saw in Example 5.2(2) that V is a reducible FG-module, and has an FG-submodule $W = \mathrm{sp}\,(v_1 + v_2 + v_3)$. Let $\mathscr{B}$ be the basis $v_1 + v_2 + v_3, v_1, v_2$ of V. Then

$$[1]_{\mathscr{B}} = \begin{pmatrix} 1 & 0 & 0 \\ \hline 0 & 1 & 0 \\ 0 & 0 & 1 \end{pmatrix}, [a]_{\mathscr{B}} = \begin{pmatrix} 1 & 0 & 0 \\ \hline 0 & 0 & 1 \\ 1 & -1 & -1 \end{pmatrix},$$

$$[a^2]_{\mathscr{B}} = \begin{pmatrix} 1 & 0 & 0 \\ \hline 1 & -1 & -1 \\ 0 & 1 & 0 \end{pmatrix}.$$

This reducible representation gives us two other representations: at the 'top left' we have the trivial representation and at the 'bottom right' we have the representation which is given by

$$1 \to \begin{pmatrix} 1 & 0 \\ 0 & 1 \end{pmatrix}, \; a \to \begin{pmatrix} 0 & 1 \\ -1 & -1 \end{pmatrix}, \; a^2 \to \begin{pmatrix} -1 & -1 \\ 1 & 0 \end{pmatrix}.$$

(2) Let $G = D_8$ and let $V = F^2$ be the 2-dimensional FG-module described in Example 4.5(1). Thus $G = \langle a, b \rangle$, and for all $(\lambda, \mu) \in V$ we have

$$(\lambda, \mu)a = (-\mu, \lambda), \quad (\lambda, \mu)b = (\lambda, -\mu).$$

We claim that V is an irreducible FG-module. To see this, suppose that there is an FG-submodule U which is not equal to V. Then $\dim U \leqslant 1$, so $U = \mathrm{sp}\,((\alpha, \beta))$ for some $\alpha, \beta \in F$. As U is an FG-module, $(\alpha, \beta)b$ is a scalar multiple of (α, β), and hence either $\alpha = 0$ or $\beta = 0$. Since $(\alpha, \beta)a$ is also a scalar multiple of (α, β), this forces $\alpha = \beta = 0$, so $U = \{0\}$. Consequently V is irreducible, as claimed.

Summary of Chapter 5

1. If V is an FG-module, and W is a subspace of V which is itself an FG-module, then W is an FG-submodule of V.

2. The FG-module V is irreducible if it is non-zero and the only FG-submodules are $\{0\}$ and V.

Exercises for Chapter 5

1. Let $G = C_2 = \langle a: a^2 = 1 \rangle$, and let $V = F^2$. For $(\alpha, \beta) \in V$, define
   $(\alpha, \beta)1 = (\alpha, \beta)$ and $(\alpha, \beta)a = (\beta, \alpha)$. Verify that V is an FG-module
   and find all the FG-submodules of V.

2. Let ρ and σ be equivalent representations of the group G over F.
   Prove that if ρ is reducible then σ is reducible.

3. Which of the four representations of D_{12} defined in Exercise 3.5 are
   irreducible?

4. Define the permutations $a, b, c \in S_6$ by

$$a = (1\ 2\ 3),\ b = (4\ 5\ 6),\ c = (2\ 3)(4\ 5),$$

   and let $G = \langle a, b, c \rangle$.
   (a) Check that

$$a^3 = b^3 = c^2 = 1,\ ab = ba,$$

$$c^{-1}ac = a^{-1}\ \text{and}\ c^{-1}bc = b^{-1}.$$

   Deduce that G has order 18.
   (b) Suppose that ε and η are complex cube roots of unity. Prove
   that there is a representation ρ of G over $\mathbb{C}$ such that

$$a\rho = \begin{pmatrix} \varepsilon & 0 \\ 0 & \varepsilon^{-1} \end{pmatrix},\ b\rho = \begin{pmatrix} \eta & 0 \\ 0 & \eta^{-1} \end{pmatrix},\ c\rho = \begin{pmatrix} 0 & 1 \\ 1 & 0 \end{pmatrix}.$$

   (c) For which values of ε, η is ρ faithful?
   (d) For which values of ε, η is ρ irreducible?

5. Let $G = C_{13}$. Find a $\mathbb{C}G$-module which is neither reducible nor
   irreducible.

6

Group algebras

The group algebra of a finite group G is a vector space of dimension $|G|$ which also carries extra structure involving the product operation on G. In a sense, group algebras are the source of all you need to know about representation theory. In particular, the ultimate goal of representation theory – that of understanding all the representations of finite groups – would be achieved if group algebras could be fully analysed. Group algebras are therefore of great interest.

After defining the group algebra of G, we shall use it to construct an important faithful representation, known as the regular representation of G, which will be explored in greater detail later on.

The group algebra of G

Let G be a finite group whose elements are $g_1, \ldots, g_n$, and let F be $\mathbb{R}$ or $\mathbb{C}$.

We define a vector space over F with $g_1, \ldots, g_n$ as a basis, and we call this vector space FG. Take as the elements of FG all expressions of the form

$$\lambda_1 g_1 + \ldots + \lambda_n g_n \quad (\text{all } \lambda_i \in F).$$

The rules for addition and scalar multiplication in FG are the natural ones: namely, if

$$u = \sum_{i=1}^{n} \lambda_i g_i \text{ and } v = \sum_{i=1}^{n} \mu_i g_i$$

are elements of FG, and $\lambda \in F$, then

$$u + v = \sum_{i=1}^{n} (\lambda_i + \mu_i) g_i \text{ and } \lambda u = \sum_{i=1}^{n} (\lambda \lambda_i) g_i.$$

With these rules, FG is a vector space over F of dimension n, with basis $g_1, \ldots, g_n$. The basis $g_1, \ldots, g_n$ is called the *natural* basis of FG.

6.1 Example

Let $G = C_3 = \langle a: a^3 = e \rangle$. (To avoid confusion with the element 1 of F, we write e for the identity element of G, in this example.) The vector space $\mathbb{C}G$ contains

$$u = e - a + 2a^2 \text{ and } v = \tfrac{1}{2}e + 5a.$$

We have

$$u + v = \tfrac{3}{2}e + 4a + 2a^2, \tfrac{1}{3}u = \tfrac{1}{3}e - \tfrac{1}{3}a + \tfrac{2}{3}a^2.$$

Sometimes we write elements of FG in the form

$$\sum_{g \in G} \lambda_g g \quad (\lambda_g \in F).$$

Now, FG carries more structure than that of a vector space – we can use the product operation on G to define multiplication in FG as follows:

$$\left(\sum_{g \in G} \lambda_g g \right) \left(\sum_{h \in G} \mu_h h \right) = \sum_{g, h \in G} \lambda_g \mu_h (gh)$$

$$= \sum_{g \in G} \sum_{h \in G} (\lambda_h \mu_{h^{-1}g}) g$$

where all $\lambda_g, \mu_h \in F$.

6.2 Example

If $G = C_3$ and u, v are the elements of $\mathbb{C}G$ which appear in Example 6.1, then

$$uv = (e - a + 2a^2)(\tfrac{1}{2}e + 5a)$$

$$= \tfrac{1}{2}e + 5a - \tfrac{1}{2}a - 5a^2 + a^2 + 10a^3$$

$$= \tfrac{21}{2}e + \tfrac{9}{2}a - 4a^2.$$

6.3 Definition
The vector space FG, with multiplication defined by

$$\left(\sum_{g \in G} \lambda_g g\right)\left(\sum_{h \in G} \mu_h h\right) = \sum_{g,h \in G} \lambda_g \mu_h (gh)$$

$(\lambda_g, \mu_h \in F)$, is called the *group algebra* of G over F.

The group algebra FG contains an identity for multiplication, namely the element $1e$ (where 1 is the identity of F and e is the identity of G). We write this element simply as 1.

6.4 Proposition
Multiplication in FG satisfies the following properties, for all $r, s, t \in FG$ and $\lambda \in F$:

(1) $rs \in FG$;
(2) $r(st) = (rs)t$;
(3) $r1 = 1r = r$;
(4) $(\lambda r)s = \lambda(rs) = r(\lambda s)$;
(5) $(r + s)t = rt + st$;
(6) $r(s + t) = rs + rt$;
(7) $r0 = 0r = 0$.

Proof (1) It follows immediately from the definition of rs that $rs \in FG$.

(2) Let

$$r = \sum_{g \in G} \lambda_g g, \; s = \sum_{g \in G} \mu_g g, \; t = \sum_{g \in G} \nu_g g,$$

$(\lambda_g, \mu_g, \nu_g \in F)$. Then

$$(rs)t = \sum_{g,h,k \in G} \lambda_g \mu_h \nu_k (gh)k$$

$$= \sum_{g,h,k \in G} \lambda_g \mu_h \nu_k g(hk)$$

$$= r(st).$$

We leave the proofs of the other equations as easy exercises. ∎

In fact, any vector space equipped with a multiplication satisfying properties (1)–(7) of Proposition 6.4 is called an *algebra*. We shall be concerned only with group algebras, but it is worth pointing out that the axioms for an algebra mean that it is both a vector space and a ring.

The regular FG-module

We now use the group algebra to define an important FG-module.

Let $V = FG$, so that V is a vector space of dimension n over F, where $n = |G|$. For all $u, v \in V$, $\lambda \in F$ and $g, h \in G$, we have

$$vg \in V,$$

$$v(gh) = (vg)h,$$

$$v1 = v,$$

$$(\lambda v)g = \lambda(vg),$$

$$(u + v)g = ug + vg,$$

by parts (1), (2), (3), (4) and (5) of Proposition 6.4, respectively. Therefore V is an FG-module.

6.5 Definition
Let G be a finite group and F be $\mathbb{R}$ or $\mathbb{C}$. The vector space FG, with the natural multiplication vg ($v \in FG$, $g \in G$), is called the *regular FG*-module.

The representation $g \to [g]_{\mathscr{B}}$ obtained by taking $\mathscr{B}$ to be the natural basis of FG is called the *regular representation* of G over F.

Note that the regular FG-module has dimension equal to $|G|$.

6.6 Proposition
The regular FG-module is faithful.

Proof Suppose that $g \in G$ and $vg = v$ for all $v \in FG$. Then $1g = 1$, so $g = 1$, and the result follows. ∎

6.7 Example
Let $G = C_3 = \langle a: a^3 = e \rangle$. The elements of FG have the form

$$\lambda_1 e + \lambda_2 a + \lambda_3 a^2 \quad (\lambda_i \in F).$$

We have

$$(\lambda_1 e + \lambda_2 a + \lambda_3 a^2)e = \lambda_1 e + \lambda_2 a + \lambda_3 a^2,$$

$$(\lambda_1 e + \lambda_2 a + \lambda_3 a^2)a = \lambda_3 e + \lambda_1 a + \lambda_2 a^2,$$

$$(\lambda_1 e + \lambda_2 a + \lambda_3 a^2)a^2 = \lambda_2 e + \lambda_3 a + \lambda_1 a^2.$$

By taking matrices relative to the basis e, a, a^2 of FG, we obtain the regular representation of G:

$$e \rightarrow \begin{pmatrix} 1 & 0 & 0 \\ 0 & 1 & 0 \\ 0 & 0 & 1 \end{pmatrix}, \ a \rightarrow \begin{pmatrix} 0 & 1 & 0 \\ 0 & 0 & 1 \\ 1 & 0 & 0 \end{pmatrix}, \ a^2 \rightarrow \begin{pmatrix} 0 & 0 & 1 \\ 1 & 0 & 0 \\ 0 & 1 & 0 \end{pmatrix}.$$

FG acts on an *FG*-module

You will remember that an FG-module is a vector space over F, together with a multiplication vg for $v \in V$ and $g \in G$ (and the multiplication satisfies various axioms). Now, it is sometimes helpful to extend the definition of the multiplication so that we have an element vr of V for all elements r in the group algebra FG. This is done in the following natural way.

6.8 Definition
Suppose that V is an FG-module, and that $v \in V$ and $r \in FG$; say $r = \sum_{g \in G} \mu_g g \ (\mu_g \in F)$. Define vr by

$$vr = \sum_{g \in G} \mu_g(vg).$$

6.9 Examples
(1) Let V be the permutation module for S_4, as described in Example 4.9. If

$$r = \lambda(1 \ 2) + \mu(1 \ 3 \ 4) \quad (\lambda, \mu \in F)$$

then

$$v_1 r = \lambda v_1(1 \ 2) + \mu v_1(1 \ 3 \ 4) = \lambda v_2 + \mu v_3,$$

$$v_2 r = \lambda v_1 + \mu v_2,$$

$$(2v_1 + v_2)r = \lambda v_1 + (2\lambda + \mu)v_2 + 2\mu v_3.$$

(2) If V is the regular FG-module, then for all $v \in V$ and $r \in FG$, the element vr is simply the product of v and r as elements of the group algebra, given by Definition 6.3.

Compare the next result with Proposition 6.4.

6.10 Proposition
Suppose that V is an FG-module. Then the following properties hold for all $u, v \in V$, all $\lambda \in F$ and all $r, s \in FG$:

(1) $vr \in V$;
(2) $v(rs) = (vr)s$;
(3) $v1 = v$;
(4) $(\lambda v)r = \lambda(vr) = v(\lambda r)$;
(5) $(u + v)r = ur + vr$;
(6) $v(r + s) = vr + vs$;
(7) $v0 = 0r = 0$.

Proof All parts except (2) are straightforward, and we leave them to you. We shall give a proof of part (2), assuming the other parts.

Let $v \in V$, and let $r, s \in FG$ with

$$r = \sum_{g \in G} \lambda_g g, \ s = \sum_{h \in G} \mu_h h.$$

Then

$$v(rs) = v\left(\sum_{g,h} \lambda_g \mu_h(gh)\right)$$

$$= \sum_{g,h} \lambda_g \mu_h(v(gh)) \qquad \text{by (4) and (6)}$$

$$= \sum_{g,h} \lambda_g \mu_h((vg)h)$$

$$= \left(\sum_g \lambda_g(vg)\right)\left(\sum_h \mu_h h\right) \qquad \text{by (4), (5), (6)}$$

$$= (vr)s. \qquad\qquad\qquad\qquad\qquad \blacksquare$$

Summary of Chapter 6

1. The group algebra FG of G over F consists of all linear combinations of elements of G, and has a natural multiplication defined on it.

2. The vector space FG, with the natural multiplication vg ($v \in FG$, $g \in G$) is the regular FG-module.

3. The regular FG-module is faithful.

Exercises for Chapter 6

1. Suppose that $G = D_8 = \langle a, b: a^4 = b^2 = 1, b^{-1}ab = a^{-1} \rangle$.
   (a) Let x and y be the following elements of $\mathbb{C}G$:

   $$x = a + 2a^2, \; y = b + ab - a^2.$$

   Calculate xy, yx and x^2.
   (b) Let $z = b + a^2b$. Show that $zg = gz$ for all g in G. Deduce that $zr = rz$ for all r in $\mathbb{C}G$.

2. Work out matrices for the regular representation of $C_2 \times C_2$ over F.

3. Let $G = C_2$. For r and s in $\mathbb{C}G$, does $rs = 0$ imply that $r = 0$ or $s = 0$?

4. Assume that G is a finite group, say $G = \{g_1, \ldots, g_n\}$, and write c for the element $\sum_{i=1}^{n} g_i$ of $\mathbb{C}G$.
   (a) Prove that $ch = hc = c$ for all h in G.
   (b) Deduce that $c^2 = |G|c$.
   (c) Let $\vartheta: \mathbb{C}G \to \mathbb{C}G$ be the linear transformation sending v to vc for all v in $\mathbb{C}G$. What is the matrix $[\vartheta]_{\mathscr{B}}$, where $\mathscr{B}$ is the basis $g_1, \ldots, g_n$ *of* $\mathbb{C}G$?

5. If V is an FG-module, prove from the definition that

   $$0r = 0 \text{ for all } r \in FG, \text{ and}$$

   $$v0 = 0 \text{ for all } v \in V,$$

   where the symbol 0 is used for the zero elements of V and FG.
   Show that for every finite group G, with $|G| > 1$, there exists an FG-module V and elements $v \in V$, $r \in FG$ such that $vr = 0$, but neither v nor r is 0.

6. Suppose that $G = D_6 = \langle a, b : a^3 = b^2 = 1, b^{-1}ab = a^{-1} \rangle$, and let $\omega = e^{2\pi i/3}$. Prove that the 2-dimensional subspace W of $\mathbb{C}G$, defined by

$$W = \mathrm{sp}\,(1 + \omega^2 a + \omega a^2, \; b + \omega^2 ab + \omega a^2 b),$$

is an irreducible $\mathbb{C}G$-submodule of the regular $\mathbb{C}G$-module.

7
FG-homomorphisms

For groups and vector spaces, the 'structure-preserving' functions are, respectively, group homomorphisms and linear transformations. The analogous functions for *FG*-modules are called *FG*-homomorphisms, and we introduce these in this chapter.

FG-homomorphisms

7.1 Definition
Let V and W be *FG*-modules. A function $\vartheta\colon V \to W$ is said to be an *FG-homomorphism* if ϑ is a linear transformation and

$$(vg)\vartheta = (v\vartheta)g \quad \text{for all } v \in V,\, g \in G.$$

In other words, if ϑ sends v to w then it sends vg to wg.

Note that if G is a finite group and $\vartheta\colon V \to W$ is an *FG*-homomorphism, then for all $v \in V$ and $r = \sum_{g \in G} \lambda_g g \in FG$, we have

$$(vr)\vartheta = (v\vartheta)r$$

since

$$(vr)\vartheta = \sum_{g \in G} \lambda_g(vg)\vartheta = \sum_{g \in G} \lambda_g(v\vartheta)g = (v\vartheta)r.$$

The next result shows that *FG*-homomorphisms give rise to *FG*-submodules in a natural way.

7.2 Proposition
Let V and W be FG-modules and let $\vartheta\colon V \to W$ be an FG-homomorphism. Then $\operatorname{Ker}\vartheta$ is an FG-submodule of V, and $\operatorname{Im}\vartheta$ is an FG-submodule of W.

61

Proof First note that $\text{Ker}\,\vartheta$ is a subspace of V and $\text{Im}\,\vartheta$ is a subspace of W, since ϑ is a linear transformation.

Let $v \in \text{Ker}\,\vartheta$ and $g \in G$. Then

$$(vg)\vartheta = (v\vartheta)g = 0g = 0,$$

so $vg \in \text{Ker}\,\vartheta$. Therefore $\text{Ker}\,\vartheta$ is an FG-submodule of V.

Now let $w \in \text{Im}\,\vartheta$, so that $w = v\vartheta$ for some $v \in V$. For all $g \in G$,

$$wg = (v\vartheta)g = (vg)\vartheta \in \text{Im}\,\vartheta,$$

and so $\text{Im}\,\vartheta$ is an FG-submodule of W. $\blacksquare$

7.3 Examples

(1) If $\vartheta\colon V \to W$ is defined by $v\vartheta = 0$ for all $v \in V$, then ϑ is an FG-homomorphism, and $\text{Ker}\,\vartheta = V$, $\text{Im}\,\vartheta = \{0\}$.

(2) Let $\lambda \in F$, and define $\vartheta\colon V \to V$ by $v\vartheta = \lambda v$ for all $v \in V$. Then ϑ is an FG-homomorphism. Provided $\lambda \neq 0$, we have $\text{Ker}\,\vartheta = \{0\}$, $\text{Im}\,\vartheta = V$.

(3) Suppose that G is a subgroup of S_n. Let $V = \text{sp}\,(v_1, \ldots, v_n)$ be the permutation module for G over F (see Definition 4.10), and let $W = \text{sp}\,(w)$ be the trivial FG-module (see Definition 4.8). We construct an FG-homomorphism ϑ from V to W. Define

$$\vartheta\colon \sum_{i=1}^{n} \lambda_i v_i \to \left(\sum_{i=1}^{n} \lambda_i\right) w \quad (\lambda_i \in F).$$

Thus $v_i\vartheta = w$ for all i. Then ϑ is a linear transformation, and for all $v = \sum \lambda_i v_i \in V$ and all $g \in G$, we have

$$(vg)\vartheta = \left(\sum \lambda_i v_{ig}\right)\vartheta = \left(\sum \lambda_i\right) w,$$

and

$$(v\vartheta)g = \left(\sum \lambda_i\right) wg = \left(\sum \lambda_i\right) w.$$

Therefore ϑ is an FG-homomorphism. Here,

$$\text{Ker}\,\vartheta = \left\{\sum_{i=1}^{n} \lambda_i v_i \colon \sum_{i=1}^{n} \lambda_i = 0\right\},$$

$$\text{Im}\,\vartheta = W.$$

By Proposition 7.2, $\mathrm{Ker}\,\vartheta$ is an *FG*-submodule of the permutation module V.

Isomorphic *FG*-modules

7.4 Definition

Let V and W be *FG*-modules. We call a function $\vartheta\colon V \to W$ an *FG-isomorphism* if ϑ is an *FG*-homomorphism and ϑ is invertible. If there is such an *FG*-isomorphism, then we say that V and W are *isomorphic FG*-modules and write $V \cong W$.

In the next result, we check that if $V \cong W$ then $W \cong V$.

7.5 Proposition

If $\vartheta\colon V \to W$ is an FG-isomorphism, then the inverse $\vartheta^{-1}\colon W \to V$ is also an FG-isomorphism.

Proof Certainly ϑ^{-1} is an invertible linear transformation, so we need only show that ϑ^{-1} is an *FG*-homomorphism. For $w \in W$ and $g \in G$,

$$((w\vartheta^{-1})g)\vartheta = ((w\vartheta^{-1})\vartheta)g \quad \text{as } \vartheta \text{ is an } FG\text{-homomorphism}$$

$$= wg$$

$$= ((wg)\vartheta^{-1})\vartheta.$$

Hence $(w\vartheta^{-1})g = (wg)\vartheta^{-1}$, as required. ∎

Suppose that $\vartheta\colon V \to W$ is an *FG*-isomorphism. Then we may use ϑ and ϑ^{-1} to translate back and forth between the isomorphic *FG*-modules V and W, and prove that V and W share the same structural properties. We list some examples below:

(1) $\dim V = \dim W$ (since $v_1, \ldots, v_n$ is a basis of V if and only if $v_1\vartheta, \ldots, v_n\vartheta$ is a basis of W);

(2) V is irreducible if and only if W is irreducible (since X is an *FG*-submodule of V if and only if $X\vartheta$ is an *FG*-submodule of W);

(3) V contains a trivial *FG*-submodule if and only if W contains a trivial *FG*-submodule (since X is a trivial *FG*-submodule of V if and only if $X\vartheta$ is a trivial *FG*-submodule of W).

Just as we often regard isomorphic groups as being identical, we frequently disdain to distinguish between isomorphic FG-modules. For the moment, though, we continue simply to emphasize the similarity between isomorphic FG-modules. In the next result, we show that isomorphic FG-modules correspond to equivalent representations.

7.6 Theorem

Suppose that V is an FG-module with basis $\mathscr{B}$, and W is an FG-module with basis $\mathscr{B}'$. Then V and W are isomorphic if and only if the representations

$$\rho: g \rightarrow [g]_{\mathscr{B}} \text{ and } \sigma: g \rightarrow [g]_{\mathscr{B}'}$$

are equivalent.

Proof We first establish the following fact:

(7.7)   The FG-modules V and W are isomorphic if and only if there are a basis $\mathscr{B}_1$ of V and a basis $\mathscr{B}_2$ of W such that

$$[g]_{\mathscr{B}_1} = [g]_{\mathscr{B}_2} \quad \text{for all } g \in G.$$

To see this, suppose first that ϑ is an FG-isomorphism from V to W, and let $v_1, \ldots, v_n$ be a basis $\mathscr{B}_1$ of V; then $v_1\vartheta, \ldots, v_n\vartheta$ is a basis $\mathscr{B}_2$ of W. Let $g \in G$. Since $(v_ig)\vartheta = (v_i\vartheta)g$ for each i, it follows that $[g]_{\mathscr{B}_1} = [g]_{\mathscr{B}_2}$.

Conversely, suppose that $v_1, \ldots, v_n$ is a basis $\mathscr{B}_1$ of V and $w_1, \ldots, w_n$ is a basis $\mathscr{B}_2$ of W such that $[g]_{\mathscr{B}_1} = [g]_{\mathscr{B}_2}$ for all $g \in G$. Let ϑ be the invertible linear transformation from V to W for which $v_i\vartheta = w_i$ for all i. Let $g \in G$. Since $[g]_{\mathscr{B}_1} = [g]_{\mathscr{B}_2}$, we deduce that $(v_ig)\vartheta = (v_i\vartheta)g$ for all i, and hence ϑ is an FG-isomorphism. This completes the proof of (7.7).

Now assume that V and W are isomorphic FG-modules. By (7.7), there are a basis $\mathscr{B}_1$ of V and a basis $\mathscr{B}_2$ of W such that $[g]_{\mathscr{B}_1} = [g]_{\mathscr{B}_2}$ for all $g \in G$. Define a representation ϕ of G by $\phi: g \rightarrow [g]_{\mathscr{B}_1}$. Then by Theorem 4.12(1), ϕ is equivalent to both ρ and σ. Hence ρ and σ are equivalent.

Conversely, suppose that ρ and σ are equivalent. Then by Theorem 4.12(2), there is a basis $\mathscr{B}''$ of V such that $g\sigma = [g]_{\mathscr{B}''}$ for all $g \in G$;

that is, $[g]_{\mathscr{B}'} = [g]_{\mathscr{B}''}$ for all $g \in G$. Therefore V and W are isomorphic FG-modules, by (7.7). ∎

7.8 Example

Let $G = \langle a: a^3 = 1 \rangle$, a cyclic group of order 3, and let W denote the regular FG-module. Then $1, a, a^2$ is a basis of W; call it $\mathscr{B}'$. We have

$$[1]_{\mathscr{B}'} = \begin{pmatrix} 1 & 0 & 0 \\ 0 & 1 & 0 \\ 0 & 0 & 1 \end{pmatrix}, \quad [a]_{\mathscr{B}'} = \begin{pmatrix} 0 & 1 & 0 \\ 0 & 0 & 1 \\ 1 & 0 & 0 \end{pmatrix},$$

$$[a^2]_{\mathscr{B}'} = \begin{pmatrix} 0 & 0 & 1 \\ 1 & 0 & 0 \\ 0 & 1 & 0 \end{pmatrix}.$$

Compare the FG-module V defined in Example 4.11, with basis v_1, v_2, v_3 such that

$$v_1 a = v_2, \; v_2 a = v_3, \; v_3 a = v_1.$$

Writing $\mathscr{B}$ for the basis v_1, v_2, v_3 of V, we have

$$[g]_{\mathscr{B}} = [g]_{\mathscr{B}'} \quad \text{for all } g \in G.$$

According to (7.7), the FG-modules V and W are therefore isomorphic. Indeed, the function

$$\vartheta: \lambda_1 v_1 + \lambda_2 v_2 + \lambda_3 v_3 \rightarrow \lambda_1 1 + \lambda_2 a + \lambda_3 a^2 \quad (\lambda_i \in F)$$

is an FG-isomorphism from V to W.

7.9 Example

Let $G = D_8 = \langle a, b: a^4 = b^2 = 1, b^{-1}ab = a^{-1} \rangle$. In Example 3.4(1) we encountered two equivalent representations ρ and σ of G, where

$$a\rho = \begin{pmatrix} 0 & 1 \\ -1 & 0 \end{pmatrix}, \; b\rho = \begin{pmatrix} 1 & 0 \\ 0 & -1 \end{pmatrix}$$

and

$$a\sigma = \begin{pmatrix} i & 0 \\ 0 & -i \end{pmatrix}, \; b\sigma = \begin{pmatrix} 0 & 1 \\ 1 & 0 \end{pmatrix}.$$

Let V be the $\mathbb{C}G$-module with basis v_1, v_2 for which

$$v_1 a = v_2, \ v_1 b = v_1,$$

$$v_2 a = -v_1, \ v_2 b = -v_2$$

(see Example 4.5(1)), and, in a similar way, let W be the $\mathbb{C}G$-module with basis w_1, w_2 for which

$$w_1 a = i w_1, \qquad w_1 b = w_2,$$
$$w_2 a = -i w_2, \qquad w_2 b = w_1$$

Thus, if we write $\mathscr{B}$ for the basis v_1, v_2 of V and $\mathscr{B}'$ for the basis w_1, w_2 of W, then for all $g \in G$ we have

$$\rho: g \rightarrow [g]_{\mathscr{B}} \text{ and } \sigma: g \rightarrow [g]_{\mathscr{B}'}.$$

According to Theorem 7.6, the $\mathbb{C}G$-modules V and W are isomorphic, since ρ and σ are equivalent. To verify this directly, let $\vartheta: V \rightarrow W$ be the invertible linear transformation such that

$$\vartheta: v_1 \rightarrow w_1 + w_2,$$

$$v_2 \rightarrow i w_1 - i w_2.$$

Then $(v_j a)\vartheta = (v_j \vartheta)a$ and $(v_j b)\vartheta = (v_j \vartheta)b$ for $j = 1$, 2, and hence ϑ is a $\mathbb{C}G$-isomorphism from V to W. (Compare Example 3.4(1).)

Direct sums

We conclude the chapter with a discussion of direct sums of FG-modules, and we show that these give rise to FG-homomorphisms.

Let V be an FG-module, and suppose that

$$V = U \oplus W,$$

where U and W are FG-submodules of V. Let $u_1, \ldots, u_m$ be a basis $\mathscr{B}_1$ of U, and $w_1, \ldots, w_n$ be a basis $\mathscr{B}_2$ of W. Then by (2.9), $u_1, \ldots, u_m, w_1, \ldots, w_n$ is a basis $\mathscr{B}$ of V, and for $g \in G$,

$$[g]_{\mathscr{B}} = \left(\begin{array}{c|c} [g]_{\mathscr{B}_1} & 0 \\ \hline 0 & [g]_{\mathscr{B}_2} \end{array} \right).$$

More generally, if $V = U_1 \oplus \ldots \oplus U_r$, a direct sum of FG-submodules U_i, and $\mathscr{B}_i$ is a basis of U_i, then we can amalgamate $\mathscr{B}_1, \ldots, \mathscr{B}_r$ to obtain a basis $\mathscr{B}$ of V, and for $g \in G$,

$$(7.10) \qquad [g]_{\mathscr{B}} = \begin{pmatrix} [g]_{\mathscr{B}_1} & & 0 \\ & \ddots & \\ 0 & & [g]_{\mathscr{B}_r} \end{pmatrix}.$$

The next result shows that direct sums give rise naturally to *FG*-homomorphisms.

7.11 Proposition

Let V be an FG-module, and suppose that

$$V = U_1 \oplus \ldots \oplus U_r$$

where each U_i is an FG-submodule of V. For $v \in V$, we have $v = u_1 + \ldots + u_r$ for unique vectors $u_i \in U_i$, and we define $\pi_i : V \to V$ $(1 \leqslant i \leqslant r)$ by setting

$$v\pi_i = u_i.$$

Then each π_i is an FG-homomorphism, and is also a projection of V.

Proof Clearly π_i is a linear transformation; and π_i is an *FG*-homomorphism, since for $v \in V$ with $v = u_1 + \ldots + u_r$ ($u_j \in U_j$ for all j), and $g \in G$, we have

$$(vg)\pi_i = (u_1 g + \ldots + u_r g)\pi_i = u_i g = (v\pi_i)g.$$

Also,

$$v\pi_i^2 = u_i \pi_i = u_i = v\pi_i,$$

so $\pi_i^2 = \pi_i$. Thus π_i is a projection (see Definition 2.30). ∎

We now present a technical result concerning sums of irreducible *FG*-modules which will be used at a later stage.

7.12 Proposition

Let V be an FG-module, and suppose that

$$V = U_1 + \ldots + U_r,$$

where each U_i is an irreducible FG-submodule of V. Then V is a direct sum of some of the FG-submodules U_i.

Proof The idea is to choose as many as we can of the FG-submodules $U_1, \ldots, U_r$ so that the sum of our chosen FG-submodules is direct. To this end, choose a subset $Y = \{W_1, \ldots, W_s\}$ of $\{U_1, \ldots, U_r\}$ which has the properties that

$$W_1 + \ldots + W_s \text{ is direct (i.e. equal to } W_1 \oplus \ldots \oplus W_s), \text{ but}$$

$$W_1 + \ldots + W_s + U_i \text{ is not direct, if } U_i \notin Y.$$

Let

$$W = W_1 + \ldots + W_s.$$

We claim that $U_i \subseteq W$ for all i. If $U_i \in Y$ this is clear, so assume that $U_i \notin Y$. Then $W + U_i$ is not a direct sum, so $W \cap U_i \neq \{0\}$. But $W \cap U_i$ is an FG-submodule of U_i, and U_i is irreducible; therefore $W \cap U_i = U_i$, and so $U_i \subseteq W$, as claimed.

Since $U_i \subseteq W$ for all i with $1 \leqslant i \leqslant r$, we have $V = W = W_1 \oplus \ldots \oplus W_s$, as required. ∎

Finally, we remark that if $V_1, \ldots, V_r$ are FG-modules, then we can make the external direct sum $V_1 \oplus \ldots \oplus V_r$ (see Chapter 2) into an FG-module by defining

$$(v_1, \ldots, v_r)g = (v_1 g, \ldots, v_r g)$$

for all $v_i \in V_i$ ($1 \leqslant i \leqslant r$) and all $g \in G$.

Summary of Chapter 7

1. If V and W are FG-modules and $\vartheta\colon V \to W$ is a linear transformation which satisfies

$$(vg)\vartheta = (v\vartheta)g$$

for all $v \in V$, $g \in G$, then ϑ is an FG-homomorphism.

2. Kernels and images of FG-homomorphisms are FG-modules.

3. Isomorphic FG-modules correspond to equivalent representations.

Exercises for Chapter 7

1. Let U, V and W be FG-modules, and let $\vartheta\colon U \to V$ and $\phi\colon V \to W$ be FG-homomorphisms. Prove that $\vartheta\phi\colon U \to W$ is an FG-homomorphism.

2. Let G be the subgroup of S_5 which is generated by $(1\,2\,3\,4\,5)$. Prove that the permutation module for G over F is isomorphic to the regular FG-module.

3. Assume that V is an FG-module. Prove that the subset

$$V_0 = \{v \in V \colon vg = v \text{ for all } g \in G\}$$

is an FG-submodule of V. Show that the function

$$\vartheta \colon v \to \sum_{g \in G} vg \quad (v \in V)$$

is an FG-homomorphism from V to V_0. Is it necessarily surjective?

4. Suppose that V and W are isomorphic FG-modules. Define the FG-submodules V_0 and W_0 of V and W as in Exercise 3. Prove that V_0 and W_0 are isomorphic FG-modules.

5. Let G be the subgroup of S_4 which is generated by $(1\,2)$ and $(3\,4)$. Is the permutation module for G over F isomorphic to the regular FG-module?

6. Let $G = C_2 = \langle x \colon x^2 = 1 \rangle$.
   (a) Show that the function

   $$\vartheta \colon \alpha 1 + \beta x \to (\alpha - \beta)(1 - x) \quad (\alpha, \beta \in F)$$

   is an FG-homomorphism from the regular FG-module to itself.
   (b) Prove that $\vartheta^2 = 2\vartheta$.
   (c) Find a basis $\mathscr{B}$ of FG such that

   $$[\vartheta]_{\mathscr{B}} = \begin{pmatrix} 2 & 0 \\ 0 & 0 \end{pmatrix}.$$

8
Maschke's Theorem

We now come to our first major result in representation theory, namely Maschke's Theorem. A consequence of this theorem is that every FG-module is a direct sum of irreducible FG-submodules, where as usual $F = \mathbb{R}$ or $\mathbb{C}$. (The assumption on F is important – see Example 8.2(2) below.) This essentially reduces representation theory to the study of irreducible FG-modules.

Maschke's Theorem

8.1 Maschke's Theorem
Let G be a finite group, let F be $\mathbb{R}$ or $\mathbb{C}$, and let V be an FG-module. If U is an FG-submodule of V, then there is an FG-submodule W of V such that

$$V = U \oplus W.$$

Before proving Maschke's Theorem, we illustrate it with some examples.

8.2 Examples
(1) Let $G = S_3$ and let $V = \text{sp}(v_1, v_2, v_3)$ be the permutation module for G over F (see Definition 4.10). Put

$$u = v_1 + v_2 + v_3 \text{ and } U = \text{sp}(u).$$

Then U is an FG-submodule of V, since $ug = u$ for all $g \in G$.

There are many *subspaces* W of V such that $V = U \oplus W$, for instance $\text{sp}(v_2, v_3)$ and $\text{sp}(v_1, v_2 - 2v_3)$. But there is, in fact, only one FG-*submodule* W of V with $V = U \oplus W$. We shall find this W in an example after proving Maschke's Theorem (but you may like to look for it yourself now).

(2) The conclusion of Maschke's Theorem can fail if F is not $\mathbb{R}$ or $\mathbb{C}$. For example, let p be a prime number, let $G = C_p = \langle a: a^p = 1 \rangle$, and take F to be the field of integers modulo p. Check that the function

$$a^j \to \begin{pmatrix} 1 & 0 \\ j & 1 \end{pmatrix} \quad (j = 0, 1, \ldots, p-1)$$

is a representation from G to $GL(2, F)$. The corresponding FG-module is $V = \text{sp}(v_1, v_2)$, where, for $0 \leqslant j \leqslant p-1$,

$$v_1 a^j = v_1,$$

$$v_2 a^j = jv_1 + v_2.$$

Clearly, $U = \text{sp}(v_1)$ is an FG-submodule of V. But there is no FG-submodule W such that $V = U \oplus W$, since U is the only 1-dimensional FG-submodule of V, as can easily be seen.

Proof of Maschke's Theorem 8.1 We are given U, an FG-submodule of the FG-module V. Choose any *subspace* W_0 of V such that

$$V = U \oplus W_0.$$

(There are many choices for W_0 – simply take a basis $v_1, \ldots, v_m$ of U, extend it to a basis $v_1, \ldots, v_n$ of V, and let $W_0 = \text{sp}(v_{m+1}, \ldots, v_n)$.)

For $v \in V$, we have $v = u + w$ for unique vectors $u \in U$ and $w \in W_0$, and we define $\phi: V \to V$ by setting $v\phi = u$. By Proposition 2.29, ϕ is a projection of V with kernel W_0 and image U.

We aim to modify the projection ϕ to create an FG-homomorphism from V to V with image U. To this end, define $\vartheta: V \to V$ by

(8.3) $$v\vartheta = \frac{1}{|G|} \sum_{g \in G} v g \phi g^{-1} \quad (v \in V).$$

It is clear that ϑ is an endomorphism of V and $\text{Im}\,\vartheta \subseteq U$.

We show first that ϑ is an FG-homomorphism. For $v \in V$ and $x \in G$,

$$(vx)\vartheta = \frac{1}{|G|} \sum_{g \in G} (vx) g \phi g^{-1}.$$

As g runs over the elements of G, so does $h = xg$. Hence

$$(vx)\vartheta = \frac{1}{|G|}\sum_{h \in G} vh\phi h^{-1}x$$

$$= \left(\frac{1}{|G|}\sum_{h \in G} vh\phi h^{-1}\right)x$$

$$= (v\vartheta)x.$$

Thus ϑ is an FG-homomorphism.

Next, we prove that $\vartheta^2 = \vartheta$. First note that for $u \in U$, $g \in G$, we have $ug \in U$, and so $(ug)\phi = ug$. Using this,

$$(8.4) \qquad u\vartheta = \frac{1}{|G|}\sum_{g \in G} ug\phi g^{-1} = \frac{1}{|G|}\sum_{g \in G}(ug)g^{-1} = \frac{1}{|G|}\sum_{g \in G} u = u.$$

Now let $v \in V$. Then $v\vartheta \in U$, so by (8.4) we have $(v\vartheta)\vartheta = v\vartheta$. Consequently $\vartheta^2 = \vartheta$, as claimed.

We have now established that $\vartheta: V \to V$ is a projection and an FG-homomorphism. Moreover, (8.4) shows that $\operatorname{Im}\vartheta = U$. Let $W = \operatorname{Ker}\vartheta$. Then W is an FG-submodule of V by Proposition 7.2, and $V = U \oplus W$ by Proposition 2.32.

This completes the proof of Maschke's Theorem. ∎

8.5 Example

Let $G = S_3$ and let $V = \operatorname{sp}(v_1, v_2, v_3)$ be the permutation module, with submodule $U = \operatorname{sp}(v_1 + v_2 + v_3)$, as in Example 8.2(1). We use the proof of Maschke's Theorem to find an FG-submodule W of V such that $V = U \oplus W$.

First, let $W_0 = \operatorname{sp}(v_1, v_2)$. Then $V = U \oplus W_0$ (but of course W_0 is not an FG-submodule). The projection ϕ onto U is given by

$$\phi: v_1 \to 0, \ v_2 \to 0, \ v_3 \to v_1 + v_2 + v_3.$$

Check now that the FG-homomorphism ϑ given by (8.3) is

$$\vartheta: v_i \to \tfrac{1}{3}(v_1 + v_2 + v_3) \quad (i = 1, 2, 3).$$

The required FG-submodule W is then $\operatorname{Ker}\vartheta$, so

$$W = \operatorname{sp}(v_1 - v_2, v_2 - v_3).$$

(In fact, $W = \{\sum\lambda_i v_i: \sum\lambda_i = 0\}$, the FG-submodule constructed in Example 7.3(3).)

Note that if $\mathscr{B}$ is the basis $v_1 + v_2 + v_3$, v_1, v_2 of V, then for all $g \in G$, the matrix $[g]_{\mathscr{B}}$ has the form

$$[g]_{\mathscr{B}} = \begin{pmatrix} \blacksquare & 0 & 0 \\ \blacksquare & \blacksquare & \blacksquare \\ \blacksquare & \blacksquare & \blacksquare \end{pmatrix}.$$

The zeros reflect the fact that U is an FG-submodule of V (see (5.4)). If instead we use $v_1 + v_2 + v_3$, $v_1 - v_2$, $v_2 - v_3$ as a basis $\mathscr{B}'$, then we get

$$[g]_{\mathscr{B}'} = \begin{pmatrix} \blacksquare & 0 & 0 \\ 0 & \blacksquare & \blacksquare \\ 0 & \blacksquare & \blacksquare \end{pmatrix},$$

because $\text{sp}\,(v_1 - v_2, v_2 - v_3)$ is also an FG-submodule of V.

This example illustrates the matrix version of Maschke's Theorem: for an arbitrary finite group G, if we can choose a basis $\mathscr{B}$ of an FG-module V such that $[g]_{\mathscr{B}}$ has the form

$$\begin{pmatrix} * & 0 \\ \hline * & * \end{pmatrix}$$

for all $g \in G$ (see (5.4)), then we can find a basis $\mathscr{B}'$ such that $[g]_{\mathscr{B}'}$ has the form

$$\begin{pmatrix} * & 0 \\ \hline 0 & * \end{pmatrix}$$

for all $g \in G$.

To put this another way, suppose that ρ is a reducible representation of a finite group G over F of degree n. Then we know that ρ is equivalent to a representation of the form

$$g \to \left(\begin{array}{c|c} X_g & 0 \\ \hline Y_g & Z_g \end{array} \right) \quad (g \in G),$$

for some matrices X_g, Y_g, Z_g, where X_g is $k \times k$ with $0 < k < n$.

Maschke's Theorem asserts further that ρ is equivalent to a representation of the form

$$g \rightarrow \left(\begin{array}{c|c} A_g & 0 \\ \hline 0 & B_g \end{array} \right),$$

where A_g is also a $k \times k$ matrix.

Consequences of Maschke's Theorem

We now use Maschke's Theorem to show that every non-zero FG-module is a direct sum of irreducible FG-submodules. (By an irreducible FG-submodule, we simply mean an FG-submodule which is an irreducible FG-module.)

8.6 Definition
An FG-module V is said to be *completely reducible* if $V = U_1 \oplus \ldots \oplus U_r$, where each U_i is an irreducible FG-submodule of V.

8.7 Theorem
If G is a finite group and $F = \mathbb{R}$ or $\mathbb{C}$, then every non-zero FG-module is completely reducible.

Proof Let V be a non-zero FG-module. The proof goes by induction on dim V. The result is true if dim $V = 1$, since V is irreducible in this case.

If V is irreducible then the result holds, so suppose that V is reducible. Then V has an FG-submodule U not equal to $\{0\}$ or V. By Maschke's Theorem, there is an FG-submodule W such that $V = U \oplus W$. Since dim $U <$ dim V and dim $W <$ dim V, we have, by induction,

$$U = U_1 \oplus \ldots \oplus U_r, \quad W = W_1 \oplus \ldots \oplus W_s,$$

where each U_i and W_j is an irreducible FG-module. Then by (2.10),

$$V = U_1 \oplus \ldots \oplus U_r \oplus W_1 \oplus \ldots \oplus W_s,$$

a direct sum of irreducible FG-modules.   ∎

Another useful consequence of Maschke's Theorem is the next proposition.

8.8 Proposition
Let V be an FG-module, where $F = \mathbb{R}$ or $\mathbb{C}$ and G is a finite group. Suppose that U is an FG-submodule of V. Then there exists a surjective FG-homomorphism from V onto U.

Proof By Maschke's Theorem, there is an FG-submodule W of V such that $V = U \oplus W$. Then the function $\pi: V \to U$ which is defined by

$$\pi: u + w \to u \quad (u \in U, w \in W)$$

is an FG-homomorphism onto U, by Proposition 7.11. ∎

Theorem 8.7 tells us that every non-zero FG-module is a direct sum of irreducible FG-modules. Thus, in order to understand FG-modules, we may concentrate upon the irreducible FG-modules. We begin our study of these in the next chapter.

Summary of Chapter 8

Assume that G is a finite group and $F = \mathbb{R}$ or $\mathbb{C}$.

1. Maschke's Theorem says that for every FG-submodule U of an FG-module V, there is an FG-submodule W with

$$V = U \oplus W.$$

2. Every non-zero FG-module V is a direct sum of irreducible FG-modules:

$$V = U_1 \oplus \ldots \oplus U_r.$$

Exercises for Chapter 8

1. Let $G = \langle x: x^3 = 1 \rangle \cong C_3$, and let V be the 2-dimensional $\mathbb{C}G$-module with basis v_1, v_2, where

$$v_1 x = v_2, \, v_2 x = -v_1 - v_2.$$

(This *is* a $\mathbb{C}G$-module, by Exercise 3.2.)
Express V as a direct sum of irreducible $\mathbb{C}G$-submodules.

2. If $G = C_2 \times C_2$, express the group algebra $\mathbb{R}G$ as a direct sum of 1-dimensional $\mathbb{R}G$-submodules.

3. Find a group G, a $\mathbb{C}G$-module V and a $\mathbb{C}G$-homomorphism $\vartheta\colon V \to V$ such that $V \neq \operatorname{Ker}\vartheta \oplus \operatorname{Im}\vartheta$.

4. Let G be a finite group and let $\rho\colon G \to \mathrm{GL}\,(2,\,\mathbb{C})$ be a representation of G. Suppose that there are elements $g,\ h$ in G such that the matrices $g\rho$ and $h\rho$ do not commute. Prove that ρ is irreducible.

   (You may care to revisit Example 5.5(2) and Exercises 5.1, 5.3, 5.4, 6.6 in the light of this result.)

5. Suppose that G is the infinite group

$$\left\{ \begin{pmatrix} 1 & 0 \\ n & 1 \end{pmatrix}\colon n \in \mathbb{Z} \right\}$$

   and let V be the $\mathbb{C}G$-module $\mathbb{C}^2$, with the natural multiplication by elements of G (so that for $v \in V,\ g \in G$, the vector vg is just the product of the row vector v with the matrix g).
      Show that V is not completely reducible.

   (This shows that Maschke's Theorem fails for infinite groups – compare Example 8.2(2).)

6. *An alternative proof of Maschke's Theorem for $\mathbb{C}G$-modules.*

   Let V be a $\mathbb{C}G$-module with basis $v_1, \ldots, v_n$ and suppose that U is a $\mathbb{C}G$-submodule of V. Define a complex inner product $(\ ,\ )$ on V as follows (see (14.2) for the definition of a complex inner product): for $\lambda_i,\ \mu_j \in \mathbb{C}$,

$$\left(\sum_{i=1}^{n} \lambda_i v_i,\ \sum_{j=1}^{n} \mu_j v_j \right) = \sum_{i=1}^{n} \lambda_i \bar{\mu}_i.$$

   Define another complex inner product $[\ ,\ ]$ on V by

$$[u,\,v] = \sum_{x\in G}(ux,\,vx) \quad (u,\,v \in V).$$

   (1) Verify that $[\ ,\ ]$ is a complex inner product, which satisfies
   $$[ug,\,vg] = [u,\,v] \quad \text{for all } u,\,v \in V \text{ and } g \in G.$$

   (2) Suppose that U is a $\mathbb{C}G$-submodule of V, and define
   $$U^{\perp} = \{v \in V\colon [u,\,v] = 0 \text{ for all } u \in U\}.$$

   Show that $U^{\perp}$ is a $\mathbb{C}G$-submodule of V.

(3) Deduce Maschke's Theorem. (Hint: it is a standard property of complex inner products that $V = U \oplus U^\perp$ for all subspaces U of V.)

7. Prove that for every finite simple group G, there exists a faithful irreducible $\mathbb{C}G$-module.

9

Schur's Lemma

Schur's Lemma is a basic result concerning irreducible modules. Though simple in both statement and proof, Schur's Lemma is fundamental to representation theory, and we give an immediate application by determining all the irreducible representations of finite abelian groups.

Schur's Lemma concerns $\mathbb{C}G$-modules rather than $\mathbb{R}G$-modules, and since much of the ensuing theory depends on it, we shall deal with $\mathbb{C}G$-modules for the remainder of the book (except in Chapter 23).

Throughout, G denotes a finite group.

Schur's Lemma

9.1 Schur's Lemma

Let V and W be irreducible $\mathbb{C}G$-modules.

(1) *If $\vartheta: V \to W$ is a $\mathbb{C}G$-homomorphism, then either ϑ is a $\mathbb{C}G$-isomorphism, or $v\vartheta = 0$ for all $v \in V$.*

(2) *If $\vartheta: V \to V$ is a $\mathbb{C}G$-isomorphism, then ϑ is a scalar multiple of the identity endomorphism 1_V.*

Proof (1) Suppose that $v\vartheta \neq 0$ for some $v \in V$. Then $\operatorname{Im} \vartheta \neq \{0\}$. As $\operatorname{Im} \vartheta$ is a $\mathbb{C}G$-submodule of W by Proposition 7.2, and W is irreducible, we have $\operatorname{Im} \vartheta = W$. Also by Proposition 7.2, $\operatorname{Ker} \vartheta$ is a $\mathbb{C}G$-submodule of V; as $\operatorname{Ker} \vartheta \neq V$ and V is irreducible, $\operatorname{Ker} \vartheta = \{0\}$. Thus ϑ is invertible, and hence is a $\mathbb{C}G$-isomorphism.

(2) By (2.26), the endomorphism ϑ has an eigenvalue $\lambda \in \mathbb{C}$, and so $\operatorname{Ker}(\vartheta - \lambda 1_V) \neq \{0\}$. Thus $\operatorname{Ker}(\vartheta - \lambda 1_V)$ is a non-zero $\mathbb{C}G$-submodule

78

of V. Since V is irreducible, $\mathrm{Ker}\,(\vartheta - \lambda 1_V) = V$. Therefore

$$v(\vartheta - \lambda 1_V) = 0 \quad \text{for all } v \in V.$$

That is, $\vartheta = \lambda 1_V$, as required. ∎

Part (2) of Schur's Lemma has the following converse.

9.2 Proposition
Let V be a non-zero $\mathbb{C}G$-module, and suppose that every $\mathbb{C}G$-homo-morphism from V to V is a scalar multiple of 1_V. Then V is irreducible.

Proof Suppose that V is reducible, so that V has a $\mathbb{C}G$-submodule U not equal to $\{0\}$ or V. By Maschke's Theorem, there is a $\mathbb{C}G$-submodule W of V such that

$$V = U \oplus W.$$

Then the projection $\pi: V \to V$ defined by $(u + w)\pi = u$ for all $u \in U$, $w \in W$ is a $\mathbb{C}G$-homomorphism (see Proposition 7.11), and is not a scalar multiple of 1_V, which is a contradiction. Hence V is irreducible. ∎

We next interpret Schur's Lemma and its converse in terms of representations.

9.3 Corollary
Let $\rho: G \to \mathrm{GL}\,(n, \mathbb{C})$ be a representation of G. Then ρ is irreducible if and only if every $n \times n$ matrix A which satisfies

$$(g\rho)A = A(g\rho) \quad \text{for all } g \in G$$

has the form $A = \lambda I_n$ with $\lambda \in \mathbb{C}$.

Proof As in Theorem 4.4(1), regard $\mathbb{C}^n$ as a $\mathbb{C}G$-module by defining $vg = v(g\rho)$ for all $v \in \mathbb{C}^n$, $g \in G$.

Let A be an $n \times n$ matrix over $\mathbb{C}$. The endomorphism $v \to vA$ of $\mathbb{C}^n$ is a $\mathbb{C}G$-homomorphism if and only if

$$(vg)A = (vA)g \quad \text{for all } v \in \mathbb{C}^n, g \in G;$$

that is, if and only if

$$(g\rho)A = A(g\rho) \quad \text{for all } g \in G.$$

The result now follows from Schur's Lemma 9.1 and Proposition 9.2.

∎

9.4 Examples

(1) Let $G = C_3 = \langle a: a^3 = 1 \rangle$, and let $\rho: G \to GL(2, \mathbb{C})$ be the representation for which

$$a\rho = \begin{pmatrix} 0 & 1 \\ -1 & -1 \end{pmatrix}$$

(see Exercise 3.2). Since the matrix

$$\begin{pmatrix} 0 & 1 \\ -1 & -1 \end{pmatrix}$$

commutes with all $g\rho$ $(g \in G)$, Corollary 9.3 implies that ρ is reducible.

(2) Let $G = D_{10} = \langle a, b: a^5 = b^2 = 1, b^{-1}ab = a^{-1} \rangle$, and let $\omega = e^{2\pi i/5}$. Check that there is a representation $\rho: G \to GL(2, \mathbb{C})$ for which

$$a\rho = \begin{pmatrix} \omega & 0 \\ 0 & \omega^{-1} \end{pmatrix}, \quad b\rho = \begin{pmatrix} 0 & 1 \\ 1 & 0 \end{pmatrix}.$$

Assume that the matrix

$$A = \begin{pmatrix} \alpha & \beta \\ \gamma & \delta \end{pmatrix}$$

commutes with both $a\rho$ and $b\rho$. The fact that $(a\rho)A = A(a\rho)$ forces $\beta = \gamma = 0$; and then $(b\rho)A = A(b\rho)$ gives $\alpha = \delta$. Hence

$$A = \begin{pmatrix} \alpha & 0 \\ 0 & \alpha \end{pmatrix} = \alpha I.$$

Consequently ρ is irreducible, by Corollary 9.3.

Representation theory of finite abelian groups

Let G be a finite abelian group, and let V be an irreducible $\mathbb{C}G$-module. Pick $x \in G$. Since G is abelian,

$$vgx = vxg \quad \text{for all } g \in G,$$

and hence the endomorphism $v \to vx$ of V is a $\mathbb{C}G$-homomorphism. By Schur's Lemma 9.1(2), this endomorphism is a scalar multiple of the identity 1_V, say $\lambda_x 1_V$. Thus

$$vx = \lambda_x v \quad \text{for all } v \in V.$$

This implies that every subspace of V is a $\mathbb{C}G$-submodule. As V is irreducible, we deduce that dim $V = 1$. Thus we have proved

9.5 Proposition
If G is a finite abelian group, then every irreducible $\mathbb{C}G$-module has dimension 1.

The next result is a major structure theorem for finite abelian groups. We shall not prove it here, but refer you to Chapter 9 of the book of J. B. Fraleigh listed in the Bibliography.

9.6 Theorem
Every finite abelian group is isomorphic to a direct product of cyclic groups.

We shall determine the irreducible representations of all direct products

$$C_{n_1} \times C_{n_2} \times \ldots \times C_{n_r}$$

where $n_1, \ldots, n_r$ are positive integers. By Theorem 9.6, this covers the irreducible representations of all finite abelian groups.

Let $G = C_{n_1} \times \ldots \times C_{n_r}$, and for $1 \leqslant i \leqslant r$, let c_i be a generator for C_{n_i}. Write

$$g_i = (1, \ldots, c_i, \ldots, 1) \quad (c_i \text{ in } i\text{th position}).$$

Then

$$G = \langle g_1, \ldots, g_r \rangle, \text{ with } g_i^{n_i} = 1 \text{ and } g_i g_j = g_j g_i \text{ for all } i, j.$$

Now let $\rho: G \to \mathrm{GL}(n, \mathbb{C})$ be an irreducible representation of G

over $\mathbb{C}$. Then $n = 1$ by Proposition 9.5, so for $1 \leqslant i \leqslant r$, there exists $\lambda_i \in \mathbb{C}$ such that

$$g_i \rho = (\lambda_i)$$

(where of course (λ_i) is a 1×1 matrix). As g_i has order n_i, we have $\lambda_i^{n_i} = 1$; that is, λ_i is an n_ith root of unity. Also, the values $\lambda_1, \ldots, \lambda_r$ determine ρ, since for $g \in G$, we have $g = g_1^{i_1} \ldots g_r^{i_r}$ for some integers $i_1, \ldots, i_r$, and then

(9.7)
$$g\rho = (g_1^{i_1} \ldots g_r^{i_r})\rho = (\lambda_1^{i_1} \ldots \lambda_r^{i_r}).$$

For a representation ρ of G satisfying (9.7) for all $i_1, \ldots, i_r$, write

$$\rho = \rho_{\lambda_1, \ldots, \lambda_r}.$$

Conversely, given any n_ith roots of unity λ_i $(1 \leqslant i \leqslant r)$, the function

$$g_1^{i_1} \ldots g_r^{i_r} \to (\lambda_1^{i_1} \ldots \lambda_r^{i_r})$$

is a representation of G. There are $n_1 n_2 \ldots n_r$ such representations, and no two of them are equivalent.

We have proved the following theorem.

9.8 Theorem
Let G be the abelian group $C_{n_1} \times \ldots \times C_{n_r}$. The representations $\rho_{\lambda_1, \ldots, \lambda_r}$ of G constructed above are irreducible and have degree 1. There are $|G|$ of these representations, and every irreducible representation of G over $\mathbb{C}$ is equivalent to precisely one of them.

9.9 Examples
(1) Let $G = C_n = \langle a: a^n = 1 \rangle$, and put $\omega = e^{2\pi i/n}$. The n irreducible representations of G over $\mathbb{C}$ are ρ_{ω^j} $(0 \leqslant j \leqslant n - 1)$, where

$$a^k \rho_{\omega^j} = (\omega^{jk}) \quad (0 \leqslant k \leqslant n - 1).$$

(2) The four irreducible $\mathbb{C}G$-modules for $G = C_2 \times C_2 = \langle g_1, g_2 \rangle$ are V_1, V_2, V_3, V_4, where V_i is a 1-dimensional space with basis v_i $(i = 1, 2, 3, 4)$ and

$$v_1 g_1 = v_1, \qquad v_1 g_2 = v_2;$$
$$v_2 g_1 = v_2, \qquad v_2 g_2 = -v_2;$$
$$v_3 g_1 = -v_3, \qquad v_3 g_2 = v_3;$$
$$v_4 g_1 = -v_4, \qquad v_4 g_2 = -v_4.$$

Diagonalization

Let $H = \langle g \rangle$ be a cyclic group of order n, and let V be a non-zero $\mathbb{C}H$-module. By Theorem 8.7,

$$V = U_1 \oplus \ldots \oplus U_r,$$

a direct sum of irreducible $\mathbb{C}H$-submodules U_i of V. Each U_i has dimension 1, by Proposition 9.5; let u_i be a vector spanning U_i. Put $\omega = e^{2\pi i/n}$. Then for each i, there exists an integer m_i such that

$$u_i g = \omega^{m_i} u_i.$$

Thus if $\mathscr{B}$ is the basis $u_1, \ldots, u_r$ of V, then

(9.10)
$$[g]_{\mathscr{B}} = \begin{pmatrix} \omega^{m_1} & & 0 \\ & \ddots & \\ 0 & & \omega^{m_r} \end{pmatrix}.$$

The following useful result is an immediate consequence of this.

9.11 Proposition
Let G be a finite group and V a $\mathbb{C}G$-module. If $g \in G$, then there is a basis $\mathscr{B}$ of V such that the matrix $[g]_{\mathscr{B}}$ is diagonal. If g has order n, then the entries on the diagonal of $[g]_{\mathscr{B}}$ are nth roots of unity.

Proof Let $H = \langle g \rangle$. As V is also a $\mathbb{C}H$-module, the result follows from (9.10). ∎

Some further applications of Schur's Lemma

Our next application concerns an important subspace of the group algebra $\mathbb{C}G$.

9.12 Definition
Let G be a finite group. The *centre* of the group algebra $\mathbb{C}G$, written $Z(\mathbb{C}G)$, is defined by

$$Z(\mathbb{C}G) = \{z \in \mathbb{C}G: zr = rz \text{ for all } r \in \mathbb{C}G\}.$$

Using (2.5), it is easy to check that $Z(\mathbb{C}G)$ is a subspace of $\mathbb{C}G$.

For abelian groups G, the centre $Z(\mathbb{C}G)$ is the whole group algebra. For arbitrary groups G, we shall see that $Z(\mathbb{C}G)$ plays a crucial role in the study of representations of G (for example, its dimension is equal to the number of irreducible representations of G – see Chapter 15).

9.13 Example

The elements 1 and $\sum_{g \in G} g$ lie in $Z(\mathbb{C}G)$. Indeed, if H is any normal subgroup of G, then

$$\sum_{h \in H} h \in Z(\mathbb{C}G).$$

To see this, write $z = \sum_{h \in H} h$. Then for all $g \in G$,

$$g^{-1}zg = \sum_{h \in H} g^{-1}hg = \sum_{h \in H} h = z,$$

and so $zg = gz$. Consequently $zr = rz$ for all $r \in \mathbb{C}G$.

For example, if $G = D_6 = \langle a, b \colon a^3 = b^2 = 1, \ b^{-1}ab = a^{-1} \rangle$, then $\{1\}$, $\langle a \rangle$ and G are normal subgroups of G, so the elements

$$1, \ 1 + a + a^2 \text{ and } 1 + a + a^2 + b + ab + a^2 b$$

lie in $Z(\mathbb{C}G)$. We shall see later that these elements in fact form a basis of $Z(\mathbb{C}G)$.

We use Schur's Lemma to prove the following important property of the elements of $Z(\mathbb{C}G)$.

9.14 Proposition

Let V be an irreducible $\mathbb{C}G$-module, and let $z \in Z(\mathbb{C}G)$. Then there exists $\lambda \in \mathbb{C}$ such that

$$vz = \lambda v \quad \text{for all } v \in V.$$

Proof For all $r \in \mathbb{C}G$ and $v \in V$, we have

$$vrz = vzr,$$

and hence the function $v \to vz$ is a $\mathbb{C}G$-homomorphism from V to V. By Schur's Lemma 9.1(2), this $\mathbb{C}G$-homomorphism is equal to $\lambda 1_V$ for some $\lambda \in \mathbb{C}$, and the result follows. ∎

Some elements of the centre of $\mathbb{C}G$ are provided by the centre of G, which we now define.

9.15 Definition

The *centre* of G, written $Z(G)$, is defined by

$$Z(G) = \{z \in G : zg = gz \text{ for all } g \in G\}.$$

Clearly $Z(G)$ is a normal subgroup of G, and is a subset of $Z(\mathbb{C}G)$.

Although we have seen in Proposition 6.6 that for every finite group G there is a faithful $\mathbb{C}G$-module, it is not necessarily the case that there is a faithful *irreducible* $\mathbb{C}G$-module. Indeed, the following result shows that the existence of a faithful irreducible $\mathbb{C}G$-module imposes a strong restriction on the structure of G.

9.16 Proposition

If there exists a faithful irreducible $\mathbb{C}G$-module, then $Z(G)$ is cyclic.

Proof Let V be a faithful irreducible $\mathbb{C}G$-module. If $z \in Z(G)$ then z lies in $Z(\mathbb{C}G)$, and hence by Proposition 9.14, there exists $\lambda_z \in \mathbb{C}$ such that

$$vz = \lambda_z v \quad \text{for all } v \in V.$$

Since V is faithful, the function

$$z \to \lambda_z \quad (z \in Z(G))$$

is an injective homomorphism from $Z(G)$ into the multiplicative group $\mathbb{C}^*$ of non-zero complex numbers. Therefore $Z(G) \cong \{\lambda_z : z \in Z(G)\}$, which, being a finite subgroup of $\mathbb{C}^*$, is cyclic (see Exercise 1.7). ∎

We remark that the converse of Proposition 9.16 is false, since in Exercise 25.6, we give an example of a group G such that $Z(G)$ is cyclic but there exists no faithful irreducible $\mathbb{C}G$-module.

9.17 Example

If G is an abelian group, then $G = Z(G)$, and so by Proposition 9.16, there is no faithful irreducible $\mathbb{C}G$-module unless G is cyclic. For example, $C_2 \times C_2$ has no faithful irreducible representation (compare Example 9.9(2)).

The irreducible representations of non-abelian groups are more difficult to construct than those of abelian groups. In particular, they do not all have degree 1, as is shown by the following converse to Proposition 9.5.

9.18 Proposition
Suppose that G is a finite group such that every irreducible $\mathbb{C}G$-module has dimension 1. Then G is abelian.

Proof By Theorem 8.7, we can write

$$\mathbb{C}G = V_1 \oplus \ldots \oplus V_n,$$

where each V_i is an irreducible $\mathbb{C}G$-submodule of the regular $\mathbb{C}G$-module $\mathbb{C}G$. Then dim $V_i = 1$ for all i, since we are assuming that all irreducible $\mathbb{C}G$-modules have dimension 1. For $1 \leqslant i \leqslant n$, let v_i be a vector spanning V_i. Then $v_1, \ldots, v_n$ is a basis of $\mathbb{C}G$; call it $\mathscr{B}$. For all $x, y \in G$, the matrices $[x]_{\mathscr{B}}$ and $[y]_{\mathscr{B}}$ are diagonal, and hence they commute. Since the representation

$$g \rightarrow [g]_{\mathscr{B}} \quad (g \in G)$$

of G is faithful (see Proposition 6.6), we deduce that x and y commute. Hence G is abelian, as required.   ∎

Summary of Chapter 9

1. Schur's Lemma states that every $\mathbb{C}G$-homomorphism between irreducible $\mathbb{C}G$-modules is either zero or a $\mathbb{C}G$-isomorphism. Also, the only $\mathbb{C}G$-homomorphisms from an irreducible $\mathbb{C}G$-module to itself are scalar multiples of the identity.

2. The centre $Z(\mathbb{C}G)$ of the group algebra $\mathbb{C}G$ consists of those elements which commute with all elements of $\mathbb{C}G$. The elements of $Z(\mathbb{C}G)$ act as scalar multiples of the identity on all irreducible $\mathbb{C}G$-modules.

3. All irreducible $\mathbb{C}G$-modules for a finite abelian group G have dimension 1, and there are precisely $|G|$ of them.

Exercises for Chapter 9

1. Write down the irreducible representations over $\mathbb{C}$ of the groups C_2, C_3 and $C_2 \times C_2$.

2. Let $G = C_4 \times C_4$.

   (a) Find a non-trivial irreducible representation ρ of G such that $g^2\rho = (1)$ for all $g \in G$.

   (b) Prove that there is no irreducible representation σ of G such that $g\sigma = (-1)$ for all elements g of order 2 in G.

3. Let G be the finite abelian group $C_{n_1} \times \ldots \times C_{n_r}$. Prove that G has a faithful representation of degree r. Can G have a faithful representation of degree less than r?

4. Suppose that $G = D_8 = \langle a, b: a^4 = b^2 = 1, b^{-1}ab = a^{-1} \rangle$. Check that there is a representation ρ of G over $\mathbb{C}$ such that

$$a\rho = \begin{pmatrix} -7 & 10 \\ -5 & 7 \end{pmatrix}, \quad b\rho = \begin{pmatrix} -5 & 6 \\ -4 & 5 \end{pmatrix}.$$

Find all 2×2 matrices M such that $M(g\rho) = (g\rho)M$ for all $g \in G$. Hence determine whether or not ρ is irreducible.

Do the same for the representation σ of G, where

$$a\sigma = \begin{pmatrix} 5 & -6 \\ 4 & -5 \end{pmatrix}, \quad b\sigma = \begin{pmatrix} -5 & 6 \\ -4 & 5 \end{pmatrix}.$$

5. Show that if V is an irreducible $\mathbb{C}G$-module, then there exists $\lambda \in \mathbb{C}$ such that

$$v\left(\sum_{g \in G} g\right) = \lambda v \quad \text{for all } v \in V.$$

6. Let $G = D_6 = \langle a, b: a^3 = b^2 = 1, b^{-1}ab = a^{-1} \rangle$. Write $\omega = e^{2\pi i/3}$, and let W be the irreducible $\mathbb{C}G$-submodule of the regular $\mathbb{C}G$-module defined by

$$W = \text{sp}\,(1 + \omega^2 a + \omega a^2,\, b + \omega^2 ab + \omega a^2 b)$$

(see Exercise 6.6).

   (a) Show that $a + a^{-1} \in Z(\mathbb{C}G)$.

   (b) Find $\lambda \in \mathbb{C}$ such that

$$w(a + a^{-1}) = \lambda w$$

for all $w \in W$. (Compare Proposition 9.14.)

7. Which of the following groups have a faithful irreducible representation?

   (a) C_n (n a positive integer);

   (b) D_8;

   (c) $C_2 \times D_8$;

   (d) $C_3 \times D_8$.

10
Irreducible modules and the group algebra

Let G be a finite group and $\mathbb{C}G$ be the group algebra of G over $\mathbb{C}$. Consider $\mathbb{C}G$ as the regular $\mathbb{C}G$-module. By Theorem 8.7, we can write

$$\mathbb{C}G = U_1 \oplus \ldots \oplus U_r$$

where each U_i is an irreducible $\mathbb{C}G$-module. We shall show in this chapter that *every* irreducible $\mathbb{C}G$-module is isomorphic to one of the $\mathbb{C}G$-modules $U_1, \ldots, U_r$. As a consequence, there are only finitely many non-isomorphic irreducible $\mathbb{C}G$-modules (a result which has already been established for abelian groups in Theorem 9.8). Also, in theory, to find all irreducible $\mathbb{C}G$-modules, it is sufficient to decompose $\mathbb{C}G$ as a direct sum of irreducible $\mathbb{C}G$-submodules. However, this is not really a practical way of finding the irreducible $\mathbb{C}G$-modules, unless G is a small group.

Irreducible submodules of $\mathbb{C}G$

We begin with another consequence of Maschke's Theorem.

10.1 Proposition
Let V and W be $\mathbb{C}G$-modules and let $\vartheta: V \to W$ be a $\mathbb{C}G$-homomorphism. Then there is a $\mathbb{C}G$-submodule U of V such that $V = \operatorname{Ker}\vartheta \oplus U$ and $U \cong \operatorname{Im}\vartheta$.

Proof Since $\operatorname{Ker}\vartheta$ is a $\mathbb{C}G$-submodule of V by Proposition 7.2, there is by Maschke's Theorem a $\mathbb{C}G$-submodule U of V such that $V = \operatorname{Ker}\vartheta \oplus U$. Define a function $\overline{\vartheta}: U \to \operatorname{Im}\vartheta$ by

$$u\overline{\vartheta} = u\vartheta \quad (u \in U).$$

We show that $\overline{\vartheta}$ is a $\mathbb{C}G$-isomorphism from U to $\operatorname{Im}\vartheta$. Clearly $\overline{\vartheta}$ is a $\mathbb{C}G$-homomorphism, since ϑ is a $\mathbb{C}G$-homomorphism. If $u \in \operatorname{Ker}\overline{\vartheta}$ then $u \in \operatorname{Ker}\vartheta \cap U = \{0\}$; hence $\operatorname{Ker}\overline{\vartheta} = \{0\}$. Now let $w \in \operatorname{Im}\vartheta$; so $w = v\vartheta$ for some $v \in V$. Write $v = k + u$ with $k \in \operatorname{Ker}\vartheta$, $u \in U$. Then

$$w = v\vartheta = k\vartheta + u\vartheta = u\vartheta = u\overline{\vartheta}.$$

Therefore $\operatorname{Im}\overline{\vartheta} = \operatorname{Im}\vartheta$. We have now established that $\overline{\vartheta}\colon U \to \operatorname{Im}\vartheta$ is an invertible $\mathbb{C}G$-homomorphism. Thus $U \cong \operatorname{Im}\vartheta$, as required.

10.2 Proposition
Let V be a $\mathbb{C}G$-module, and write

$$V = U_1 \oplus \ldots \oplus U_s,$$

a direct sum of irreducible $\mathbb{C}G$-submodules U_i. If U is any irreducible $\mathbb{C}G$-submodule of V, then $U \cong U_i$ for some i.

Proof For $u \in U$, we have $u = u_1 + \ldots + u_s$ for unique vectors $u_i \in U_i$ $(1 \leqslant i \leqslant s)$. Define $\pi_i \colon U \to U_i$ by setting $u\pi_i = u_i$. Choosing i such that $u_i \neq 0$ for some $u \in U$, we have $\pi_i \neq 0$.

Now π_i is a $\mathbb{C}G$-homomorphism (see Proposition 7.11). As U and U_i are irreducible, and $\pi_i \neq 0$, Schur's Lemma 9.1(1) implies that π_i is a $\mathbb{C}G$-isomorphism. Therefore $U \cong U_i$. ∎

Of course it can happen that U is an irreducible $\mathbb{C}G$-submodule of $U_1 \oplus \ldots \oplus U_s$ (each U_i irreducible) without U being equal to any U_i, as the following example shows.

10.3 Example
Let G be any group and let V be a 2-dimensional $\mathbb{C}G$-module, with basis v_1, v_2, such that $vg = v$ for all $v \in V$ and $g \in G$. Then

$$V = U_1 \oplus U_2,$$

where $U_1 = \operatorname{sp}(v_1)$ and $U_2 = \operatorname{sp}(v_2)$ are irreducible $\mathbb{C}G$-submodules. However, $U = \operatorname{sp}(v_1 + v_2)$ is an irreducible $\mathbb{C}G$-submodule which is not equal to U_1 or U_2.

10.4 Definitions
(1) If V is a $\mathbb{C}G$-module and U is an irreducible $\mathbb{C}G$-module, then we say that U is a *composition factor* of V if V has a $\mathbb{C}G$-submodule which is isomorphic to U.

(2) Two $\mathbb{C}G$-modules V and W are said to have a *common composition factor* if there is an irreducible $\mathbb{C}G$-module which is a composition factor of both V and W.

We now come to the main result of the chapter, which shows that every irreducible $\mathbb{C}G$-module is a composition factor of the regular $\mathbb{C}G$-module.

10.5 Theorem
Regard $\mathbb{C}G$ as the regular $\mathbb{C}G$-module, and write

$$\mathbb{C}G = U_1 \oplus \ldots \oplus U_r,$$

a direct sum of irreducible $\mathbb{C}G$-submodules. Then every irreducible $\mathbb{C}G$-module is isomorphic to one of the $\mathbb{C}G$-modules U_i.

Proof Let W be an irreducible $\mathbb{C}G$-module, and choose a non-zero vector $w \in W$. Observe that $\{wr: r \in \mathbb{C}G\}$ is a $\mathbb{C}G$-submodule of W; since W is irreducible, it follows that

(10.6) $$W = \{wr: r \in \mathbb{C}G\}.$$

Now define $\vartheta: \mathbb{C}G \to W$ by

$$r\vartheta = wr \quad (r \in \mathbb{C}G).$$

Clearly ϑ is a linear transformation, and $\operatorname{Im} \vartheta = W$ by (10.6). Moreover, ϑ is a $\mathbb{C}G$-homomorphism, since for $r, s \in \mathbb{C}G$,

$$(rs)\vartheta = w(rs) = (wr)s = (r\vartheta)s.$$

By Proposition 10.1, there is a $\mathbb{C}G$-submodule U of $\mathbb{C}G$ such that

$$\mathbb{C}G = U \oplus \operatorname{Ker} \vartheta \text{ and } U \cong \operatorname{Im} \vartheta = W.$$

As W is irreducible, so is U. By Proposition 10.2 we have $U \cong U_i$ for some i; then $W \cong U_i$, and the result is proved. ∎

Theorem 10.5 shows that there is a finite set of irreducible $\mathbb{C}G$-modules such that every irreducible $\mathbb{C}G$-module is isomorphic to one of them. We record this fact in the following corollary.

10.7 Corollary
If G is a finite group, then there are only finitely many non-isomorphic irreducible $\mathbb{C}G$-modules.

According to Theorem 10.5, to find all the irreducible $\mathbb{C}G$-modules we need only decompose the regular $\mathbb{C}G$-module as a direct sum of irreducible $\mathbb{C}G$-submodules. We now do this for a couple of examples; however, this is not a practical method for studying $\mathbb{C}G$-modules in general.

10.8 Examples

(1) Let $G = C_3 = \langle a : a^3 = 1 \rangle$, and write $\omega = e^{2\pi i/3}$. Define $v_0, v_1, v_2 \in \mathbb{C}G$ by

$$v_0 = 1 + a + a^2,$$
$$v_1 = 1 + \omega^2 a + \omega a^2,$$
$$v_2 = 1 + \omega a + \omega^2 a^2,$$

and let $U_i = \mathrm{sp}\,(v_i)$ for $i = 0, 1, 2$. Then $v_1 a = a + \omega^2 a^2 + \omega 1 = \omega v_1$, and similarly

$$v_i a = \omega^i v_i \quad \text{for } i = 0, 1, 2.$$

Hence U_i is a $\mathbb{C}G$-submodule of $\mathbb{C}G$ for $i = 0, 1, 2$.

It is easy to check that v_0, v_1, v_2 is a basis of $\mathbb{C}G$, and hence

$$\mathbb{C}G = U_0 \oplus U_1 \oplus U_2,$$

a direct sum of irreducible $\mathbb{C}G$-submodules U_i. By Theorem 10.5, every irreducible $\mathbb{C}G$-module is isomorphic to U_0, U_1 or U_2. The irreducible representation of G corresponding to U_i is the representation ρ_{ω^i} of Example 9.9(1).

(2) Let $G = D_6 = \langle a, b : a^3 = b^2 = 1, \ b^{-1}ab = a^{-1} \rangle$. We decompose $\mathbb{C}G$ as a direct sum of irreducible $\mathbb{C}G$-submodules. Let $\omega = e^{2\pi i/3}$ and define

$$v_0 = 1 + a + a^2, \qquad w_0 = bv_0 \quad (= b + ba + ba^2),$$
$$v_1 = 1 + \omega^2 a + \omega a^2, \qquad w_1 = bv_1,$$
$$v_2 = 1 + \omega a + \omega^2 a^2, \qquad w_2 = bv_2.$$

As in (1) above, $v_i a = \omega^i v_i$ for $i = 0, 1, 2$, and so $\mathrm{sp}\,(v_i)$ and $\mathrm{sp}\,(w_i)$ are $\mathbb{C}\langle a \rangle$-modules. Next, note that

$$v_0 b = w_0, \quad w_0 b = v_0,$$
$$v_1 b = w_2, \quad w_1 b = v_2,$$
$$v_2 b = w_1, \quad w_2 b = v_1.$$

Therefore, $\mathrm{sp}\,(v_0, w_0)$, $\mathrm{sp}\,(v_1, w_2)$ and $\mathrm{sp}\,(v_2, w_1)$ are $\mathbb{C}\langle b\rangle$-modules, and hence are $\mathbb{C}G$-submodules of $\mathbb{C}G$. By the argument in Example 5.5(2), the $\mathbb{C}G$-submodules $U_3 = \mathrm{sp}\,(v_1, w_2)$ and $U_4 = \mathrm{sp}\,(v_2, w_1)$ are irreducible. However, $\mathrm{sp}(v_0, w_0)$ is reducible, as $U_1 = \mathrm{sp}(v_0 + w_0)$ and $U_2 = \mathrm{sp}(v_0 - w_0)$ are $\mathbb{C}G$-submodules.

Now v_0, v_1, v_2, w_0, w_1, w_2 is a basis of $\mathbb{C}G$, and hence

$$\mathbb{C}G = U_1 \oplus U_2 \oplus U_3 \oplus U_4,$$

a direct sum of irreducible $\mathbb{C}G$-submodules. Note that U_1 is the trivial $\mathbb{C}G$-module, and U_1 is not isomorphic to U_2, the other 1-dimensional U_i. But $U_3 \cong U_4$ (there is a $\mathbb{C}G$-isomorphism sending $v_1 \to w_1$, $w_2 \to v_2$).

We conclude from Theorem 10.5 that there are exactly three non-isomorphic irreducible $\mathbb{C}G$-modules, namely U_1, U_2 and U_3. Correspondingly, every irreducible representation of D_6 over $\mathbb{C}$ is equivalent to precisely one of the following:

$$\rho_1\colon a \to (1),\ b \to (1);$$

$$\rho_2\colon a \to (1),\ b \to (-1);$$

$$\rho_3\colon a \to \begin{pmatrix} \omega & 0 \\ 0 & \omega^{-1} \end{pmatrix},\ b \to \begin{pmatrix} 0 & 1 \\ 1 & 0 \end{pmatrix}.$$

Summary of Chapter 10

1. Every irreducible $\mathbb{C}G$-module occurs as a composition factor of the regular $\mathbb{C}G$-module.

2. There are only finitely many non-isomorphic irreducible $\mathbb{C}G$-modules.

Exercises for Chapter 10

1. Let G be a finite group. Find a $\mathbb{C}G$-submodule of $\mathbb{C}G$ which is isomorphic to the trivial $\mathbb{C}G$-module. Is there only one such $\mathbb{C}G$-submodule?

2. Let $G = C_4$. Express $\mathbb{C}G$ as a direct sum of irreducible $\mathbb{C}G$-submodules. (Hint: copy the method of Example 10.8(1).)

3. Let $G = D_8 = \langle a, b: a^4 = b^2 = 1, b^{-1}ab = a^{-1} \rangle$. Find a 1-dimen-sional $\mathbb{C}G$-submodule, $\mathrm{sp}(u_1)$ say, of $\mathbb{C}G$ such that

$$u_1 a = u_1, \; u_1 b = -u_1.$$

Find also 1-dimensional $\mathbb{C}G$-submodules, $\mathrm{sp}(u_2)$ and $\mathrm{sp}(u_3)$, such that

$$u_2 a = -u_2, \quad u_2 b = u_2, \text{ and}$$

$$u_3 a = -u_3, \quad u_3 b = -u_3.$$

4. Use the method of Example 10.8(2) to find all the irreducible representations of D_8 over $\mathbb{C}$.

5. Suppose that V is a non-zero $\mathbb{C}G$-module such that $V = U_1 \oplus U_2$, where U_1 and U_2 are isomorphic $\mathbb{C}G$-modules. Show that there is a $\mathbb{C}G$-submodule U of V which is not equal to U_1 or U_2, but is isomorphic to both of them.

6. Let $G = Q_8 = \langle a, b: a^4 = 1, b^2 = a^2, b^{-1}ab = a^{-1} \rangle$, and let V be the $\mathbb{C}G$-module given in Example 4.5(2). Thus V has basis v_1, v_2 and

$$v_1 a = iv_1, \qquad v_1 b = v_2,$$

$$v_2 a = -iv_2, \quad v_2 b = -v_1.$$

Show that V is irreducible, and find a $\mathbb{C}G$-submodule of $\mathbb{C}G$ which is isomorphic to V.

11

More on the group algebra

We now go further into the structure of the group algebra $\mathbb{C}G$ of a finite group G. As in Chapter 10, we write

$$\mathbb{C}G = U_1 \oplus \ldots \oplus U_r,$$

a direct sum of irreducible $\mathbb{C}G$-modules U_i. In Theorem 10.5 we proved that *every* irreducible $\mathbb{C}G$-module U is isomorphic to one of the U_i. The question arises: how many of the U_i are isomorphic to U? There is an elegant and significant answer to this question: the number is precisely $\dim U$ (see Theorem 11.9).

Our proof of Theorem 11.9 is based on a study of the vector space of $\mathbb{C}G$-homomorphisms from one $\mathbb{C}G$-module to another.

The space of $\mathbb{C}G$-homomorphisms

11.1 Definition
Let V and W be $\mathbb{C}G$-modules. We write $\mathrm{Hom}_{\mathbb{C}G}(V, W)$ for the set of all $\mathbb{C}G$-homomorphisms from V to W.

Define addition and scalar multiplication on $\mathrm{Hom}_{\mathbb{C}G}(V, W)$ as follows: for $\vartheta, \phi \in \mathrm{Hom}_{\mathbb{C}G}(V, W)$ and $\lambda \in \mathbb{C}$, define $\vartheta + \phi$ and $\lambda\vartheta$ by

$$v(\vartheta + \phi) = v\vartheta + v\phi,$$

$$v(\lambda\vartheta) = \lambda(v\vartheta)$$

for all $v \in V$. Then $\vartheta + \phi, \lambda\vartheta \in \mathrm{Hom}_{\mathbb{C}G}(V, W)$. With these definitions, it is easily checked that $\mathrm{Hom}_{\mathbb{C}G}(V, W)$ is a vector space over $\mathbb{C}$.

We begin our study of the vector space $\mathrm{Hom}_{\mathbb{C}G}(V, W)$ with an easy consequence of Schur's Lemma.

11.2 Proposition
Suppose that V and W are irreducible $\mathbb{C}G$-modules. Then

$$\dim\left(\operatorname{Hom}_{\mathbb{C}G}(V, W)\right) = \begin{cases} 1, & \text{if } V \cong W, \\ 0, & \text{if } V \not\cong W. \end{cases}$$

Proof If $V \not\cong W$ then this is immediate from Schur's Lemma 9.1(1).

Now suppose that $V \cong W$, and let $\vartheta: V \to W$ be a $\mathbb{C}G$-isomorphism. If $\phi \in \operatorname{Hom}_{\mathbb{C}G}(V, W)$, then $\phi\vartheta^{-1}$ is a $\mathbb{C}G$-isomorphism from V to V, so by Schur's Lemma 9.1(2), there exists $\lambda \in \mathbb{C}$ such that

$$\phi\vartheta^{-1} = \lambda 1_V.$$

Then $\phi = \lambda\vartheta$, and so $\operatorname{Hom}_{\mathbb{C}G}(V, W) = \{\lambda\vartheta: \lambda \in \mathbb{C}\}$, a 1-dimensional space. ∎

For the next result, recall the definition of a composition factor of a $\mathbb{C}G$-module from 10.4.

11.3 Proposition
Let V and W be $\mathbb{C}G$-modules, and suppose that $\operatorname{Hom}_{\mathbb{C}G}(V, W) \neq \{0\}$. Then V and W have a common composition factor.

Proof Let ϑ be a non-zero element of $\operatorname{Hom}_{\mathbb{C}G}(V, W)$. Then $V = \operatorname{Ker}\vartheta \oplus U$ for some non-zero $\mathbb{C}G$-module U, by Maschke's Theorem. Let X be an irreducible $\mathbb{C}G$-submodule of U. Since $X\vartheta \neq \{0\}$, Schur's Lemma 9.1(1) implies that $X\vartheta \cong X$. Therefore X is a common composition factor of V and W. ∎

The next few results show how to calculate the dimension of $\operatorname{Hom}_{\mathbb{C}G}(V, W)$ in general. The key step is the following proposition.

11.4 Proposition
Let V, V_1, V_2 and W, W_1, W_2 be $\mathbb{C}G$-modules. Then

(1) $\dim\left(\operatorname{Hom}_{\mathbb{C}G}(V, W_1 \oplus W_2)\right) =$

   $\dim\left(\operatorname{Hom}_{\mathbb{C}G}(V, W_1)\right) + \dim\left(\operatorname{Hom}_{\mathbb{C}G}(V, W_2)\right),$

(2) $\dim\left(\operatorname{Hom}_{\mathbb{C}G}(V_1 \oplus V_2, W)\right) =$

   $\dim\left(\operatorname{Hom}_{\mathbb{C}G}(V_1, W)\right) + \dim\left(\operatorname{Hom}_{\mathbb{C}G}(V_2, W)\right).$

Proof (1) Define the functions $\pi_1: W_1 \oplus W_2 \to W_1$ and $\pi_2: W_1 \oplus W_2 \to W_2$ by

$$(w_1 + w_2)\pi_1 = w_1, \quad (w_1 + w_2)\pi_2 = w_2$$

for all $w_1 \in W_1$, $w_2 \in W_2$. By Proposition 7.11, π_1 and π_2 are $\mathbb{C}G$-homomorphisms. If $\vartheta \in \mathrm{Hom}_{\mathbb{C}G}(V, W_1 \oplus W_2)$, then $\vartheta\pi_1 \in \mathrm{Hom}_{\mathbb{C}G}(V, W_1)$ and $\vartheta\pi_2 \in \mathrm{Hom}_{\mathbb{C}G}(V, W_2)$ (see Exercise 7.1).

We now define a function f from $\mathrm{Hom}_{\mathbb{C}G}(V, W_1 \oplus W_2)$ to the (external) direct sum of $\mathrm{Hom}_{\mathbb{C}G}(V, W_1)$ and $\mathrm{Hom}_{\mathbb{C}G}(V, W_2)$ by

$$f: \vartheta \to (\vartheta\pi_1, \vartheta\pi_2) \quad (\vartheta \in \mathrm{Hom}_{\mathbb{C}G}(V, W_1 \oplus W_2)).$$

Clearly f is a linear transformation. We show that f is invertible.

Given $\phi_i \in \mathrm{Hom}_{\mathbb{C}G}(V, W_i)$ $(i = 1, 2)$, the function

$$\phi: v \to v\phi_1 + v\phi_2 \quad (v \in V)$$

lies in $\mathrm{Hom}_{\mathbb{C}G}(V, W_1 \oplus W_2)$, and the image of ϕ under f is (ϕ_1, ϕ_2). Hence f is surjective.

If $\vartheta \in \mathrm{Ker}\, f$, then $v\vartheta\pi_1 = 0$ and $v\vartheta\pi_2 = 0$ for all $v \in V$, so $v\vartheta = v\vartheta(\pi_1 + \pi_2) = 0$. Therefore $\vartheta = 0$, so $\mathrm{Ker}\, f = \{0\}$ and f is injective.

We have established that f is an invertible linear transformation from $\mathrm{Hom}_{\mathbb{C}G}(V, W_1 \oplus W_2)$ to $\mathrm{Hom}_{\mathbb{C}G}(V, W_1) \oplus \mathrm{Hom}_{\mathbb{C}G}(V, W_2)$. Consequently these two vector spaces have equal dimensions, and (1) follows.

(2) For $\vartheta \in \mathrm{Hom}_{\mathbb{C}G}(V_1 \oplus V_2, W)$, define $\vartheta_{V_i}: V_i \to W$ $(i = 1, 2)$ to be the restriction of ϑ to V_i; that is, ϑ_{V_i} is the function

$$v_i\vartheta_{V_i} = v_i\vartheta \quad (v_i \in V_i).$$

Then $\vartheta_{V_i} \in \mathrm{Hom}_{\mathbb{C}G}(V_i, W)$ for $i = 1, 2$.

Now let h be the function from $\mathrm{Hom}_{\mathbb{C}G}(V_1 \oplus V_2, W)$ to $\mathrm{Hom}_{\mathbb{C}G}(V_1, W) \oplus \mathrm{Hom}_{\mathbb{C}G}(V_2, W)$ which is given by

$$h: \vartheta \to (\vartheta_{V_1}, \vartheta_{V_2}) \quad (\vartheta \in \mathrm{Hom}_{\mathbb{C}G}(V_1 \oplus V_2, W)).$$

Clearly h is an injective linear transformation. Given $\phi_i \in \mathrm{Hom}_{\mathbb{C}G}(V_i, W)$ $(i = 1, 2)$, the function

$$\phi: v_1 + v_2 \to v_1\phi_1 + v_2\phi_2 \quad (v_i \in V_i \text{ for } i = 1, 2)$$

lies in $\mathrm{Hom}_{\mathbb{C}G}(V_1 \oplus V_2, W)$ and has image (ϕ_1, ϕ_2) under h. Hence h is surjective. We have shown that h is an invertible linear transformation, and (2) follows. ∎

Now suppose that we have $\mathbb{C}G$-modules V, W, V_i, W_j ($1 \leqslant i \leqslant r$, $1 \leqslant j \leqslant s$). By an obvious induction using Proposition 11.4, we have

(11.5)   (1) $\dim(\text{Hom}_{\mathbb{C}G}(V, W_1 \oplus \ldots \oplus W_s))$

$$= \sum_{j=1}^{s} \dim(\text{Hom}_{\mathbb{C}G}(V, W_j)),$$

(2) $\dim(\text{Hom}_{\mathbb{C}G}(V_1 \oplus \ldots \oplus V_r, W))$

$$= \sum_{i=1}^{r} \dim(\text{Hom}_{\mathbb{C}G}(V_i, W)).$$

These in turn imply

(3) $\dim(\text{Hom}_{\mathbb{C}G}(V_1 \oplus \ldots \oplus V_r, W_1 \oplus \ldots \oplus W_s))$

$$= \sum_{i=1}^{r} \sum_{j=1}^{s} \dim(\text{Hom}_{\mathbb{C}G}(V_i, W_j)).$$

By applying (3) when all V_i and W_j are irreducible, and using Proposition 11.2, we can find $\dim(\text{Hom}_{\mathbb{C}G}(V, W))$ in general. In the following corollary we single out the case where one of the $\mathbb{C}G$-modules is irreducible.

11.6 Corollary
Let V be a $\mathbb{C}G$-module with

$$V = U_1 \oplus \ldots \oplus U_s,$$

where each U_i is an irreducible $\mathbb{C}G$-module. Let W be any irreducible $\mathbb{C}G$-module. Then the dimensions of $\text{Hom}_{\mathbb{C}G}(V, W)$ and $\text{Hom}_{\mathbb{C}G}(W, V)$ are both equal to the number of $\mathbb{C}G$-modules U_i such that $U_i \cong W$.

Proof By (11.5),

$$\dim(\text{Hom}_{\mathbb{C}G}(V, W)) = \sum_{i=1}^{s} \dim(\text{Hom}_{\mathbb{C}G}(U_i, W)), \text{ and}$$

$$\dim(\text{Hom}_{\mathbb{C}G}(W, V)) = \sum_{i=1}^{s} \dim(\text{Hom}_{\mathbb{C}G}(W, U_i)).$$

And by Proposition 11.2,

$$\dim(\mathrm{Hom}_{\mathbb{C}G}(U_i, W)) = \dim(\mathrm{Hom}_{\mathbb{C}G}(W, U_i)) = \begin{cases} 1, & \textit{if } U_i \cong W, \\ 0, & \textit{if } U_i \ncong W. \end{cases}$$

The result follows. ∎

11.7 Example
For $G = D_6$, we saw in Example 10.8(2) that

$$\mathbb{C}G = U_1 \oplus U_2 \oplus U_3 \oplus U_4,$$

a direct sum of irreducible $\mathbb{C}G$-submodules, with $U_3 \cong U_4$ but U_3 not isomorphic to U_1 or U_2. Thus by Corollary 11.6, we have

$$\dim(\mathrm{Hom}_{\mathbb{C}G}(\mathbb{C}G, U_3)) = \dim(\mathrm{Hom}_{\mathbb{C}G}(U_3, \mathbb{C}G)) = 2.$$

You are asked in Exercise 11.5 to find bases for these two vector spaces of $\mathbb{C}G$-homomorphisms.

The next proposition investigates the space of $\mathbb{C}G$-homomorphisms from the regular $\mathbb{C}G$-module to any other $\mathbb{C}G$-module. When combined with Corollary 11.6, it will give the main result of this chapter.

11.8 Proposition
If U is a $\mathbb{C}G$-module, then

$$\dim(\mathrm{Hom}_{\mathbb{C}G}(\mathbb{C}G, U)) = \dim U.$$

Proof Let $d = \dim U$. Choose a basis $u_1, \ldots, u_d$ of U. For $1 \leq i \leq d$, define $\phi_i : \mathbb{C}G \to U$ by

$$r\phi_i = u_i r \quad (r \in \mathbb{C}G).$$

Then $\phi_i \in \mathrm{Hom}_{\mathbb{C}G}(\mathbb{C}G, U)$ since for all $r, s \in \mathbb{C}G$,

$$(rs)\phi_i = u_i(rs) = (u_i r)s = (r\phi_i)s.$$

We shall prove that $\phi_1, \ldots, \phi_d$ is a basis of $\mathrm{Hom}_{\mathbb{C}G}(\mathbb{C}G, U)$.
  Suppose that $\phi \in \mathrm{Hom}_{\mathbb{C}G}(\mathbb{C}G, U)$. Then

$$1\phi = \lambda_1 u_1 + \ldots + \lambda_d u_d$$

for some $\lambda_i \in \mathbb{C}$. Since ϕ is a $\mathbb{C}G$-homomorphism, for all $r \in \mathbb{C}G$ we have

$$r\phi = (1r)\phi = (1\phi)r$$

$$= \lambda_1 u_1 r + \ldots + \lambda_d u_d r$$

$$= r(\lambda_1 \phi_1 + \ldots + \lambda_d \phi_d).$$

Hence $\phi = \lambda_1 \phi_1 + \ldots + \lambda_d \phi_d$. Therefore $\phi_1, \ldots, \phi_d$ span $\text{Hom}_{\mathbb{C}G}(\mathbb{C}G, U)$.

Now assume that

$$\lambda_1 \phi_1 + \ldots + \lambda_d \phi_d = 0 \quad (\lambda_i \in \mathbb{C}).$$

Evaluating both sides at the identity 1, we have

$$0 = 1(\lambda_1 \phi_1 + \ldots + \lambda_d \phi_d)$$

$$= \lambda_1 u_1 + \ldots + \lambda_d u_d,$$

which forces $\lambda_i = 0$ for all i. Hence $\phi_1, \ldots, \phi_d$ is a basis of $\text{Hom}_{\mathbb{C}G}(\mathbb{C}G, U)$, which therefore has dimension d. ∎

We now come to the main theorem of the chapter, which tells us how often each irreducible $\mathbb{C}G$-module occurs in the regular $\mathbb{C}G$-module.

11.9 Theorem
Suppose that

$$\mathbb{C}G = U_1 \oplus \ldots \oplus U_r,$$

a direct sum of irreducible $\mathbb{C}G$-submodules. If U is any irreducible $\mathbb{C}G$-module, then the number of $\mathbb{C}G$-modules U_i with $U_i \cong U$ is equal to $\dim U$.

Proof By Proposition 11.8,

$$\dim U = \dim (\text{Hom}_{\mathbb{C}G}(\mathbb{C}G, U)),$$

and by Corollary 11.6, this is equal to the number of U_i with $U_i \cong U$. ∎

11.10 Example
Recall again from Example 10.8(2) that if $G = D_6$ then

$$\mathbb{C}G = U_1 \oplus U_2 \oplus U_3 \oplus U_4,$$

where U_1, U_2 are non-isomorphic 1-dimensional $\mathbb{C}G$-modules, and

U_3, U_4 are isomorphic irreducible 2-dimensional $\mathbb{C}G$-modules. This illustrates Theorem 11.9:

$$U_1 \text{ occurs once, } \dim U_1 = 1;$$

$$U_2 \text{ occurs once, } \dim U_2 = 1;$$

$$U_3 \text{ occurs twice, } \dim U_3 = 2.$$

We conclude the chapter with a significant consequence of Theorem 11.9 concerning the dimensions of all irreducible $\mathbb{C}G$-modules.

11.11 Definition
We say that the irreducible $\mathbb{C}G$-modules $V_1, \ldots, V_k$ form a *complete set of non-isomorphic irreducible $\mathbb{C}G$-modules* if every irreducible $\mathbb{C}G$-module is isomorphic to some V_i, and no two of $V_1, \ldots, V_k$ are isomorphic. (By Corollary 10.7, for any finite group G there exists a complete set of non-isomorphic irreducible $\mathbb{C}G$-modules.)

11.12 Theorem
Let $V_1, \ldots, V_k$ form a complete set of non-isomorphic irreducible $\mathbb{C}G$-modules. Then

$$\sum_{i=1}^{k} (\dim V_i)^2 = |G|.$$

Proof Let $\mathbb{C}G = U_1 \oplus \ldots \oplus U_r$, a direct sum of irreducible $\mathbb{C}G$-submodules. For $1 \leqslant i \leqslant k$, write $d_i = \dim V_i$. By Theorem 11.9, for each i, the number of $\mathbb{C}G$-modules U_j with $U_j \cong V_i$ is equal to d_i. Therefore

$$\dim \mathbb{C}G = \dim U_1 + \ldots + \dim U_r$$

$$= \sum_{i=1}^{k} d_i(\dim V_i) = \sum_{i=1}^{k} d_i^2.$$

As $\dim \mathbb{C}G = |G|$, the result follows. ∎

11.13 Example
Let G be a group of order 8, and let $d_1, \ldots, d_k$ be the dimensions of all the irreducible $\mathbb{C}G$-modules. By Theorem 11.12,

$$\sum_{i=1}^{k} d_i^2 = 8.$$

Observe that the trivial $\mathbb{C}G$-module is irreducible of dimension 1, and so $d_i = 1$ for some i. Hence the possibilities for $d_1, \ldots, d_k$ are

$$1, 1, 1, 1, 1, 1, 1, 1 \quad \text{and}$$

$$1, 1, 1, 1, 2.$$

Both these possibilities do occur: the first holds when G is an abelian group (see Proposition 9.5), and the second when $G = D_8$ (see Exercise 10.4).

We shall see later that $\dim V_i$ divides $|G|$ for all i, and this fact, combined with Theorem 11.12, is quite a powerful tool in finding the dimensions of irreducible $\mathbb{C}G$-modules.

Summary of Chapter 11

1. $\dim (\mathrm{Hom}_{\mathbb{C}G} (V_1 \oplus \ldots \oplus V_r, W_1 \oplus \ldots \oplus W_s))$

$$= \sum_{i=1}^{r} \sum_{j=1}^{s} \dim (\mathrm{Hom}_{\mathbb{C}G} (V_i, W_j)).$$

2. $\dim (\mathrm{Hom}_{\mathbb{C}G} (\mathbb{C}G, U)) = \dim U.$

3. Let $\mathbb{C}G = U_1 \oplus \ldots \oplus U_r$, a direct sum of irreducible $\mathbb{C}G$-modules, and let U be any irreducible $\mathbb{C}G$-module. Then the number of U_i with $U_i \cong U$ is equal to $\dim U.$

4. If $V_1, \ldots, V_k$ is a complete set of non-isomorphic irreducible $\mathbb{C}G$-modules, then

$$\sum_{i=1}^{k} (\dim V_i)^2 = |G|.$$

Exercises for Chapter 11

1. If G is a non-abelian group of order 6, find the dimensions of all the irreducible $\mathbb{C}G$-modules.

2. If G is a group of order 12, what are the possible degrees of all the irreducible representations of G?

   Find the degrees of the irreducible representations of D_{12}.

   (Hint: use Exercise 5.3.)

3. Let G be a finite group. Find a basis for $\mathrm{Hom}_{\mathbb{C}G} (\mathbb{C}G, \mathbb{C}G).$

4. Suppose that $G = S_n$ and V is the n-dimensional permutation module for G over $\mathbb{C}$, as defined in 4.10. If U is the trivial $\mathbb{C}G$-module, show that $\text{Hom}_{\mathbb{C}G}(V, U)$ has dimension 1.

5. Let $G = D_6$ and let $\mathbb{C}G = U_1 \oplus U_2 \oplus U_3 \oplus U_4$, a direct sum of irreducible $\mathbb{C}G$-modules, as in Example 10.8(2). Find a basis for $\text{Hom}_{\mathbb{C}G}(\mathbb{C}G, U_3)$ and a basis for $\text{Hom}_{\mathbb{C}G}(U_3, \mathbb{C}G)$.

6. Let $V_1, \ldots, V_k$ be a complete set of non-isomorphic irreducible $\mathbb{C}G$-modules, and let V, W be arbitrary $\mathbb{C}G$-modules. Assume that for $1 \le i \le k$,

$$d_i = \dim(\text{Hom}_{\mathbb{C}G}(V, V_i)) \text{ and } e_i = \dim(\text{Hom}_{\mathbb{C}G}(W, V_i)).$$

Show that $\dim(\text{Hom}_{\mathbb{C}G}(V, W)) = \sum_{i=1}^{k} d_i e_i$.

12
Conjugacy classes

We take a break from representation theory to discuss some topics in group theory which will be relevant in our further study of representations. After defining conjugacy classes, we develop enough theory to determine the conjugacy classes of dihedral, symmetric and alternating groups. At the end of the chapter we prove a result linking the conjugacy classes of a group to the structure of its group algebra.

Throughout the chapter, G is a finite group.

Conjugacy classes

12.1 Definition

Let $x, y \in G$. We say that x is *conjugate to y in G* if

$$y = g^{-1}xg \quad \text{for some } g \in G.$$

The set of all elements conjugate to x in G is

$$x^G = \{g^{-1}xg : g \in G\},$$

and is called the *conjugacy class* of x in G.

Our first result shows that two distinct conjugacy classes have no elements in common.

12.2 Proposition

If $x, y \in G$, then either $x^G = y^G$ or $x^G \cap y^G$ is empty.

Proof Suppose that $x^G \cap y^G$ is not empty, and pick $z \in x^G \cap y^G$. Then there exist $g, h \in G$ such that

$$z = g^{-1}xg = h^{-1}yh.$$

Hence $x = gh^{-1}yhg^{-1} = k^{-1}yk$, where $k = hg^{-1}$. So

$$a \in x^G \Rightarrow a = b^{-1}xb \quad \text{for some } b \in G$$

$$\Rightarrow a = b^{-1}k^{-1}ykb$$

$$\Rightarrow a = c^{-1}yc \quad \text{where } c = kb$$

$$\Rightarrow a \in y^G.$$

Therefore $x^G \subseteq y^G$. Similarly $y^G \subseteq x^G$ (using $y = kxk^{-1}$), and so $x^G = y^G$. ∎

Since every element x of G lies in the conjugacy class x^G (as $x = 1^{-1}x1$ with $1 \in G$), G is the union of its conjugacy classes and so we deduce immediately

12.3 Corollary
Every group is a union of conjugacy classes, and distinct conjugacy classes are disjoint.

Another way of seeing this Corollary 12.3 is to observe that conjugacy is an equivalence relation, and that the conjugacy classes are the equivalence classes.

12.4 Definition
If $G = x_1^G \cup \ldots \cup x_l^G$, where the conjugacy classes $x_1^G, \ldots, x_l^G$ are distinct, then we call $x_1, \ldots, x_l$ *representatives* of the conjugacy classes of G.

12.5 Examples
(1) For every group G, $1^G = \{1\}$ is a conjugacy class of G.

(2) Let $G = D_6 = \langle a, b : a^3 = b^2 = 1, b^{-1}ab = a^{-1} \rangle$. The elements of G are $1, a, a^2, b, ab, a^2b$. Since $g^{-1}ag$ is a or a^2 for every $g \in G$, and $b^{-1}ab = a^2$, we have

$$a^G = \{a, a^2\}.$$

Also, $a^{-i}ba^i = a^{-2i}b$ for all integers i, so

$$b^G = \{b, ab, a^2b\}.$$

Thus the conjugacy classes of G are

$$\{1\}, \{a, a^2\}, \{b, ab, a^2b\}.$$

(3) If G is abelian then $g^{-1}xg = x$ for all $x, g \in G$, and so $x^G = \{x\}$. Hence every conjugacy class of G consists of just one element.

The next proposition is often useful when calculating conjugacy classes.

12.6 Proposition
Let $x, y \in G$. If x is conjugate to y in G, then x^n is conjugate to y^n in G for every integer n, and x and y have the same order.

Proof Observe that for $a, b \in G$, we have

$$g^{-1}abg = (g^{-1}ag)(g^{-1}bg).$$

Hence $g^{-1}x^n g = (g^{-1}xg)^n$. Suppose that x is conjugate to y in G, so that $y = g^{-1}xg$ for some $g \in G$. Then $y^n = g^{-1}x^n g$ and therefore x^n is conjugate to y^n in G. Let x have order m. Then $y^m = g^{-1}x^m g = 1$, and for $0 < r < m$, $y^r = g^{-1}x^r g \neq 1$, so y also has order m. $\blacksquare$

Conjugacy class sizes

The next theorem determines the sizes of the conjugacy classes in G in terms of certain subgroups which we now define.

12.7 Definition
Let $x \in G$. The *centralizer of x in G*, written $C_G(x)$, is the set of elements of G which commute with x; that is,

$$C_G(x) = \{g \in G\colon xg = gx\}.$$

(So also $C_G(x) = \{g \in G\colon g^{-1}xg = x\}$.)

It is easy to check that $C_G(x)$ is a subgroup of G (Exercise 12.1). Observe that $x \in C_G(x)$ and indeed, $\langle x \rangle \subseteq C_G(x)$ for all $x \in G$.

12.8 Theorem
Let $x \in G$. Then the size of the conjugacy class x^G is given by

$$|x^G| = |G\colon C_G(x)| = |G|/|C_G(x)|.$$

In particular, $|x^G|$ divides $|G|$.

Proof Observe first that for $g, h \in G$, we have

$$g^{-1}xg = h^{-1}xh \Leftrightarrow hg^{-1}x = xhg^{-1}$$

$$\Leftrightarrow hg^{-1} \in C_G(x)$$

$$\Leftrightarrow C_G(x)g = C_G(x)h.$$

By dint of this, we may define an injective function f from x^G to the set of right cosets of $C_G(x)$ in G by

$$f: g^{-1}xg \to C_G(x)g \quad (g \in G).$$

Clearly f is surjective. Hence f is a bijection, proving that $|x^G| = |G:C_G(x)|$. ∎

Before summarizing our results on conjugacy classes, we make the observation that

(12.9) $\qquad |x^G| = 1 \Leftrightarrow g^{-1}xg = x \quad$ for all $g \in G$

$$\Leftrightarrow x \in Z(G),$$

where $Z(G)$ is the centre of G, as defined in 9.15.

We have now proved all parts of the following result.

12.10 The Class Equation

Let $x_1, \ldots, x_l$ be representatives of the conjugacy classes of G. Then

$$|G| = |Z(G)| + \sum_{x_i \notin Z(G)} |x_i^G|,$$

where $|x_i^G| = |G:C_G(x_i)|$, and both $|Z(G)|$ and $|x_i^G|$ divide $|G|$.

Conjugacy classes of dihedral groups

We illustrate the use of Theorem 12.8 by finding the conjugacy classes of all dihedral groups.

Let $G = D_{2n}$, the dihedral group of order $2n$. Thus

$$G = \langle a, b: a^n = b^2 = 1, b^{-1}ab = a^{-1} \rangle.$$

In finding the conjugacy classes of G, it is convenient to consider separately the cases where n is odd and where n is even.

(1) *n odd*

First consider a^i $(1 \leqslant i \leqslant n - 1)$. Since $C_G(a^i)$ contains $\langle a \rangle$,

$$|G: C_G(a^i)| \leqslant |G: \langle a \rangle| = 2.$$

Also $b^{-1}a^i b = a^{-i}$, so $\{a^i, a^{-i}\} \subseteq (a^i)^G$. As n is odd, $a^i \neq a^{-i}$, and so $|(a^i)^G| \geq 2$. Using Theorem 12.8, we have

$$2 \geq |G: C_G(a^i)| = |(a^i)^G| \geq 2.$$

Hence equality holds here, and

$$C_G(a^i) = \langle a \rangle, \ (a^i)^G = \{a^i, a^{-i}\}.$$

Next, $C_G(b)$ contains $\{1, b\}$; and as $b^{-1}a^i b = a^{-i}$, no element a^i or $a^i b$ (with $1 \leq i \leq n - 1$) commutes with b. Thus

$$C_G(b) = \{1, b\}.$$

Therefore by Theorem 12.8, $|b^G| = n$. Since all the elements a^i have been accounted for, b^G must consist of the remaining n elements of G. That is,

$$b^G = \{b, ab, \ldots, a^{n-1}b\}.$$

We have shown

(12.11) *The dihedral group D_{2n} (n odd) has precisely $\frac{1}{2}(n+3)$ conjugacy classes:*

$$\{1\}, \ \{a, a^{-1}\}, \ldots, \{a^{(n-1)/2}, a^{-(n-1)/2}\}, \ \{b, ab, \ldots, a^{n-1}b\}.$$

(2) *n even*

Write $n = 2m$. As $b^{-1}a^m b = a^{-m} = a^m$, the centralizer of a^m in G contains both a and b, and hence $C_G(a^m) = G$. Therefore the conjugacy class of a^m in G is just $\{a^m\}$. As in case (1), $(a^i)^G = \{a^i, a^{-i}\}$ for $1 \leq i \leq m - 1$.

For every integer j,

$$a^j ba^{-j} = a^{2j}b, \ a^j(ab)a^{-j} = a^{2j+1}b.$$

It follows that

$$b^G = \{a^{2j}b: 0 \leq j \leq m - 1\}, \ (ab)^G = \{a^{2j+1}b: 0 \leq j \leq m - 1\}.$$

Hence

(12.12) *The dihedral group D_{2n} (n even, $n = 2m$) has precisely $m + 3$ conjugacy classes:*

$$\{1\}, \ \{a^m\}, \ \{a, a^{-1}\}, \ldots, \{a^{m-1}, a^{-m+1}\},$$
$$\{a^{2j}b: 0 \leq j \leq m - 1\}, \ \{a^{2j+1}b: 0 \leq j \leq m - 1\}.$$

Conjugacy classes of S_n

We shall later need to know the conjugacy classes of the symmetric group S_n. Our first observation is simple but crucial.

12.13 Proposition
Let x be a k-cycle $(i_1 \, i_2 \ldots i_k)$ in S_n, and let $g \in S_n$. Then $g^{-1}xg$ is the k-cycle $(i_1 g \, i_2 g \ldots i_k g)$.

Proof Write $A = \{i_1, \ldots, i_k\}$. For $i_r \in A$,

$$i_r g(g^{-1}xg) = i_r xg = i_{r+1} g \text{ (or } i_1 g \text{ if } r = k).$$

Also, for $1 \le i \le n$ and $i \notin A$,

$$ig(g^{-1}xg) = ixg = ig.$$

Hence $g^{-1}(i_1 \, i_2 \ldots i_k)g = (i_1 g \, i_2 g \ldots i_k g)$, as required. ∎

Now consider an arbitrary permutation $x \in S_n$. Write

$$x = (a_1 \ldots a_{k_1})(b_1 \ldots b_{k_2}) \ldots (c_1 \ldots c_{k_s}),$$

a product of disjoint cycles, with $k_1 \ge k_2 \ge \ldots \ge k_s$. By Proposition 12.13, for $g \in S_n$ we have

(12.14)

$$g^{-1}xg = g^{-1}(a_1 \ldots a_{k_1})gg^{-1}(b_1 \ldots b_{k_2})g \ldots g^{-1}(c_1 \ldots c_{k_s})g$$

$$= (a_1 g \ldots a_{k_1} g)(b_1 g \ldots b_{k_2} g) \ldots (c_1 g \ldots c_{k_s} g).$$

We call $(k_1, \ldots, k_s)$ the *cycle-shape* of x, and note that x and $g^{-1}xg$ have the same cycle-shape. On the other hand, given any two permutations x, y of the same cycle-shape, say

$$x = (a_1 \ldots a_{k_1}) \ldots (c_1 \ldots c_{k_s}),$$

$$y = (a_1' \ldots a_{k_1}') \ldots (c_1' \ldots c_{k_s}'),$$

(products of disjoint cycles), there exists $g \in S_n$ sending $a_1 \to a_1', \ldots, c_{k_s} \to c_{k_s}'$, and so by (12.14),

$$g^{-1}xg = y.$$

We have proved the following result.

12.15 Theorem

For $x \in S_n$, the conjugacy class x^{S_n} of x in S_n consists of all permutations in S_n which have the same cycle-shape as x.

12.16 Examples

(1) The conjugacy classes of S_3 are

Class	Cycle-shape
{1}	(1)
{(1 2), (1 3), (2 3)}	(2)
{(1 2 3), (1 3 2)}	(3)

(2) The conjugacy class of (1 2)(3 4) in S_4 consists of all the elements of cycle-shape (2, 2) and is

$$\{(1\ 2)(3\ 4), (1\ 3)(2\ 4), (1\ 4)(2\ 3)\}.$$

(3) There are precisely five conjugacy classes of S_4, with representatives (see Definition 12.4):

$$1, (1\ 2), (1\ 2\ 3), (1\ 2)(3\ 4), (1\ 2\ 3\ 4).$$

To calculate the sizes of the conjugacy classes, we simply count the number of 2-cycles, 3-cycles, and so on. The number of 2-cycles is equal to the number of pairs that can be chosen from $\{1, 2, 3, 4\}$, which is $\binom{4}{2} = 6$. (The notation $\binom{n}{r}$ means the binomial coefficient $n!/(r!(n-r)!)$.) The number of 3-cycles is 4×2 (4 for the choice of fixed point and 2 because there are two 3-cycles fixing a given point). Similarly, there are three elements of cycle-shape (2, 2) and there are six 4-cycles. Thus for $G = S_4$, the conjugacy class representatives g, the conjugacy class sizes $|g^G|$ and the centralizer orders $|C_G(g)|$ (obtained using Theorem 12.8) are as follows:

Representative	g	1	(1 2)	(1 2 3)	(1 2)(3 4)	(1 2 3 4)
Class size	$|g^G|$	1	6	8	3	6
	$|C_G(g)|$	24	4	3	8	4

We check our arithmetic by noting that

$$|S_4| = 1 + 6 + 8 + 3 + 6.$$

(4) Similarly, the corresponding table for $G = S_5$ is

Rep. g	1	(1 2)	(1 2 3)	(1 2)(3 4)	(1 2 3 4)	(1 2 3)(4 5)	(1 2 3 4 5)
$\lvert g^G \rvert$	1	10	20	15	30	20	24
$\lvert C_G(g) \rvert$	120	12	6	8	4	6	5

Conjugacy classes of A_n

Given an even permutation $x \in A_n$, we have seen in Theorem 12.15 that the conjugacy class x^{S_n} consists of all permutations in S_n which have the same cycle-shape as x. The conjugacy class x^{A_n} of x in A_n, given by

$$x^{A_n} = \{g^{-1}xg : g \in A_n\},$$

is of course contained in x^{S_n}; however, x^{A_n} might not be equal to x^{S_n}. For an easy example where equality does not hold, consider $x = (1\ 2\ 3) \in A_3$; here $x^{A_3} = \{x\}$, while $x^{S_3} = \{x, x^{-1}\}$.

The next result determines precisely when x^{A_n} and x^{S_n} are equal, and what happens when equality fails.

12.17 Proposition
Let $x \in A_n$ with $n > 1$.
 (1) *If x commutes with some odd permutation in S_n, then $x^{S_n} = x^{A_n}$.*
 (2) *If x does not commute with any odd permutation in S_n then x^{S_n} splits into two conjugacy classes in A_n of equal size, with representatives x and $(1\ 2)^{-1}x(1\ 2)$.*

Proof (1) Assume that x commutes with an odd permutation g. Let $y \in x^{S_n}$, so that $y = h^{-1}xh$ for some $h \in S_n$. If h is even then $y \in x^{A_n}$; and if h is odd then $gh \in A_n$ and

$$y = h^{-1}xh = h^{-1}g^{-1}xgh = (gh)^{-1}x(gh),$$

so again $y \in x^{A_n}$. Thus $x^{S_n} \subseteq x^{A_n}$, and so $x^{S_n} = x^{A_n}$.

(2) Assume that x does not commute with any odd permutation. Then

$$C_{S_n}(x) = C_{A_n}(x).$$

Hence by Theorem 12.8,

$$|x^{A_n}| = |A_n : C_{A_n}(x)| = \tfrac{1}{2}|S_n : C_{A_n}(x)| \quad (\text{as } |A_n| = \tfrac{1}{2}|S_n|)$$
$$= \tfrac{1}{2}|S_n : C_{S_n}(x)| = \tfrac{1}{2}|x^{S_n}|.$$

Next, we observe that

$$\{h^{-1}xh : h \text{ is odd}\} = ((1\ 2)^{-1}x(1\ 2))^{A_n}$$

since every odd permutation has the form $(1\ 2)a$ for some $a \in A_n$. Now

$$x^{S_n} = \{h^{-1}xh : h \text{ is even}\} \cup \{h^{-1}xh : h \text{ is odd}\}$$
$$= x^{A_n} \cup ((1\ 2)^{-1}x(1\ 2))^{A_n}.$$

Since $|x^{A_n}| = \tfrac{1}{2}|x^{S_n}|$, the conjugacy classes x^{A_n} and $((1\ 2)^{-1}x(1\ 2))^{A_n}$ must be disjoint and of equal size, as we wished to show. ∎

12.18 Examples

(1) We find the conjugacy classes of A_4. The elements of A_4 are the identity, together with the permutations of cycle-shapes $(2, 2)$ and (3). Since $(1\ 2)(3\ 4)$ commutes with the odd permutation $(1\ 2)$, Proposition 12.17 implies that

$$(1\ 2)(3\ 4)^{A_4} = (1\ 2)(3\ 4)^{S_4} = \{(1\ 2)(3\ 4), (1\ 3)(2\ 4), (1\ 4)(2\ 3)\}.$$

However, the 3-cycle $(1\ 2\ 3)$ commutes with no odd permutation: for if $g^{-1}(1\ 2\ 3)g = (1\ 2\ 3)$ then $(1\ 2\ 3) = (1g\ 2g\ 3g)$ by Proposition 12.13, so g is 1, $(1\ 2\ 3)$ or $(1\ 3\ 2)$, an even permutation. Hence by Proposition 12.17, $(1\ 2\ 3)^{S_4}$ splits into two conjugacy classes in A_4 of size 4, with representatives $(1\ 2\ 3)$ and $(1\ 2)^{-1}(1\ 2\ 3)(1\ 2) = (1\ 3\ 2)$.

Thus the conjugacy classes of A_4 are

Representative	1	$(1\ 2)(3\ 4)$	$(1\ 2\ 3)$	$(1\ 3\ 2)$
Class size	1	3	4	4
Centralizer order	12	4	3	3

(2) We find the conjugacy classes of A_5. The non-identity even permutations in S_5 are those of cycle-shapes (3), $(2, 2)$ and (5). The elements $(1\ 2\ 3)$ and $(2\ 3)(4\ 5)$ commute with the odd permutation $(4\ 5)$; but $(1\ 2\ 3\ 4\ 5)$ commutes with no odd permutation. (Check this by using the argument in (1) above.) Hence by Proposition 12.17, the

conjugacy classes of A_5 are represented by 1, $(1\,2\,3)$, $(1\,2)(3\,4)$, $(1\,2\,3\,4\,5)$ and $(1\,2)^{-1}(1\,2\,3\,4\,5)(1\,2) = (1\,3\,4\,5\,2)$. Using Proposition 12.17(2), we see that the class sizes and centralizer orders are as follows:

Representative	1	$(1\,2\,3)$	$(1\,2)(3\,4)$	$(1\,2\,3\,4\,5)$	$(1\,3\,4\,5\,2)$
Class size	1	20	15	12	12
Centralizer order	60	3	4	5	5

Normal subgroups

Normal subgroups are related to conjugacy classes by the following elementary result.

12.19 Proposition
Let H be a subgroup of G. Then $H \lhd G$ if and only if H is a union of conjugacy classes of G.

Proof If H is a union of conjugacy classes, then

$$h \in H, g \in G \Rightarrow g^{-1}hg \in H,$$

so $g^{-1}Hg \subseteq H$. Thus $H \lhd G$.

Conversely, if $H \lhd G$ then for all $h \in H$, $g \in G$, we have $g^{-1}hg \in H$, and so $h^G \subseteq H$. Therefore

$$H = \bigcup_{h \in H} h^G,$$

and so H is a union of conjugacy classes of G. ∎

12.20 Example
We find all the normal subgroups of S_4. Let $H \lhd S_4$. Then by Proposition 12.19, H is a union of conjugacy classes of S_4. As we saw in Example 12.16(3), these conjugacy classes have sizes 1, 6, 8, 3, 6. Since $|H|$ divides 24 by Lagrange's Theorem, and $1 \in H$, there are just four possibilities:

$$|H| = 1, 1+3, 1+8+3 \text{ or } 1+6+8+3+6.$$

In the first case $H = \{1\}$, in the last case $H = S_4$, and in the third case $H = A_4$. In the case where $|H| = 1 + 3$, we have

$$H = 1^{S_4} \cup (1\ 2)(3\ 4)^{S_4} = \{1, (1\ 2)(3\ 4), (1\ 3)(2\ 4), (1\ 4)(2\ 3)\}.$$

This is easily checked to be a subgroup of S_4; we write it as V_4 (V stands for 'Viergruppe', meaning 'four-group').

We have now shown that S_4 has exactly four normal subgroups:

$$\{1\}, S_4, A_4 \text{ and } V_4 = \{1, (1\ 2)(3\ 4), (1\ 3)(2\ 4), (1\ 4)(2\ 3)\}.$$

The centre of a group algebra

In this final section we link the conjugacy classes of the group G to the centre of the group algebra $\mathbb{C}G$. Recall from Definition 9.12 that the centre of $\mathbb{C}G$ is

$$Z(\mathbb{C}G) = \{z \in \mathbb{C}G: zr = rz \text{ for all } r \in \mathbb{C}G\}.$$

We know that $Z(\mathbb{C}G)$ is a subspace of the vector space $\mathbb{C}G$. There is a convenient basis for this subspace which can be described in terms of the conjugacy classes of G.

12.21 Definition
Let $C_1, \ldots, C_l$ be the distinct conjugacy classes of G. For $1 \leqslant i \leqslant l$, define

$$\bar{C}_i = \sum_{g \in C_i} g \in \mathbb{C}G.$$

The elements $\bar{C}_1, \ldots, \bar{C}_l$ of $\mathbb{C}G$ are called *class sums*.

12.22 Proposition
The class sums $\bar{C}_1, \ldots, \bar{C}_l$ form a basis of $Z(\mathbb{C}G)$.

Proof First we show that each $\bar{C}_i$ belongs to $Z(\mathbb{C}G)$. Let C_i consist of the r distinct conjugates $y_1^{-1}gy_1, \ldots, y_r^{-1}gy_r$ of an element g, so

$$\bar{C}_i = \sum_{j=1}^{r} y_j^{-1}gy_j.$$

For all $h \in G$,

$$h^{-1}\bar{C}_i h = \sum_{j=1}^{r} h^{-1}y_j^{-1}gy_j h.$$

As j runs from 1 to r, the elements $h^{-1}y_j^{-1}gy_jh$ run through C_i, since

$$h^{-1}y_j^{-1}gy_jh = h^{-1}y_k^{-1}gy_kh \Leftrightarrow y_j^{-1}gy_j = y_k^{-1}gy_k.$$

Hence

$$\sum_{j=1}^{r}h^{-1}y_j^{-1}gy_jh = \bar{C_i},$$

and so $h^{-1}\bar{C_i}h = \bar{C_i}$. That is,

$$\bar{C_i}h = h\bar{C_i}.$$

Therefore each $\bar{C_i}$ commutes with all $h \in G$, hence with all $\sum_{h \in G}\lambda_h h \in \mathbb{C}G$, and so $\bar{C_i} \in Z(\mathbb{C}G)$.

Next, observe that $\bar{C_1}, \ldots, \bar{C_l}$ are linearly independent: for if $\sum_{i=1}^{l}\lambda_i\bar{C_i} = 0$ ($\lambda_i \in \mathbb{C}$), then all $\lambda_i = 0$ as the classes $C_1, \ldots, C_l$ are pairwise disjoint by Corollary 12.3.

It remains to show that $\bar{C_1}, \ldots, \bar{C_l}$ span $Z(\mathbb{C}G)$. Let $r = \sum_{g \in G}\lambda_g g \in Z(\mathbb{C}G)$. For $h \in G$, we have $rh = hr$, so $h^{-1}rh = r$. That is,

$$\sum_{g \in G}\lambda_g h^{-1}gh = \sum_{g \in G}\lambda_g g.$$

So for every $h \in G$, the coefficient λ_g of g is equal to the coefficient $\lambda_{h^{-1}gh}$ of $h^{-1}gh$. That is to say, the function $g \to \lambda_g$ is constant on conjugacy classes of G. It follows that $r = \sum_{i=1}^{l}\lambda_i\bar{C_i}$ where λ_i is the coefficient λ_{g_i} for some $g_i \in C_i$. This completes the proof. ∎

12.23 Examples

(1) From Example 12.16(1), a basis for $Z(\mathbb{C}S_3)$ is

$$1, (1\ 2) + (1\ 3) + (2\ 3), (1\ 2\ 3) + (1\ 3\ 2).$$

(2) From (12.12), a basis for $Z(\mathbb{C}D_8)$ is

$$1, a^2, a + a^3, b + a^2b, ab + a^3b.$$

Summary of Chapter 12

1. Every group is a union of conjugacy classes, and distinct conjugacy classes are disjoint.

2. For an element x of a group G, the centralizer $C_G(x)$ is the set of

elements of G which commute with x. It is a subgroup of G, and the number of elements in the conjugacy class x^G is equal to $|G:C_G(x)|$.

3. The conjugacy classes of S_n correspond to the cycle-shapes of permutations in S_n.

4. If $x \in A_n$ then $x^{S_n} = x^{A_n}$ if and only if x commutes with some odd permutation in S_n.

5. The class sums in $\mathbb{C}G$ form a basis for the centre of $\mathbb{C}G$.

Exercises for Chapter 12

1. If G is a group and $x \in G$, show that $C_G(x)$ is a subgroup of G which contains $Z(G)$.

2. Let G be a finite group and suppose that $g \in G$ and $z \in Z(G)$. Prove that the conjugacy classes g^G and $(gz)^G$ have the same size.

3. Let $G = S_n$.
   (a) Prove that $|(1\ 2)^G| = \binom{n}{2}$ and find $C_G((1\ 2))$. Verify that your solution satisfies Theorem 12.8.
   (b) Show that $|(1\ 2\ 3)^G| = 2\binom{n}{3}$ and $|(1\ 2)(3\ 4)^G| = 3\binom{n}{4}$.
   (c) Now let $n = 6$. Show that

$$|(1\ 2\ 3)(4\ 5\ 6)^G| = 40 \text{ and } |(1\ 2)(3\ 4)(5\ 6)^G| = 15,$$

   and find the sizes of the other conjugacy classes of S_6. (There are 11 conjugacy classes in all.)

4. What are the cycle-shapes of those permutations $x \in A_6$ for which $x^{A_6} \neq x^{S_6}$?

5. Show that A_5 is a simple group. (Hint: use the method of Example 12.20.)

6. Find the conjugacy classes of the quaternion group Q_8. Give a basis of the centre of the group algebra $\mathbb{C}Q_8$.

7. Let p be a prime number, and let n be a positive integer. Suppose that G is a group of order p^n.
   (a) Use the Class Equation 12.10 to show that $Z(G) \neq \{1\}$.
   (b) Suppose that $n \geq 3$ and that $|Z(G)| = p$. Prove that G has a conjugacy class of size p.

13
Characters

Suppose that $\rho\colon G \to \mathrm{GL}(n, \mathbb{C})$ is a representation of the finite group G. With each $n \times n$ matrix $g\rho$ $(g \in G)$ we associate the complex number given by adding all the diagonal entries of the matrix, and call this number $\chi(g)$. The function $\chi\colon G \to \mathbb{C}$ is called the character of the representation ρ. Characters of representations have many remarkable properties, and they are the fundamental tools for performing calculations in representation theory. For example, we shall show later that two representations have the same character if and only if they are equivalent. Moreover, basic problems, such as deciding whether or not a given representation is irreducible, can be resolved by doing some easy arithmetic with the character of the representation. These facts are surprising, since from the definition of a representation $\rho\colon G \to \mathrm{GL}(n, \mathbb{C})$, it appears that we must keep track of all the n^2 entries in each matrix $g\rho$, whereas the character records just one number for each matrix.

The theory of characters will occupy a considerable portion of the rest of the book. In this chapter we present some basic properties and examples.

The trace of a matrix

13.1 Definition
If $A = (a_{ij})$ is an $n \times n$ matrix, then the *trace* of A, written $\mathrm{tr}\, A$, is given by

$$\mathrm{tr}\, A = \sum_{i=1}^{n} a_{ii}.$$

That is, the trace of A is the sum of the diagonal entries of A.

117

13.2 Proposition

Let $A = (a_{ij})$ and $B = (b_{ij})$ be $n \times n$ matrices. Then

$$\text{tr}\,(A + B) = \text{tr}\,A + \text{tr}\,B, \text{ and}$$

$$\text{tr}\,(AB) = \text{tr}\,(BA).$$

Moreover, if T is an invertible $n \times n$ matrix, then

$$\text{tr}\,(T^{-1}AT) = \text{tr}\,A.$$

Proof The *ii*-entry of $A + B$ is $a_{ii} + b_{ii}$, and the *ii*-entry of AB is $\sum_{j=1}^{n} a_{ij}b_{ji}$. Therefore

$$\text{tr}\,(A + B) = \sum_{i=1}^{n}(a_{ii} + b_{ii}) = \sum_{i=1}^{n} a_{ii} + \sum_{i=1}^{n} b_{ii} = \text{tr}\,A + \text{tr}\,B,$$

and

$$\text{tr}\,(AB) = \sum_{i=1}^{n} \sum_{j=1}^{n} a_{ij}b_{ji} = \sum_{j=1}^{n} \sum_{i=1}^{n} b_{ji}a_{ij} = \text{tr}\,(BA).$$

For the last part,

$$\text{tr}\,(T^{-1}AT) = \text{tr}\,((T^{-1}A)T)$$

$$= \text{tr}\,(T(T^{-1}A)) \quad \text{(by the second part)}$$

$$= \text{tr}\,A. \qquad \blacksquare$$

Notice that, unlike the determinant function, the trace function is not multiplicative; that is, $\text{tr}\,(AB)$ need not equal $(\text{tr}\,A)(\text{tr}\,B)$.

Characters

13.3 Definition

Suppose that V is a $\mathbb{C}G$-module with a basis $\mathscr{B}$. Then the *character* of V is the function $\chi \colon G \to \mathbb{C}$ defined by

$$\chi(g) = \text{tr}\,[g]_{\mathscr{B}} \quad (g \in G).$$

The character of V does not depend on the basis $\mathscr{B}$, since if $\mathscr{B}$ and $\mathscr{B}'$ are bases of V, then

$$[g]_{\mathscr{B}'} = T^{-1}[g]_{\mathscr{B}}T$$

for some invertible matrix T (see (2.24)), and so by Proposition 13.2,

$$\operatorname{tr}[g]_{\mathscr{B}'} = \operatorname{tr}[g]_{\mathscr{B}} \quad \text{for all } g \in G.$$

Naturally enough, we define the character of a representation $\rho\colon G \to \operatorname{GL}(n, \mathbb{C})$ to be the character χ of the corresponding $\mathbb{C}G$-module $\mathbb{C}^n$, namely

$$\chi(g) = \operatorname{tr}(g\rho) \quad (g \in G).$$

13.4 Definition

We say that χ is a *character* of G if χ is the character of some $\mathbb{C}G$-module. Further, χ is an *irreducible* character of G if χ is the character of an irreducible $\mathbb{C}G$-module; and χ is *reducible* if it is the character of a reducible $\mathbb{C}G$-module.

You will have noticed that we are writing characters as functions on the left. That is, we write $\chi(g)$ and not $g\chi$.

13.5 Proposition

(1) *Isomorphic $\mathbb{C}G$-modules have the same character.*
(2) *If x and y are conjugate elements of the group G, then*

$$\chi(x) = \chi(y)$$

for all characters χ of G.

Proof (1) Suppose that V and W are isomorphic $\mathbb{C}G$-modules. Then by (7.7), there are a basis $\mathscr{B}_1$ of V and a basis $\mathscr{B}_2$ of W such that

$$[g]_{\mathscr{B}_1} = [g]_{\mathscr{B}_2} \quad \text{for all } g \in G.$$

Consequently $\operatorname{tr}[g]_{\mathscr{B}_1} = \operatorname{tr}[g]_{\mathscr{B}_2}$ for all $g \in G$, and so V and W have the same character.

(2) Assume that x and y are conjugate elements of G, so that $x = g^{-1}yg$ for some $g \in G$. Let V be a $\mathbb{C}G$-module, and let $\mathscr{B}$ be a basis of V. Then

$$[x]_{\mathscr{B}} = [g^{-1}yg]_{\mathscr{B}} = [g]_{\mathscr{B}}^{-1}[y]_{\mathscr{B}}[g]_{\mathscr{B}}.$$

Hence by Proposition 13.2, we have $\operatorname{tr}[x]_{\mathscr{B}} = \operatorname{tr}[y]_{\mathscr{B}}$. Therefore $\chi(x) = \chi(y)$, where χ is the character of V. ∎

The result corresponding to Proposition 13.5(1) for representations is that equivalent representations have the same character.

Later, we shall prove an astonishing converse of Proposition 13.5(1): if two $\mathbb{C}G$-modules have the same character, then they are isomorphic.

13.6 Examples

(1) Let $G = D_8 = \langle a, b: a^4 = b^2 = 1, b^{-1}ab = a^{-1} \rangle$, and let $\rho: G \to GL(2, \mathbb{C})$ be the representation for which

$$a\rho = \begin{pmatrix} 0 & 1 \\ -1 & 0 \end{pmatrix}, \; b\rho = \begin{pmatrix} 1 & 0 \\ 0 & -1 \end{pmatrix}$$

(see Example 3.2(1)). Let χ be the character of this representation. The following table records g, $g\rho$ and $\chi(g)$ as g runs through G. (We obtain $\chi(g)$ by adding the two entries on the diagonal of $g\rho$.)

g	1	a	a^2	a^3
$g\rho$	$\begin{pmatrix} 1 & 0 \\ 0 & 1 \end{pmatrix}$	$\begin{pmatrix} 0 & 1 \\ -1 & 0 \end{pmatrix}$	$\begin{pmatrix} -1 & 0 \\ 0 & -1 \end{pmatrix}$	$\begin{pmatrix} 0 & -1 \\ 1 & 0 \end{pmatrix}$
$\chi(g)$	2	0	-2	0

g	b	ab	a^2b	a^3b
$g\rho$	$\begin{pmatrix} 1 & 0 \\ 0 & -1 \end{pmatrix}$	$\begin{pmatrix} 0 & -1 \\ -1 & 0 \end{pmatrix}$	$\begin{pmatrix} -1 & 0 \\ 0 & 1 \end{pmatrix}$	$\begin{pmatrix} 0 & 1 \\ 1 & 0 \end{pmatrix}$
$\chi(g)$	0	0	0	0

(2) Let $G = S_3$, and take V to be the 3-dimensional permutation module for G over $\mathbb{C}$ (see Definition 4.10). Let $\mathscr{B}$ be the natural basis of V; thus $\mathscr{B}$ is the basis v_1, v_2, v_3, where $v_ig = v_{ig}$ for $1 \leqslant i \leqslant 3$ and all $g \in G$. The matrices $[g]_{\mathscr{B}}$ $(g \in G)$ are given by Exercise 4.1. We record these matrices, together with the character χ of V.

g	1	$(1\,2)$	$(1\,3)$
$[g]_{\mathscr{B}}$	$\begin{pmatrix} 1 & 0 & 0 \\ 0 & 1 & 0 \\ 0 & 0 & 1 \end{pmatrix}$	$\begin{pmatrix} 0 & 1 & 0 \\ 1 & 0 & 0 \\ 0 & 0 & 1 \end{pmatrix}$	$\begin{pmatrix} 0 & 0 & 1 \\ 0 & 1 & 0 \\ 1 & 0 & 0 \end{pmatrix}$
$\chi(g)$	3	1	1

g	$(2\,3)$	$(1\,2\,3)$	$(1\,3\,2)$
$[g]_{\mathscr{B}}$	$\begin{pmatrix} 1 & 0 & 0 \\ 0 & 0 & 1 \\ 0 & 1 & 0 \end{pmatrix}$	$\begin{pmatrix} 0 & 1 & 0 \\ 0 & 0 & 1 \\ 1 & 0 & 0 \end{pmatrix}$	$\begin{pmatrix} 0 & 0 & 1 \\ 1 & 0 & 0 \\ 0 & 1 & 0 \end{pmatrix}$
$\chi(g)$	1	0	0

(3) Let $G = C_3 = \langle a: a^3 = 1 \rangle$. By Theorem 9.8, G has just three irreducible characters χ_1, χ_2, χ_3, with values

g	1	a	a^2
$\chi_1(g)$	1	1	1
$\chi_2(g)$	1	ω	ω^2
$\chi_3(g)$	1	ω^2	ω

where $\omega = e^{2\pi i/3}$.

(4) Let $G = D_6 = \langle a,\ b: a^3 = b^2 = 1,\ b^{-1}ab = a^{-1} \rangle$ (so $G \cong S_3$). In Example 10.8(2), we found a complete set of non-isomorphic irreducible $\mathbb{C}G$-modules U_1, U_2, U_3. Thus if χ_i is the character of U_i for $1 \leqslant i \leqslant 3$, then the irreducible characters of G are χ_1, χ_2 and χ_3. The values of these characters on the elements of G can be calculated from the corresponding representations ρ_1, ρ_2, ρ_3 given in Example 10.8(2), and they are as follows:

g	1	a	a^2	b	ab	a^2b
$\chi_1(g)$	1	1	1	1	1	1
$\chi_2(g)$	1	1	1	-1	-1	-1
$\chi_3(g)$	2	-1	-1	0	0	0

Notice that in all the above examples, the characters given take few distinct values. This reflects the fact that by Proposition 13.5(2), every character is constant on conjugacy classes of G. Moreover, it is much quicker to write down the single complex number $\chi(g)$ for the group element g than to record the matrix which corresponds to g. Nevertheless, the character encapsulates a great deal of information about the representation. This will become clear as the theory of characters develops.

13.7 Definition

If χ is the character of the $\mathbb{C}G$-module V, then the dimension of V is called the *degree* of χ.

13.8 Examples

(1) In Example 13.6(1) we gave a character of D_8 of degree 2; in 13.6(2) we gave a character of S_3 of degree 3; and in 13.6(4) we saw that the irreducible characters of D_6 have degrees 1, 1 and 2.

(2) If V is any 1-dimensional $\mathbb{C}G$-module, then for each $g \in G$ there is a complex number λ_g such that

$$vg = \lambda_g v \quad \text{for all } v \in V.$$

The character χ of V is given by

$$\chi(g) = \lambda_g \quad (g \in G)$$

and χ has degree 1. Characters of degree 1 are called *linear* characters; they are, of course, irreducible characters.

Observe that Theorem 9.8 gives all the irreducible characters of finite abelian groups; in particular, they are all linear characters.

Every linear character of G is a homomorphism from G to the multiplicative group of non-zero complex numbers. In fact, these are the only non-zero characters of G which are homomorphisms (see Exercise 13.4).

(3) The character of the trivial $\mathbb{C}G$-module (see Definition 4.8(1)) is a linear character, called the *trivial* character of G. We denote it by 1_G. Thus

$$1_G: g \rightarrow 1 \quad \text{for all } g \in G.$$

Given any group G, we therefore know at least one of the irreducible characters of G, namely the trivial character. Finding all the irreducible characters is usually difficult.

The values of a character

The next result gives information about the complex numbers $\chi(g)$, where χ is a character of G and $g \in G$.

13.9 Proposition

Let χ be the character of a $\mathbb{C}G$-module V. Suppose that $g \in G$ and g has order m. Then

(1) $\chi(1) = \dim V$;

(2) $\chi(g)$ *is a sum of* mth *roots of unity*;

(3) $\chi(g^{-1}) = \overline{\chi(g)}$;

(4) $\chi(g)$ *is a real number if* g *is conjugate to* g^{-1}.

Proof (1) Let $n = \dim V$, and let $\mathscr{B}$ be a basis of V. Then the matrix $[1]_{\mathscr{B}}$ of the identity element 1 relative to $\mathscr{B}$ is equal to I_n, the $n \times n$ identity matrix. Consequently

$$\chi(1) = \text{tr}\,[1]_{\mathscr{B}} = \text{tr}\,I_n = n,$$

and so $\chi(1) = \dim V$.

(2) By Proposition 9.11 there is a basis $\mathscr{B}$ of V such that

$$[g]_{\mathscr{B}} = \begin{pmatrix} \omega_1 & & 0 \\ & \ddots & \\ 0 & & \omega_n \end{pmatrix}$$

where each ω_i is an mth root of unity. Therefore

$$\chi(g) = \omega_1 + \ldots + \omega_n,$$

a sum of mth roots of unity.

(3) We have

$$[g^{-1}]_{\mathscr{B}} = \begin{pmatrix} \omega_1^{-1} & & 0 \\ & \ddots & \\ 0 & & \omega_n^{-1} \end{pmatrix}$$

and so $\chi(g^{-1}) = \omega_1^{-1} + \ldots + \omega_n^{-1}$. Every complex mth root of unity ω satisfies $\omega^{-1} = \overline{\omega}$, since for all real ϑ,

$$(e^{i\vartheta})^{-1} = e^{-i\vartheta},$$

which is the complex conjugate of $e^{i\vartheta}$. Therefore

$$\chi(g^{-1}) = \overline{\omega}_1 + \ldots + \overline{\omega}_n = \overline{\chi(g)}.$$

(4) If g is conjugate to g^{-1} then $\chi(g) = \chi(g^{-1})$ by Proposition 13.5(2). Also $\chi(g^{-1}) = \overline{\chi(g)}$ by (3), and so $\chi(g) = \overline{\chi(g)}$; that is, $\chi(g)$ is real. ∎

When the element g of G has order 2, we can be much more specific about the possibilities for $\chi(g)$:

13.10 Corollary
Let χ be a character of G, and let g be an element of order 2 in G.
Then $\chi(g)$ is an integer, and

$$\chi(g) \equiv \chi(1) \bmod 2.$$

Proof By Proposition 13.9, we have

$$\chi(g) = \omega_1 + \ldots + \omega_n,$$

where $n = \chi(1)$ and each ω_i is a square root of unity. Then each ω_i is
$+1$ or -1. Suppose r of them are $+1$, and s are -1, so that

$$\chi(g) = r - s, \quad \text{and } \chi(1) = r + s.$$

Certainly then, $\chi(g) \in \mathbb{Z}$, and since $r - s = r + s - 2s \equiv r + s \bmod 2$,
we have $\chi(g) \equiv \chi(1) \bmod 2$.   ∎

   Our next result gives the first inkling of the importance of characters,
showing that we can determine the kernel of a representation just from
knowledge of its character.

13.11 Theorem
Let $\rho\colon G \to \mathrm{GL}(n, \mathbb{C})$ be a representation of G, and let χ be the
character of ρ.
   (1) For $g \in G$,

$$|\chi(g)| = \chi(1) \Leftrightarrow g\rho = \lambda I_n \quad \text{for some } \lambda \in \mathbb{C}.$$

   (2) $\operatorname{Ker} \rho = \{g \in G \colon \chi(g) = \chi(1)\}$.

Proof (1) Let $g \in G$, and suppose that g has order m. If $g\rho = \lambda I_n$ with
$\lambda \in \mathbb{C}$, then λ is an mth root of unity, and $\chi(g) = n\lambda$, so
$|\chi(g)| = n = \chi(1)$.
   Conversely, suppose that $|\chi(g)| = \chi(1)$. By Proposition 9.11, there is
a basis $\mathcal{B}$ of $\mathbb{C}^n$ such that

$$[g]_{\mathcal{B}} = \begin{pmatrix} \omega_1 & & 0 \\ & \ddots & \\ 0 & & \omega_n \end{pmatrix}$$

where each ω_i is an mth root of unity. Then

(13.12) $|\chi(g)| = |\omega_1 + \ldots + \omega_n| = \chi(1) = n.$

Note now that for any complex numbers $z_1, \ldots, z_n$, we have

$$|z_1 + \ldots + z_n| \leqslant |z_1| + \ldots + |z_n|,$$

with equality if and only if the arguments of $z_1, \ldots, z_n$ are all equal. (To see this, consider the picture

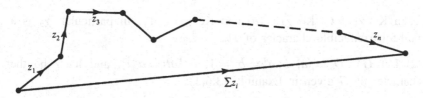

in the Argand diagram.) Since $|\omega_i| = 1$ for all i, we deduce from (13.12) that $\omega_i = \omega_j$ for all i, j. Thus

$$[g]_{\mathscr{B}} = \begin{pmatrix} \omega_1 & & 0 \\ & \ddots & \\ 0 & & \omega_1 \end{pmatrix} = \omega_1 I_n.$$

Hence for all bases $\mathscr{B}'$ of $\mathbb{C}^n$ we have $[g]_{\mathscr{B}'} = \omega_1 I_n$, and so $g\rho = \omega_1 I_n$. This completes the proof of (1).

(2) If $g \in \operatorname{Ker}\rho$ then $g\rho = I_n$, and so $\chi(g) = n = \chi(1)$.

Conversely, suppose that $\chi(g) = \chi(1)$. Then by (1), we have $g\rho = \lambda I_n$ for some $\lambda \in \mathbb{C}$. This implies that $\chi(g) = \lambda\chi(1)$, whence $\lambda = 1$. Therefore $g\rho = I_n$, and so $g \in \operatorname{Ker}\rho$. Part (2) follows. ∎

Motivated by Theorem 13.11(2), we define the kernel of a character as follows.

13.13 Definition

If χ is a character of G, then the *kernel* of χ, written $\operatorname{Ker}\chi$, is defined by

$$\operatorname{Ker}\chi = \{g \in G : \chi(g) = \chi(1)\}.$$

By Theorem 13.11(2), if ρ is a representation of G with character χ, then $\operatorname{Ker}\rho = \operatorname{Ker}\chi$. In particular, $\operatorname{Ker}\chi \lhd G$. We call χ a *faithful* character if $\operatorname{Ker}\chi = \{1\}$.

13.14 Examples

(1) According to Example 13.6(4), the irreducible characters of the group $G = D_6 = \langle a, b : a^3 = b^2 = 1, b^{-1}ab = a^{-1} \rangle$ are χ_1, χ_2, χ_3, with the following values:

g	1	a	a^2	b	ab	a^2b
$\chi_1(g)$	1	1	1	1	1	1
$\chi_2(g)$	1	1	1	-1	-1	-1
$\chi_3(g)$	2	-1	-1	0	0	0

Then $\operatorname{Ker}\chi_1 = G$, $\operatorname{Ker}\chi_2 = \langle a \rangle$ and $\operatorname{Ker}\chi_3 = \{1\}$. In particular, χ_3 is a faithful irreducible character of D_6.

(2) Let $G = D_8 = \langle a,\ b : a^4 = b^2 = 1,\ b^{-1}ab = a^{-1} \rangle$, and let χ be the character of G given in Example 13.6(1):

g	1	a	a^2	a^3	b	ab	a^2b	a^3b
$\chi(g)$	2	0	-2	0	0	0	0	0

Then $\operatorname{Ker}\chi = \{1\}$, so χ is a faithful character. And since $|\chi(a^2)| = |-2| = \chi(1)$, Theorem 13.11(1) implies that if $\rho : G \to \mathrm{GL}(2, \mathbb{C})$ is a representation with character χ, then $a^2\rho = -I$.

We next prove a result which is sometimes useful for constructing a new character from a given one. For a character χ of G, define $\bar{\chi} : G \to \mathbb{C}$ by

$$\bar{\chi}(g) = \overline{\chi(g)} \quad (g \in G).$$

Thus the values of $\bar{\chi}$ are the complex conjugates of the values of χ.

13.15 Proposition
Let χ be a character of G. Then $\bar{\chi}$ is a character of G. If χ is irreducible, then so is $\bar{\chi}$.

Proof Suppose that χ is the character of a representation $\rho : G \to \mathrm{GL}(n, \mathbb{C})$. Thus

$$\chi(g) = \operatorname{tr}(g\rho) \quad (g \in G).$$

If $A = (a_{ij})$ is an $n \times n$ matrix over $\mathbb{C}$, then we define $\bar{A}$ to be the $n \times n$ matrix $(\bar{a}_{ij})$. Observe that if $A = (a_{ij})$ and $B = (b_{ij})$ are $n \times n$ matrices over $\mathbb{C}$, then

(13.16) $$\overline{(AB)} = \bar{A}\bar{B},$$

since the ij-entry of $\overline{A}\,\overline{B}$ is

$$\sum_{k=1}^{n} \overline{a}_{ik}\overline{b}_{kj},$$

which is equal to the complex conjugate of $\sum_{k=1}^{n} a_{ik}b_{kj}$, the ij-entry of AB.

It follows from (13.16) that the function $\overline{\rho}: G \to \mathrm{GL}(n, \mathbb{C})$ defined by

$$g\overline{\rho} = \overline{(g\rho)} \quad (g \in G)$$

is a representation of G. Since

$$\mathrm{tr}\,(g\overline{\rho}) = \mathrm{tr}\,\overline{(g\rho)} = \overline{\mathrm{tr}\,(g\rho)} = \overline{\chi(g)} \quad (g \in G),$$

the character of the representation $\overline{\rho}$ is $\overline{\chi}$.

It is clear that if ρ is reducible then $\overline{\rho}$ is reducible. Hence χ is irreducible if and only if $\overline{\chi}$ is irreducible. ∎

The regular character

13.17 Definition
The *regular* character of G is the character of the regular $\mathbb{C}G$-module. We write the regular character as χ_{reg}.

In Theorem 13.19, we shall express the regular character in terms of the irreducible characters of G. First we need a preliminary result.

13.18 Proposition
Let V be a $\mathbb{C}G$-module, and suppose that

$$V = U_1 \oplus \ldots \oplus U_r,$$

a direct sum of irreducible $\mathbb{C}G$-modules U_i. Then the character of V is equal to the sum of the characters of the $\mathbb{C}G$-modules $U_1, \ldots, U_r$.

Proof This is immediate from (7.10). ∎

13.19 Theorem
Let $V_1, \ldots, V_k$ be a complete set of non-isomorphic irreducible $\mathbb{C}G$-modules (see Definition 11.11), and for $i = 1, \ldots, k$ let χ_i be the character of V_i and $d_i = \chi_i(1)$. Then

$$\chi_{\mathrm{reg}} = d_1\chi_1 + \ldots + d_k\chi_k.$$

Proof By Theorem 11.9,

$$\mathbb{C}G \cong (V_1 \oplus \ldots \oplus V_1) \oplus (V_2 \oplus \ldots \oplus V_2) \oplus \ldots$$

$$\oplus (V_k \oplus \ldots \oplus V_k),$$

where for each i there are d_i factors V_i. Now the result follows from Proposition 13.18. ∎

The values of χ_{reg} on the elements of G are easily described, and are given in the next result.

13.20 Proposition
If χ_{reg} is the regular character of G, then

$$\chi_{\text{reg}}(1) = |G|, \text{ and}$$

$$\chi_{\text{reg}}(g) = 0 \text{ if } g \neq 1.$$

Proof Let $g_1, \ldots, g_n$ be the elements of G, and let $\mathscr{B}$ be the basis $g_1, \ldots, g_n$ of $\mathbb{C}G$. By Proposition 13.9(1), $\chi_{\text{reg}}(1) = \dim \mathbb{C}G = |G|$.

Now let $g \in G$ with $g \neq 1$. Then for $1 \leqslant i \leqslant n$, we have $g_i g = g_j$ for some j with $j \neq i$. Therefore the ith row of the matrix $[g]_{\mathscr{B}}$ has zeros in every place except column j; in particular, the ii-entry is zero for all i. It follows that

$$\chi_{\text{reg}}(g) = \text{tr}\,[g]_{\mathscr{B}} = 0. \qquad ∎$$

13.21 Example
We illustrate Theorem 13.19 and Proposition 13.20 for the group $G = D_6$. By Example 13.6(4), the irreducible characters of G are χ_1, χ_2, χ_3:

g	1	a	a^2	b	ab	a^2b
$\chi_1(g)$	1	1	1	1	1	1
$\chi_2(g)$	1	1	1	-1	-1	-1
$\chi_3(g)$	2	-1	-1	0	0	0

We calculate $\chi_1 + \chi_2 + 2\chi_3$:

$(\chi_1 + \chi_2 + 2\chi_3)(g)$	6	0	0	0	0	0

This is the regular character of G, by Theorem 13.19; and it takes the value $|G|$ on 1, and the value 0 on all non-identity elements of G, illustrating Proposition 13.20.

Permutation characters

In the case where G is a subgroup of the symmetric group S_n, there is an easy construction using the permutation module which produces a character of degree n, and we now describe this.

Suppose that G is a subgroup of S_n, so that G is a group of permutations of $\{1, \ldots, n\}$. The permutation module V for G over $\mathbb{C}$ has basis $v_1, \ldots, v_n$, where for all $g \in G$,

$$v_i g = v_{ig} \quad (1 \le i \le n)$$

(see Definition 4.10). Let $\mathscr{B}$ denote the basis $v_1, \ldots, v_n$. Then the ii-entry in the matrix $[g]_{\mathscr{B}}$ is 0 if $ig \ne i$, and is 1 if $ig = i$. Therefore the character π of the permutation module V is given by

$$\pi(g) = (\text{the number of } i \text{ such that } ig = i).$$

For $g \in G$, let

$$\text{fix}(g) = \{i\colon 1 \le i \le n \text{ and } ig = i\}.$$

Then

(13.22) $$\pi(g) = |\text{fix}(g)| \quad (g \in G).$$

We call π the *permutation character* of G.

13.23 Example
Let $G = S_4$. Then by Example 12.16(3), G has five conjugacy classes, with representatives

$$1, (1\,2), (1\,2\,3), (1\,2)(3\,4), (1\,2\,3\,4).$$

The permutation character π takes the values

g_i	1	$(1\,2)$	$(1\,2\,3)$	$(1\,2)(3\,4)$	$(1\,2\,3\,4)$
$\pi(g_i)$	4	2	1	0	0

13.24 Proposition

Let G be a subgroup of S_n. Then the function $v: G \to \mathbb{C}$ defined by

$$v(g) = |\text{fix}(g)| - 1 \quad (g \in G)$$

is a character of G.

Proof Let $v_1, \ldots, v_n$ be the basis of the permutation module V as above, and let

$$u = v_1 + \ldots + v_n, \text{ and } U = \text{sp}(u).$$

Observe that $ug = u$ for all $g \in G$, so U is a $\mathbb{C}G$-submodule of V. Indeed, U is isomorphic to the trivial $\mathbb{C}G$-module, so the character of U is the trivial character 1_G (see Example 13.8(3)). By Maschke's Theorem 8.1, there is a $\mathbb{C}G$-submodule W of V such that

$$V = U \oplus W.$$

Let v be the character of W. Then

$$\pi = 1_G + v,$$

so $|\text{fix}(g)| = 1 + v(g)$ for all $g \in G$, and therefore

$$v(g) = |\text{fix}(g)| - 1 \quad (g \in G). \qquad \blacksquare$$

13.25 Example

Let $G = A_4$, a subgroup of S_4. By Example 12.18(1), the conjugacy classes of G are represented by

$$1, (1\,2)(3\,4), (1\,2\,3), (1\,3\,2).$$

The values of the character v of G are

g_i	1	$(1\,2)(3\,4)$	$(1\,2\,3)$	$(1\,3\,2)$
$v(g_i)$	3	-1	0	0

Summary of Chapter 13

1. A character is obtained from a representation by taking the trace of each matrix.

2. Characters are constant on conjugacy classes.

3. Isomorphic $\mathbb{C}G$-modules have the same character.

4. For all characters χ of G, and all $g \in G$, the complex number $\chi(g)$ is a sum of roots of unity, and $\chi(g^{-1}) = \overline{\chi(g)}$.

5. The character of a representation determines the kernel of the representation.

6. The regular character χ_{reg} of G takes the value $|G|$ on the identity and the value 0 on all other elements of G.

7. If G is a subgroup of S_n, then the function ν which is given by

$$\nu(g) = |\text{fix}(g)| - 1 \quad (g \in G)$$

is a character of G.

Exercises for Chapter 13

1. Let $G = D_{12} = \langle a, b \colon a^6 = b^2 = 1, b^{-1}ab = a^{-1} \rangle$, and let ρ_1, ρ_2 be the representations of G for which

$$a\rho_1 = \begin{pmatrix} \omega & 0 \\ 0 & \omega^{-1} \end{pmatrix}, \ b\rho_1 = \begin{pmatrix} 0 & 1 \\ 1 & 0 \end{pmatrix} \quad (\text{where } \omega = e^{2\pi i/3})$$

and

$$a\rho_2 = \begin{pmatrix} -1 & 0 \\ 0 & 1 \end{pmatrix}, \ b\rho_2 = \begin{pmatrix} 1 & 0 \\ 0 & -1 \end{pmatrix}.$$

Find the characters of ρ_1 and ρ_2. Find also $\text{Ker}\,\rho_1$ and $\text{Ker}\,\rho_2$; check that your answers are consistent with Theorem 13.11.

2. Find all the irreducible characters of C_4. Write the regular character of C_4 as a linear combination of these.

3. Let χ be the character of the 7-dimensional permutation module for S_7. Find $\chi(x)$ for $x = (1\,2)$ and for $x = (1\,6)(2\,3\,5)$.

4. Prove that the only non-zero characters of G which are homomorphisms are the linear characters.

5. Assume that χ is an irreducible character of G. Suppose that $z \in Z(G)$ and that z has order m. Prove that there exists an mth root of unity $\lambda \in \mathbb{C}$ such that for all $g \in G$,

$$\chi(zg) = \lambda\chi(g).$$

6. Prove that if χ is a faithful irreducible character of the group G, then $Z(G) = \{g \in G \colon |\chi(g)| = \chi(1)\}$.

7. Let ρ be a representation of the group G over $\mathbb{C}$.

   (a) Show that $\delta: g \rightarrow \det(g\rho)$ $(g \in G)$ is a linear character of G.
   (b) Prove that $G/\operatorname{Ker}\delta$ is abelian.
   (c) Assume that $\delta(g) = -1$ for some $g \in G$. Show that G has a normal subgroup of index 2.

8. Let g be a group of order $2k$, where k is an odd integer. By considering the regular representation of G, show that G has a normal subgroup of index 2.

9. Let χ be a character of a group G, and let g be an element of order 2 in G. Show that either

   (1) $\chi(g) \equiv \chi(1) \bmod 4$, or
   (2) G has a normal subgroup of index 2.

   (Compare Corollary 13.10. Hint: use Exercise 7.)

10. Prove that if x is a non-identity element of the group G, then $\chi(x) \neq \chi(1)$ for some irreducible character χ of G.

14

Inner products of characters

We establish some significant properties of characters in this chapter, and in particular we prove the striking result (Theorem 14.21) that if two $\mathbb{C}G$-modules have the same character then they are isomorphic. Also, we describe a method for decomposing a given $\mathbb{C}G$-module as a direct sum of $\mathbb{C}G$-submodules, using characters.

The proofs rely on an inner product involving the characters of a group, and we describe this first.

Inner products

The characters of a finite group G are certain functions from G to $\mathbb{C}$. The set of all functions from G to $\mathbb{C}$ forms a vector space over $\mathbb{C}$, if we adopt the natural rules for adding functions and multiplying functions by complex numbers. That is, if ϑ, ϕ are functions from G to $\mathbb{C}$, and $\lambda \in \mathbb{C}$, then we define $\vartheta + \phi \colon G \to \mathbb{C}$ by

$$(\vartheta + \phi)(g) = \vartheta(g) + \phi(g) \quad (g \in G)$$

and we define $\lambda\vartheta \colon G \to \mathbb{C}$ by

$$\lambda\vartheta(g) = \lambda(\vartheta(g)) \quad (g \in G).$$

(We write these functions on the left to agree with our notation for characters.)

14.1 Example
Let $G = C_3 = \langle a \colon a^3 = 1 \rangle$, and suppose that $\vartheta \colon G \to \mathbb{C}$ and $\phi \colon G \to \mathbb{C}$ are given by

	1	a	a^2
ϑ	2	i	-1
ϕ	1	1	1

This means that $\vartheta(1) = 2$, $\vartheta(a) = i$, $\vartheta(a^2) = -1$ and $\phi(1) = \phi(a) = \phi(a^2) = 1$. Then $\vartheta + \phi$ and 3ϑ are given by

	1	a	a^2
$\vartheta + \phi$	3	$1 + i$	0
3ϑ	6	3i	-3

We shall often think of functions from G to $\mathbb{C}$ as row vectors, as in this example.

The vector space of all functions from G to $\mathbb{C}$ can be equipped with an inner product in a way which we shall describe shortly. The definition of an inner product on a vector space over $\mathbb{C}$ runs as follows. With every ordered pair of vectors ϑ, ϕ in the vector space, there is associated a complex number $\langle \vartheta, \phi \rangle$ which satisfies the following conditions:

(14.2) (a) $\langle \vartheta, \phi \rangle = \overline{\langle \phi, \vartheta \rangle}$ for all ϑ, ϕ;

   (b) $\langle \lambda_1 \vartheta_1 + \lambda_2 \vartheta_2, \phi \rangle = \lambda_1 \langle \vartheta_1, \phi \rangle + \lambda_2 \langle \vartheta_2, \phi \rangle$ for all $\lambda_1, \lambda_2 \in \mathbb{C}$ and all vectors $\vartheta_1, \vartheta_2, \phi$;

   (c) $\langle \vartheta, \vartheta \rangle > 0$ if $\vartheta \neq 0$.

Notice that condition (a) implies that $\langle \vartheta, \vartheta \rangle$ is always real, and that conditions (a) and (b) give

$$\langle \phi, \lambda_1 \theta_1 + \lambda_2 \vartheta_2 \rangle = \overline{\lambda}_1 \langle \phi, \vartheta_1 \rangle + \overline{\lambda}_2 \langle \phi, \vartheta_2 \rangle$$

for all $\lambda_1, \lambda_2 \in \mathbb{C}$ and all vectors $\vartheta_1, \vartheta_2, \phi$.

We now introduce an inner product on the vector space of all functions from G to $\mathbb{C}$. This will be of basic importance in our study of characters.

14.3 Definition
Suppose that ϑ and ϕ are functions from G to $\mathbb{C}$. Define

$$\langle \vartheta, \phi \rangle = \frac{1}{|G|} \sum_{g \in G} \vartheta(g)\overline{\phi(g)}.$$

It is transparent that the conditions of (14.2) hold, so $\langle\ ,\ \rangle$ is an inner product on the vector space of functions from G to $\mathbb{C}$.

14.4 Example
As in Example 14.1, suppose that $G = C_3 = \langle a: a^3 = 1\rangle$ and that ϑ and ϕ are given by

	1	a	a^2
ϑ	2	i	-1
ϕ	1	1	1

Then

$$\langle\theta, \phi\rangle = \tfrac{1}{3}(2\cdot 1 + i\cdot 1 - 1\cdot 1) = \tfrac{1}{3}(1 + i),$$

$$\langle\theta, \theta\rangle = \tfrac{1}{3}(2\cdot 2 + i\cdot\bar{i} + (-1)\cdot(-1)) = 2,$$

$$\langle\phi, \phi\rangle = \tfrac{1}{3}(1\cdot 1 + 1\cdot 1 + 1\cdot 1) = 1.$$

Inner products of characters

We can exploit the fact that characters are constant on conjugacy classes to simplify slightly the calculation of the inner product of two characters.

14.5 Proposition
Assume that G has exactly l conjugacy classes, with representatives $g_1, \ldots, g_l$. Let χ and ψ be characters of G.

(1) $\langle\chi, \psi\rangle = \langle\psi, \chi\rangle = \dfrac{1}{|G|}\displaystyle\sum_{g\in G}\chi(g)\psi(g^{-1})$, and this is a real number.

(2) $\langle\chi, \psi\rangle = \displaystyle\sum_{i=1}^{l}\dfrac{\chi(g_i)\overline{\psi(g_i)}}{|C_G(g_i)|}$.

Proof (1) We have $\overline{\psi(g)} = \psi(g^{-1})$ for all $g \in G$, by Proposition 13.9(3). Therefore

$$\langle\chi, \psi\rangle = \frac{1}{|G|}\sum_{g\in G}\chi(g)\psi(g^{-1}).$$

Since $\{g^{-1}: g \in G\} = G$, we also have

$$\langle \chi, \psi \rangle = \frac{1}{|G|} \sum_{g \in G} \chi(g^{-1}) \psi(g) = \langle \psi, \chi \rangle.$$

Since $\langle \psi, \chi \rangle = \overline{\langle \chi, \psi \rangle}$, it follows that $\langle \chi, \psi \rangle$ is real. (We shall prove later that $\langle \chi, \psi \rangle$ is, in fact, an integer.)

(2) Recall that g_i^G denotes the conjugacy class of G which contains g_i. Since characters are constant on conjugacy classes,

$$\sum_{g \in g_i^G} \chi(g)\overline{\psi(g)} = |g_i^G| \chi(g_i)\overline{\psi(g_i)}.$$

Now

$$G = \bigcup_{i=1}^{l} g_i^G \text{ and } |g_i^G| = |G|/|C_G(g_i)|,$$

by Corollary 12.3 and Theorem 12.8. Hence

$$\langle \chi, \psi \rangle = \frac{1}{|G|} \sum_{g \in G} \chi(g)\overline{\psi(g)} = \frac{1}{|G|} \sum_{i=1}^{l} \sum_{g \in g_i^G} \chi(g)\overline{\psi(g)}$$

$$= \sum_{i=1}^{l} \frac{|g_i^G|}{|G|} \chi(g_i)\overline{\psi(g_i)}$$

$$= \sum_{i=1}^{l} \frac{1}{|C_G(g_i)|} \chi(g_i)\overline{\psi(g_i)}. \qquad \blacksquare$$

14.6 Example

The alternating group A_4 has four conjugacy classes, with representatives

$$g_1 = 1, \quad g_2 = (1\,2)(3\,4), \quad g_3 = (1\,2\,3), \quad g_4 = (1\,3\,2)$$

(see Example 12.18(1)). We shall see in Chapter 18 that there are characters χ and ψ of A_4 which take the following values on the representatives g_i:

g_i $|C_G(g_i)|$	g_1 12	g_2 4	g_3 3	g_4 3
χ	1	1	ω	ω^2
ψ	4	0	ω^2	ω

(where $\omega = e^{2\pi i/3}$). Using part (2) of Proposition 14.5, we have

$$\langle \chi, \psi \rangle = \frac{1 \cdot 4}{12} + \frac{1 \cdot 0}{4} + \frac{\omega \cdot \overline{\omega}^2}{3} + \frac{\omega^2 \cdot \overline{\omega}}{3} = 0,$$

$$\langle \psi, \psi \rangle = \frac{4 \cdot 4}{12} + \frac{0 \cdot 0}{4} + \frac{\omega^2 \cdot \overline{\omega}^2}{3} + \frac{\omega \cdot \overline{\omega}}{3} = 2.$$

We advise you to check also that $\langle \chi, \chi \rangle = 1$, and to find the inner products of χ and ψ with the trivial character (which takes the value 1 on all elements of A_4).

We are now going to pave the way to proving the key fact (Theorem 14.12) that the irreducible characters of G form an orthonormal set of vectors in the vector space of functions from G to $\mathbb{C}$; that is, for distinct irreducible characters χ and ψ of G, we have $\langle \chi, \chi \rangle = 1$ and $\langle \chi, \psi \rangle = 0$.

Recall from Chapter 10 that the regular $\mathbb{C}G$-module is a direct sum of irreducible $\mathbb{C}G$-submodules, say

$$\mathbb{C}G = U_1 \oplus \ldots \oplus U_r,$$

and that every irreducible $\mathbb{C}G$-module is isomorphic to one of the $\mathbb{C}G$-modules $U_1, \ldots, U_r$. There are several ways of choosing $\mathbb{C}G$-submodules W_1 and W_2 of $\mathbb{C}G$ such that $\mathbb{C}G = W_1 \oplus W_2$ and W_1 and W_2 have no common composition factor (see Definition 10.4). For example, we may take W_1 to be the sum of those irreducible $\mathbb{C}G$-submodules U_i which are isomorphic to a given irreducible $\mathbb{C}G$-module, and then let W_2 be the sum of the remaining $\mathbb{C}G$-modules U_i. We shall investigate some consequences of writing $\mathbb{C}G$ like this; therefore, we temporarily adopt the following Hypothesis:

14.7 Hypothesis
Let $\mathbb{C}G = W_1 \oplus W_2$, where W_1 and W_2 are $\mathbb{C}G$-submodules which have no common composition factor. Write $1 = e_1 + e_2$ where $e_1 \in W_1$ and $e_2 \in W_2$.

Among other results, we shall derive a formula for e_1 in terms of the character of W_1.

We first look at the effect of applying the elements e_1 and e_2 of $\mathbb{C}G$ to W_1 and W_2.

14.8 Proposition
For all $w_1 \in W_1$ and $w_2 \in W_2$, we have

$$w_1 e_1 = w_1, \quad w_2 e_1 = 0,$$

$$w_1 e_2 = 0, \quad w_2 e_2 = w_2.$$

Proof If $w_1 \in W_1$ then the function $w_2 \to w_1 w_2$ ($w_2 \in W_2$) is clearly a $\mathbb{C}G$-homomorphism from W_2 to W_1. But W_2 and W_1 have no common composition factor, so every $\mathbb{C}G$-homomorphism from W_2 to W_1 is zero, by Proposition 11.3. Therefore $w_1 w_2 = 0$ for all $w_1 \in W_1$, $w_2 \in W_2$. Similarly $w_2 w_1 = 0$. In particular, $w_1 e_2 = w_2 e_1 = 0$. Now

$$w_1 = w_1 1 = w_1(e_1 + e_2) = w_1 e_1, \text{ and}$$

$$w_2 = w_2 1 = w_2(e_1 + e_2) = w_2 e_2,$$

and this completes the proof.   ∎

14.9 Corollary
For the elements e_1 and e_2 of $\mathbb{C}G$ which appear in Hypothesis 14.7, we have

$$e_1^2 = e_1, \ e_2^2 = e_2 \text{ and } e_1 e_2 = e_2 e_1 = 0.$$

Proof In Proposition 14.8, take $w_1 = e_1$ and $w_2 = e_2$.   ∎

Next, we evaluate e_1.

14.10. Proposition
Let χ be the character of the $\mathbb{C}G$-module W_1 which appears in Hypothesis 14.7. Then

$$e_1 = \frac{1}{|G|} \sum_{g \in G} \chi(g^{-1}) g.$$

Proof Let $x \in G$. The function

$$\vartheta: w \to w e_1 x^{-1} \quad (w \in \mathbb{C}G)$$

is an endomorphism of $\mathbb{C}G$. We shall calculate the trace of ϑ in two ways.

First, for $w_1 \in W_1$ and $w_2 \in W_2$ we have, in view of Proposition 14.8,

$$w_1 \vartheta = w_1 e_1 x^{-1} = w_1 x^{-1},$$

$$w_2 \vartheta = w_2 e_1 x^{-1} = 0.$$

Thus ϑ acts on W_1 by $w_1 \to w_1 x^{-1}$ and on W_2 by $w_2 \to 0$. By the definition of the character χ of W_1, the endomorphism $w_1 \to w_1 x^{-1}$ of W_1 has trace equal to $\chi(x^{-1})$, and of course the endomorphism $w_2 \to 0$ of W_2 has trace 0. Therefore

$$\operatorname{tr} \vartheta = \chi(x^{-1}).$$

Secondly, $e_1 \in \mathbb{C}G$, so

$$e_1 = \sum_{g \in G} \lambda_g g$$

for some $\lambda_g \in \mathbb{C}$. By Proposition 13.20, the endomorphism $w \to wgx^{-1}$ ($w \in \mathbb{C}G$) of $\mathbb{C}G$ has trace 0 if $g \neq x$ and has trace $|G|$ if $g = x$. Hence, as $\vartheta \colon w \to w \sum_{g \in G} \lambda_g gx^{-1}$, we have

$$\operatorname{tr} \vartheta = \lambda_x |G|.$$

Comparing our two expressions for $\operatorname{tr} \vartheta$, we see that for all $x \in G$,

$$\lambda_x = \chi(x^{-1})/|G|.$$

Therefore

$$e_1 = \frac{1}{|G|} \sum_{g \in G} \chi(g^{-1}) g. \qquad \blacksquare$$

14.11 Corollary

Let χ be the character of the $\mathbb{C}G$-module W_1 which appears in Hypothesis 14.7. Then

$$\langle \chi, \chi \rangle = \chi(1).$$

Proof Using the definition 6.3 of the multiplication in $\mathbb{C}G$, we deduce from Proposition 14.10 that the coefficient of 1 in e_1^2 is

$$\frac{1}{|G|^2} \sum_{g \in G} \chi(g^{-1})\chi(g) = \frac{1}{|G|} \langle \chi, \chi \rangle.$$

On the other hand, we know from Corollary 14.9 that $e_1^2 = e_1$, and the coefficient of 1 in e_1 is $\chi(1)/|G|$. Hence $\langle \chi, \chi \rangle = \chi(1)$, as required. ∎

We can now prove the main theorem concerning the inner product $\langle \, , \, \rangle$.

14.12 Theorem
Let U and V be non-isomorphic irreducible $\mathbb{C}G$-modules, with characters χ and ψ, respectively. Then

$$\langle \chi, \chi \rangle = 1, \text{ and}$$

$$\langle \chi, \psi \rangle = 0.$$

Proof Recall from Theorem 11.9 that $\mathbb{C}G$ is a direct sum of irreducible $\mathbb{C}G$-submodules, say

$$\mathbb{C}G = U_1 \oplus \ldots \oplus U_r,$$

where the number of $\mathbb{C}G$-submodules U_i which are isomorphic to U is $\dim U$. Let $m = \dim U$, and define W to be the sum of the m irreducible $\mathbb{C}G$-submodules U_i which are isomorphic to U; let X be the sum of the remaining $\mathbb{C}G$-submodules U_i. Then

$$\mathbb{C}G = W \oplus X.$$

Moreover, every composition factor of W is isomorphic to U, and no composition factor of X is isomorphic to U. In particular, W and X have no common composition factor. The character of W is $m\chi$, since W is the direct sum of m $\mathbb{C}G$-submodules, each of which has character χ.

We now apply Corollary 14.11 to the character of W, and obtain

$$\langle m\chi, m\chi \rangle = m\chi(1).$$

As $\chi(1) = \dim U = m$, this yields $\langle \chi, \chi \rangle = 1$.

Next, let Y be the sum of those $\mathbb{C}G$-submodules U_i of $\mathbb{C}G$ which are isomorphic to either U or V, and let Z be the sum of the remaining $\mathbb{C}G$-submodules U_i. Then

$$\mathbb{C}G = Y \oplus Z,$$

and Y and Z have no common composition factor. The character of Y is $m\chi + n\psi$, where $n = \dim V$. By Corollary 14.11,

$$m\chi(1) + n\psi(1) = \langle m\chi + n\psi, \, m\chi + n\psi \rangle$$
$$= m^2 \langle \chi, \chi \rangle + n^2 \langle \psi, \psi \rangle + mn(\langle \chi, \psi \rangle + \langle \psi, \chi \rangle).$$

Now $\langle \chi, \chi \rangle = \langle \psi, \psi \rangle = 1$, by the part of the theorem which we have already proved; and $\chi(1) = m$, $\psi(1) = n$. Therefore

$$\langle \chi, \psi \rangle + \langle \psi, \chi \rangle = 0.$$

By Proposition 14.5(1), $\langle \chi, \psi \rangle = \langle \psi, \chi \rangle$, and hence $\langle \chi, \psi \rangle = 0$. ∎

Applications of Theorem 14.12

Let G be a finite group, and let $V_1, \ldots, V_k$ be a complete set of non-isomorphic irreducible $\mathbb{C}G$-modules (see Definition 11.11). If χ_i is the character of V_i $(1 \leqslant i \leqslant k)$, then by Theorem 14.12, we have

$$(14.13) \qquad \langle \chi_i, \chi_j \rangle = \delta_{ij} \quad \text{for all } i, j,$$

where δ_{ij} is the Kronecker delta function (that is, δ_{ij} is 1 if $i = j$ and is 0 if $i \neq j$). In particular, this implies that the irreducible characters $\chi_1, \ldots, \chi_k$ are all distinct.

Now let V be a $\mathbb{C}G$-module. By Theorem 8.7, V is equal to a direct sum of irreducible $\mathbb{C}G$-submodules. Each of these is isomorphic to some V_i, so there are non-negative integers $d_1, \ldots, d_k$ such that

$$(14.14) \qquad V \cong (V_1 \oplus \ldots \oplus V_1) \oplus (V_2 \oplus \ldots \oplus V_2) \oplus \ldots$$
$$\oplus (V_k \oplus \ldots \oplus V_k),$$

where for each i, there are d_i factors V_i.

Therefore the character ψ of V is given by

$$(14.15) \qquad \psi = d_1\chi_1 + \ldots + d_k\chi_k.$$

Using (14.13), we obtain from this

$$(14.16) \qquad \langle \psi, \chi_i \rangle = \langle \chi_i, \psi \rangle = d_i \text{ for } 1 \leqslant i \leqslant k, \text{ and}$$
$$\langle \psi, \psi \rangle = d_1^2 + \ldots + d_k^2.$$

Summarizing, we have

14.17 Theorem

Let $\chi_1, \ldots, \chi_k$ be the irreducible characters of G. If ψ is any character of G, then

$$\psi = d_1\chi_1 + \ldots + d_k\chi_k$$

for some non-negative integers $d_1, \ldots, d_k$. Moreover,

$$d_i = \langle \psi, \chi_i \rangle \quad for \; 1 \leq i \leq k, \; and$$

$$\langle \psi, \psi \rangle = \sum_{i=1}^{k} d_i^2.$$

14.18 Example

Recall from Example 13.6(4) that the irreducible characters of $S_3 \cong D_6$ are χ_1, χ_2, χ_3, taking the following values on the conjugacy class representatives 1, (1 2), (1 2 3):

Now let ψ be the character of the 3-dimensional permutation module

| g_i | 1 | (1 2) | (1 2 3) |
$\|C_{S_3}(g_i)\|$	6	2	3
χ_1	1	1	1
χ_2	1	−1	1
χ_3	2	0	−1

for S_3. By Example 13.6(2), we know that

$$\psi(1) = 3, \; \psi(1\,2) = 1, \; \psi(1\,2\,3) = 0.$$

Therefore, by Proposition 14.5(2),

$$\langle \psi, \chi_1 \rangle = \frac{3 \cdot 1}{6} + \frac{1 \cdot 1}{2} + 0 = 1.$$

Similarly, $\langle \psi, \chi_2 \rangle = 0$ and $\langle \psi, \chi_3 \rangle = 1$. Thus by Theorem 14.17,

$$\psi = \chi_1 + \chi_3.$$

(This can of course be checked immediately by comparing the values of ψ and $\chi_1 + \chi_3$ on each conjugacy class representative.)

A more substantial calculation along these lines is given in Example 15.7.

We shall see many more applications of the important Theorem 14.17.

14.19 Definition

Suppose that ψ is a character of G, and that χ is an irreducible character of G. We say that χ is a *constituent* of ψ if $\langle \psi, \chi \rangle \neq 0$. Thus, the constituents of ψ are the irreducible characters χ_i of G for which the integer d_i in the expression $\psi = d_1 \chi_1 + \ldots + d_k \chi_k$ is non-zero.

The next result is another significant consequence of Theorem 14.12. It gives us a quick and effective method of determining whether or not a given $\mathbb{C}G$-module is irreducible.

14.20 Theorem

Let V be a $\mathbb{C}G$-module with character ψ. Then V is irreducible if and only if $\langle \psi, \psi \rangle = 1$.

Proof If V is irreducible then $\langle \psi, \psi \rangle = 1$ by Theorem 14.12.

Conversely, assume that $\langle \psi, \psi \rangle = 1$. We have

$$\psi = d_1 \chi_1 + \ldots + d_k \chi_k$$

for some non-negative integers d_i, and by (14.16),

$$1 = \langle \psi, \psi \rangle = d_1^2 + \ldots + d_k^2.$$

It follows that one of the integers d_i is 1 and the rest are zero. Then by (14.14), $V \cong V_i$ for some i, and so V is irreducible. ∎

We are now in a position to prove the remarkable result that 'a $\mathbb{C}G$-module is determined by its character'. It is this fact which motivates our study of characters in much of the rest of the book, for it means that many questions about $\mathbb{C}G$-modules can be answered using character theory.

14.21 Theorem

Suppose that V and W are $\mathbb{C}G$-modules, with characters χ and ψ, respectively. Then V and W are isomorphic if and only if $\chi = \psi$.

Proof In Proposition 13.5 we proved the elementary fact that if $V \cong W$ then $\chi = \psi$. It is the converse which is the substantial part of this theorem.

Thus, suppose that $\chi = \psi$. Again let $V_1, \ldots, V_k$ be a complete set of non-isomorphic irreducible $\mathbb{C}G$-modules with characters $\chi_1, \ldots, \chi_k$. We know by (14.14) that there are non-negative integers c_i, d_i $(1 \leqslant i \leqslant k)$ such that

$$V \cong (V_1 \oplus \ldots \oplus V_1) \oplus (V_2 \oplus \ldots \oplus V_2) \oplus \ldots$$
$$\oplus (V_k \oplus \ldots \oplus V_k)$$

with c_i factors V_i for each i, and

$$W \cong (V_1 \oplus \ldots \oplus V_1) \oplus (V_2 \oplus \ldots \oplus V_2) \oplus \ldots$$
$$\oplus (V_k \oplus \ldots \oplus V_k)$$

with d_i factors V_i for each i. By (14.16),

$$c_i = \langle \chi, \chi_i \rangle, \; d_i = \langle \psi, \chi_i \rangle \quad (1 \leqslant i \leqslant k).$$

Since $\chi = \psi$, it follows that $c_i = d_i$ for all i, and hence $V \cong W$. ∎

14.22 Example
Let $G = C_3 = \langle a : a^3 = 1 \rangle$, and let ρ_1, ρ_2, ρ_3, ρ_4 be the representations of G over $\mathbb{C}$ for which

$$a\rho_1 = \begin{pmatrix} \omega & 0 \\ 0 & \omega \end{pmatrix}, \; a\rho_2 = \begin{pmatrix} \omega & 0 \\ 0 & \omega^{-1} \end{pmatrix},$$

$$a\rho_3 = \begin{pmatrix} 0 & 1 \\ -1 & -1 \end{pmatrix}, \; a\rho_4 = \begin{pmatrix} 1 & \omega^{-1} \\ 0 & \omega \end{pmatrix}$$

($\omega = e^{2\pi i/3}$). The characters ψ_i of the representations ρ_i ($i = 1, 2, 3, 4$) are

	1	a	a^2
ψ_1	2	2ω	$2\omega^2$
ψ_2	2	-1	-1
ψ_3	2	-1	-1
ψ_4	2	$1 + \omega$	$1 + \omega^2$

Hence by Theorem 14.21, the representations ρ_2 and ρ_3 are equivalent, but there are no other equivalences among ρ_1, ρ_2, ρ_3 and ρ_4.

The next theorem is another consequence of Theorem 14.12.

14.23 Theorem
Let $\chi_1, \ldots, \chi_k$ be the irreducible characters of G. Then $\chi_1, \ldots, \chi_k$ are linearly independent vectors in the vector space of all functions from G to $\mathbb{C}$.

Proof Assume that

$$\lambda_1 \chi_1 + \ldots + \lambda_k \chi_k = 0 \quad (\lambda_i \in \mathbb{C}).$$

Then for all i, using (14.13) we have

$$0 = \langle \lambda_1 \chi_1 + \ldots + \lambda_k \chi_k, \chi_i \rangle = \lambda_i.$$

Therefore $\chi_1, \ldots, \chi_k$ are linearly independent. ∎

We now relate inner products of characters to the spaces of $\mathbb{C}G$-homomorphisms which we constructed in Chapter 11.

14.24 Theorem
Let V and W be $\mathbb{C}G$-modules with characters χ and ψ, respectively. Then

$$\dim (\mathrm{Hom}_{\mathbb{C}G} (V, W)) = \langle \chi, \psi \rangle.$$

Proof We know from (14.14) that there are non-negative integers c_i, d_i $(1 \leq i \leq k)$ such that

$$V \cong (V_1 \oplus \ldots \oplus V_1) \oplus (V_2 \oplus \ldots \oplus V_2) \oplus \ldots$$
$$\oplus (V_k \oplus \ldots \oplus V_k)$$

with c_i factors V_i for each i, and

$$W \cong (V_1 \oplus \ldots \oplus V_1) \oplus (V_2 \oplus \ldots \oplus V_2) \oplus \ldots$$
$$\oplus (V_k \oplus \ldots \oplus V_k)$$

with d_i factors V_i for each i. By Proposition 11.2, for any i, j we have

$$\dim (\mathrm{Hom}_{\mathbb{C}G} (V_i, V_j)) = \delta_{ij}.$$

Hence, using (11.5)(3) we see that

$$\dim (\mathrm{Hom}_{\mathbb{C}G} (V, W)) = \sum_{i=1}^{k} c_i d_i.$$

On the other hand,

$$\chi = \sum_{i=1}^{k} c_i\chi_i \text{ and } \psi = \sum_{i=1}^{k} d_i\chi_i$$

and so (14.13) implies that

$$\langle \chi, \psi \rangle = \sum_{i=1}^{k} c_i d_i.$$

The result follows. ∎

Decomposing $\mathbb{C}G$-modules

It is sometimes of practical importance to be able to decompose a given $\mathbb{C}G$-module into a direct sum of $\mathbb{C}G$-submodules, and we now describe a process for doing this.

Once more we adopt Hypothesis 14.7:

> $\mathbb{C}G = W_1 \oplus W_2$, where the $\mathbb{C}G$-modules W_1 and W_2 have no common composition factor; and $1 = e_1 + e_2$ with $e_1 \in W_1$, $e_2 \in W_2$.

Let V be any $\mathbb{C}G$-module. We can write $V = V_1 \oplus V_2$, where every composition factor of V_1 is a composition factor of W_1 and every composition factor of V_2 is a composition factor of W_2.

14.25 Proposition
With the above notation, for all $v_1 \in V_1$ and $v_2 \in V_2$ we have

$$v_1 e_1 = v_1, \quad v_2 e_1 = 0,$$

$$v_1 e_2 = 0, \quad v_2 e_2 = v_2.$$

Proof If $v_1 \in V_1$ then the function $w_2 \to v_1 w_2$ $(w_2 \in W_2)$ is clearly a $\mathbb{C}G$-homomorphism from W_2 to V_1. Since W_2 and V_1 have no common composition factor, we deduce the stated results just as in the proof of Proposition 14.8. ∎

14.26 Proposition
If χ is an irreducible character of G, and V is any $\mathbb{C}G$-module, then

$$V\left(\sum_{g \in G} \chi(g^{-1})g\right)$$

is equal to the sum of those $\mathbb{C}G$-submodules of V which have character χ (where for $r \in \mathbb{C}G$, we define $Vr = \{vr : v \in V\}$).

Proof Write

$$\mathbb{C}G = U_1 \oplus \ldots \oplus U_r,$$

a direct sum of irreducible $\mathbb{C}G$-submodules U_i. Let W_1 be the sum of those $\mathbb{C}G$-submodules U_i which have character χ, and let W_2 be the sum of the remaining $\mathbb{C}G$-submodules U_i. Then the character of W_1 is $m\chi$ where $m = \chi(1)$, by Theorem 11.9. Also W_1 and W_2 satisfy Hypothesis 14.7, and by Proposition 14.10, the element e_1 of W_1 is given by

$$e_1 = \frac{m}{|G|} \sum_{g \in G} \chi(g^{-1})g.$$

Let V_1 be the sum of those $\mathbb{C}G$-submodules of V which have character χ. Then Proposition 14.25 shows that $Ve_1 = V_1$. Clearly we may omit the constant multiplier $m/|G|$, so

$$V_1 = V\left(\sum_{g \in G} \chi(g^{-1})g\right). \qquad \blacksquare$$

Once the irreducible characters of our group G are known, Proposition 14.26 provides a useful practical tool for finding $\mathbb{C}G$-submodules of a given $\mathbb{C}G$-module V. The procedure is as follows:

(14.27) (1) Choose a basis $v_1, \ldots, v_n$ of V.
        (2) For each irreducible character χ of G, calculate the vectors $v_i(\sum_{g \in G}\chi(g^{-1})g)$ for $1 \leqslant i \leqslant n$, and let V_χ be the subspace of V spanned by these vectors.
        (3) Then V is the direct sum of the $\mathbb{C}G$-modules V_χ as χ runs over the irreducible characters of G. The character of V_χ is a multiple of χ.

We illustrate this method with a couple of simple examples. Some more complicated uses of the method can be found in Chapter 32.

14.28 Examples
(1) Let G be any finite group and let V be any non-zero $\mathbb{C}G$-module. Taking χ to be the trivial character of G in Proposition 14.26, we see that

$$V\left(\sum_{g\in G} g\right)$$

is the sum of all the trivial $\mathbb{C}G$-submodules of V. For example, let $G = S_n$ and let V be the permutation module, with basis $v_1, \ldots, v_n$ such that $v_i g = v_{ig}$ for all i and all $g \in G$. Then

$$V\left(\sum_{g\in G} g\right) = \text{sp}\,(v_1 + \ldots + v_n).$$

Hence V has a unique trivial $\mathbb{C}G$-submodule.

(2) Let G be the subgroup of S_4 which is generated by

$$a = (1\,2\,3\,4) \text{ and } b = (1\,2)(3\,4).$$

Then $G \cong D_8$ (compare Example 1.5). Here is a list of the irreducible characters $\chi_1, \ldots, \chi_5$ of D_8 (see Example 16.3(3)):

	1	a	a^2	a^3	b	ab	a^2b	a^3b
χ_1	1	1	1	1	1	1	1	1
χ_2	1	1	1	1	−1	−1	−1	−1
χ_3	1	−1	1	−1	1	−1	1	−1
χ_4	1	−1	1	−1	−1	1	−1	1
χ_5	2	0	−2	0	0	0	0	0

Let V be the permutation module for G, with basis v_1, v_2, v_3, v_4 such that $v_i g = v_{ig}$ for all i and all $g \in G$.

For $1 \le i \le 5$, let

$$e_i = \frac{\chi_i(1)}{8} \sum_{g\in G} \chi_i(g^{-1}) g.$$

For example, $e_5 = \frac{1}{2}(1 - a^2)$. Then

$$Ve_1 = \text{sp}\,(v_1 + v_2 + v_3 + v_4),$$

$$Ve_2 = 0,$$

$$Ve_3 = 0,$$

$$Ve_4 = \text{sp}\,(v_1 - v_2 + v_3 - v_4),$$

$$Ve_5 = \text{sp}\,(v_1 - v_3, v_2 - v_4).$$

We have

$$V = Ve_1 \oplus Ve_4 \oplus Ve_5,$$

and so we have expressed V as a direct sum of irreducible $\mathbb{C}G$-submodules whose characters are χ_1, χ_4 and χ_5, respectively.

You might like to check that

$$e_1 + \ldots + e_5 = 1,$$

$$e_i^2 = e_i \text{ for } 1 \leq i \leq 5,$$

$$e_i e_j = 0 \text{ for } i \neq j.$$

Compare these results with Corollary 14.9.

Note that the procedure described in (14.27) does *not* in general enable us to write a given $\mathbb{C}G$-module as a direct sum of irreducible $\mathbb{C}G$-submodules (since V_χ is not in general irreducible).

Summary of Chapter 14

1. The inner product of two functions ϑ, ϕ from G to $\mathbb{C}$ is given by

$$\langle \vartheta, \phi \rangle = \frac{1}{|G|} \sum_{g \in G} \vartheta(g)\overline{\phi(g)}.$$

2. The irreducible characters $\chi_1, \ldots, \chi_k$ of G form an orthonormal set; that is, $\langle \chi_i, \chi_j \rangle = \delta_{ij}$ for all i, j.

3. Every $\mathbb{C}G$-module is determined by its character.

4. If $\chi_1, \ldots, \chi_k$ are the irreducible characters of G, and ψ is any character, then

$$\psi = d_1 \chi_1 + \ldots + d_k \chi_k \text{ where } d_i = \langle \psi, \chi_i \rangle.$$

Each d_i is a non-negative integer. Also, ψ is irreducible if and only if $\langle \psi, \psi \rangle = 1$.

Exercises for Chapter 14

1. Let $G = S_4$. We shall see in Chapter 18 that G has characters χ and ψ which take the following values on the conjugacy classes:

Class representative	1	(1 2)	(1 2 3)	(1 2)(3 4)	(1 2 3 4)
$\lvert C_G(g_i)\rvert$	24	4	3	8	4
χ	3	-1	0	3	-1
ψ	3	1	0	-1	-1

   Calculate $\langle \chi, \chi \rangle$, $\langle \chi, \psi \rangle$ and $\langle \psi, \psi \rangle$. Which of χ and ψ is irreducible?

2. Let $G = Q_8 = \langle a, b : a^4 = 1, b^2 = a^2, b^{-1}ab = a^{-1} \rangle$, and let ρ_1, ρ_2, ρ_3 be the representations of G over $\mathbb{C}$ for which

$$a\rho_1 = \begin{pmatrix} i & 0 \\ 0 & -i \end{pmatrix}, b\rho_1 = \begin{pmatrix} 0 & 1 \\ -1 & 0 \end{pmatrix},$$

$$a\rho_2 = \begin{pmatrix} 0 & i \\ i & 0 \end{pmatrix}, b\rho_2 = \begin{pmatrix} 0 & -1 \\ 1 & 0 \end{pmatrix},$$

$$a\rho_3 = \begin{pmatrix} -1 & 0 \\ 0 & 1 \end{pmatrix}, b\rho_3 = \begin{pmatrix} 1 & 0 \\ 0 & -1 \end{pmatrix}.$$

   Show that ρ_1 and ρ_2 are equivalent, but ρ_3 is not equivalent to ρ_1 or ρ_2.

3. Suppose that ρ and σ are representations of G, and that for each g in G there is an invertible matrix T_g such that

$$g\sigma = T_g^{-1}(g\rho)T_g.$$

   Prove that there is an invertible matrix T such that for all g in G,

$$g\sigma = T^{-1}(g\rho)T.$$

4. Suppose that χ is a non-zero, non-trivial character of G, and that $\chi(g)$ is a non-negative real number for all g in G. Prove that χ is reducible.

5. If χ is a character of G, show that

$$\langle \chi_{\text{reg}}, \chi \rangle = \chi(1).$$

6. If π is the permutation character of S_n, prove that

$$\langle \pi, 1_{S_n} \rangle = 1.$$

(Hint: you may find Exercise 11.4 relevant.)

7. Let $\chi_1, \ldots, \chi_k$ be the irreducible characters of the group G, and suppose that

$$\psi = d_1\chi_1 + \ldots d_k\chi_k$$

is a character of G. What can you say about the integers d_i in the cases $\langle \psi, \psi \rangle = 1, 2, 3$ or 4?

8. Suppose that χ is a character of G and that for every $g \in G$, $\chi(g)$ is an even integer. Does it follow that $\chi = 2\phi$ for some character ϕ?

15
The number of irreducible characters

We devote this chapter to the theorem which states that the number of irreducible characters of a finite group is equal to the number of conjugacy classes of the group, and to some consequences of this theorem. Together with the material from Chapter 14, the theorem provides machinery for investigating characters which is used in the remainder of the book.

Throughout, G is as usual a finite group.

Class functions

15.1 Definition
A *class function* on G is a function $\psi: G \to \mathbb{C}$ such that $\psi(x) = \psi(y)$ whenever x and y are conjugate elements of G (that is, ψ is constant on conjugacy classes).

By Proposition 13.5(2), the characters of G are class functions on G. The set C of all class functions on G is a subspace of the vector space of all functions from G to $\mathbb{C}$. A basis of C is given by those functions which take the value 1 on precisely one conjugacy class and zero on all other classes. Thus, if l is the number of conjugacy classes of G, then

(15.2) $\dim C = l.$

15.3 Theorem
The number of irreducible characters of G is equal to the number of conjugacy classes of G.

152

Proof Let $\chi_1, \ldots, \chi_k$ be the irreducible characters of G, and let l be the number of conjugacy classes of G. By Theorem 14.23, $\chi_1, \ldots, \chi_k$ are linearly independent elements of C, so (15.2) implies that $k \leq l$.

In order to prove the reverse inequality $l \leq k$, we consider the regular $\mathbb{C}G$-module. If $V_1, \ldots, V_k$ is a complete set of non-isomorphic irreducible $\mathbb{C}G$-modules, we know from Theorem 8.7 that

$$\mathbb{C}G = W_1 \oplus \ldots \oplus W_k,$$

where for each i, W_i is isomorphic to a direct sum of copies of V_i. Since $\mathbb{C}G$ contains the identity element 1, we can write

$$1 = f_1 + \ldots + f_k$$

with $f_i \in W_i$ for $1 \leq i \leq k$.

Now let $z \in Z(\mathbb{C}G)$, the centre of $\mathbb{C}G$. By Proposition 9.14, for each i there exists $\lambda_i \in \mathbb{C}$ such that for all $v \in V_i$

$$vz = \lambda_i v.$$

Hence $wz = \lambda_i w$ for all $w \in W_i$, and in particular,

$$f_i z = \lambda_i f_i \quad (1 \leq i \leq k).$$

It follows that

$$z = 1z = (f_1 + \ldots + f_k)z = f_1 z + \ldots + f_k z$$
$$= \lambda_1 f_1 + \ldots + \lambda_k f_k.$$

This shows that $Z(\mathbb{C}G)$ is contained in the subspace of $\mathbb{C}G$ spanned by $f_1, \ldots, f_k$. Since $Z(\mathbb{C}G)$ has dimension l by Proposition 12.22, we deduce that $l \leq k$. This completes the proof that $k = l$. ∎

15.4 Corollary
The irreducible characters $\chi_1, \ldots, \chi_k$ of G form a basis of the vector space of all class functions on G. Indeed, if ψ is a class function, then

$$\psi = \sum_{i=1}^{k} \lambda_i \chi_i$$

where $\lambda_i = \langle \psi, \chi_i \rangle$ for $1 \leq i \leq k$.

Proof Since $\chi_1, \ldots, \chi_k$ are linearly independent, they span a subspace of C of dimension k. By (15.2), $\dim C = l$, which is equal to k by

Theorem 15.3. Hence $\chi_1, \ldots, \chi_k$ span C, and so they form a basis of C. The last part follows, using (14.13).   ∎

Corollary 15.4 has the following useful consequence.

15.5 Proposition
Suppose that $g, h \in G$. Then g is conjugate to h if and only if $\chi(g) = \chi(h)$ for all characters χ of G.

Proof If g is conjugate to h then $\chi(g) = \chi(h)$ for all characters χ of G, by Proposition 13.5(2).

Conversely, suppose that $\chi(g) = \chi(h)$ for all characters χ. Then by Corollary 15.4, $\psi(g) = \psi(h)$ for all class functions ψ on G. In particular, this is true for the class function ψ which takes the value 1 on the conjugacy class of g and takes the value 0 elsewhere. Then $\psi(g) = \psi(h) = 1$, and so g is conjugate to h.   ∎

15.6 Corollary
Suppose that $g \in G$. Then g is conjugate to g^{-1} if and only if $\chi(g)$ is real for all characters χ of G.

Proof Since $\chi(g)$ is real if and only if $\chi(g) = \chi(g^{-1})$ (see Proposition 13.9(3)), the result follows immediately from Proposition 15.5.   ∎

We conclude the chapter with an example illustrating some practical methods of expressing characters and class functions of a group as combinations of irreducible characters. As in previous examples, we regard a character χ of G as a row vector, whose k entries are the values of χ on the k conjugacy classes of G.

15.7 Example
We shall see in Section 18.4 that there is a certain group G of order 12 which has exactly six conjugacy classes with representatives $g_1, \ldots, g_6$ (where $g_1 = 1$), and six irreducible characters $\chi_1, \ldots, \chi_6$ given as follows:

g_i $|C_G(g_i)|$	g_1 12	g_2 12	g_3 6	g_4 6	g_5 4	g_6 4
χ_1	1	1	1	1	1	1
χ_2	1	-1	-1	1	i	$-i$
χ_3	1	1	1	1	-1	-1
χ_4	1	-1	-1	1	$-i$	i
χ_5	2	2	-1	-1	0	0
χ_6	2	-2	1	-1	0	0

Suppose we are given characters χ and ψ of G as follows:

	g_1	g_2	g_3	g_4	g_5	g_6
χ	3	-3	0	0	i	$-i$
ψ	4	0	0	4	0	0

Then it is easy to spot that

$$\chi = \chi_2 + \chi_6, \quad \psi = \chi_1 + \chi_2 + \chi_3 + \chi_4.$$

For example, the second entry in the row vector for χ is equal to minus the first entry. Inspecting the values of the irreducible characters χ_i, we see that χ must be a combination of χ_2, χ_4 and χ_6. The correct answer now comes quickly to mind.

In fact, given any character ϕ of G whose degree is not large compared with the degrees of the χ_i, it is not hard to use tactical guesswork to express ϕ as a combination of the irreducible characters. The reason for this is that the required coefficients are known to be non-negative integers, and the entries in the column corresponding to $g_1 = 1$ are positive integers (indeed, they are the degrees of the χ_i).

We suggest that you use the 'guesswork method' to express the following characters λ, μ of G as combinations of $\chi_1, \ldots, \chi_6$:

	g_1	g_2	g_3	g_4	g_5	g_6
λ	2	-2	-2	2	0	0
μ	4	4	1	1	0	0

How do we cope with a class function or with a more difficult character, like the following one?

	g_1	g_2	g_3	g_4	g_5	g_6
ϕ	11	3	-3	5	$-1 + 2i$	$-1 - 2i$

The answer is to use the inner product $\langle \ , \ \rangle$. We know from Corollary 15.4 that the coefficients λ_i in the expression

$$\phi = \lambda_1 \chi_1 + \ldots + \lambda_6 \chi_6$$

are given by

$$\lambda_i = \langle \phi, \chi_i \rangle \quad (1 \leqslant i \leqslant 6).$$

Using Proposition 14.5(2), we calculate these inner products:

$$\langle \phi, \chi_1 \rangle = \frac{11 \cdot 1}{12} + \frac{3 \cdot 1}{12} + \frac{-3 \cdot 1}{6} + \frac{5 \cdot 1}{6} + \frac{(-1 + 2i) \cdot 1}{4}$$
$$+ \frac{(-1 - 2i) \cdot 1}{4} = 1,$$

$$\langle \phi, \chi_2 \rangle = \frac{11 \cdot 1}{12} + \frac{3 \cdot (-1)}{12} + \frac{(-3) \cdot (-1)}{6} + \frac{5 \cdot 1}{6} + \frac{(-1 + 2i) \cdot (-i)}{4}$$
$$+ \frac{(-1 - 2i) \cdot i}{4} = 3,$$

$$\langle \phi, \chi_3 \rangle = \frac{11 \cdot 1}{12} + \frac{3 \cdot 1}{12} + \frac{-3 \cdot 1}{6} + \frac{5 \cdot 1}{6} + \frac{(-1 + 2i) \cdot (-1)}{4}$$
$$+ \frac{(-1 - 2i) \cdot (-1)}{4} = 2,$$

and similarly $\langle \phi, \chi_4 \rangle = 1$, $\langle \phi, \chi_5 \rangle = 2$ and $\langle \phi, \chi_6 \rangle = 0$. Therefore

$$\phi = \chi_1 + 3\chi_2 + 2\chi_3 + \chi_4 + 2\chi_5.$$

Summary of Chapter 15

1. The number of irreducible characters of a group is equal to the number of conjugacy classes of the group.

2. The irreducible characters $\chi_1, \ldots, \chi_k$ of G form a basis of the vector space of all class functions on G. If ψ is a class function, then

$$\psi = \sum_{i=1}^{k} \lambda_i \chi_i \text{ where } \lambda_i = \langle \psi, \chi_i \rangle.$$

Exercises for Chapter 15

1. The three irreducible characters of S_3 are χ_1, χ_2, χ_3:

	1	(1 2)	(1 2 3)
χ_1	1	1	1
χ_2	1	-1	1
χ_3	2	0	-1

Let χ be the class function on S_3 with the following values:

	1	(1 2)	(1 2 3)
χ	19	-1	-2

Express χ as a linear combination of χ_1, χ_2 and χ_3. Is χ a character of S_3?

2. Let ψ_1, ψ_2 and ψ_3 be the class functions on S_3 taking the following values:

	1	(1 2)	(1 2 3)
ψ_1	1	0	0
ψ_2	0	1	0
ψ_3	0	0	1

Express ψ_1, ψ_2 and ψ_3 as linear combinations of the irreducible characters χ_1, χ_2 and χ_3 of S_3.

3. Suppose that G is the group of order 12 in Example 15.7, with conjugacy class representatives $g_1, \ldots, g_6$ and irreducible characters $\chi_1, \ldots, \chi_6$ as in that example. Let ψ be the class function on G taking the following values:

	g_1	g_2	g_3	g_4	g_5	g_6
ψ	6	0	3	-3	$-1 - i$	$-1 + i$

Express ψ as a linear combination of $\chi_1, \ldots, \chi_6$. Is ψ a character of G?

4. Let G be a group of order 12.
   (a) Show that G cannot have exactly 9 conjugacy classes. (Hint: show that $Z(G)$ cannot have order 6.)
   (b) Using the solution to Exercise 11.2, prove that G has 4, 6 or 12 conjugacy classes. Find groups G in which each of these possibilities is realized.

16
Character tables and orthogonality relations

The irreducible characters of a finite group G are class functions, and the number of them is equal to the number of conjugacy classes of G. It is therefore convenient to record all the values of all the irreducible characters of G in a square matrix. This matrix is called the character table of G. The entries in a character table are related to each other in subtle ways, many of which are encapsulated in the orthogonality relations (Theorem 16.4). Much of the later material in the book will be devoted to understanding character tables. The motivation for this is Theorem 14.21, which tells us that every $\mathbb{C}G$-module is determined by its character. Thus, many problems in representation theory can be solved by considering characters.

Character tables

16.1 Definition
Let $\chi_1, \ldots, \chi_k$ be the irreducible characters of G and let $g_1, \ldots, g_k$ be representatives of the conjugacy classes of G. The $k \times k$ matrix whose ij-entry is $\chi_i(g_j)$ (for all i, j with $1 \leq i \leq k$, $1 \leq j \leq k$) is called the *character table* of G.

It is usual to number the irreducible characters and conjugacy classes of G so that $\chi_1 = 1_G$, the trivial character, and $g_1 = 1$, the identity element of G. Beyond this, the numbering is arbitrary. Note that in the character table, the rows are indexed by the irreducible characters of G and the columns are indexed by the conjugacy classes (or, in practice, by conjugacy class representatives).

159

16.2 Proposition
The character table of G is an invertible matrix.

Proof This follows immediately from the fact that the irreducible characters of G, and hence also the rows of the character table, are linearly independent (Theorem 14.23). ∎

16.3 Examples
(1) Let $G = D_6 = \langle a, b: a^3 = b^2 = 1, b^{-1}ab = a^{-1}\rangle$. The irreducible characters of G are given in Example 13.6(4). We take 1, a, b as representatives of the conjugacy classes of G, and then the character table of G is

	1	a	b
χ_1	1	1	1
χ_2	1	1	-1
χ_3	2	-1	0

(2) We can write down the character table of any finite abelian group using Theorem 9.8. For example, the character table of $C_2 = \langle a: a^2 = 1\rangle$ is

	1	a
χ_1	1	1
χ_2	1	-1

and the character table of $C_3 = \langle a: a^3 = 1\rangle$ is

	1	a	a^2
χ_1	1	1	1
χ_2	1	ω	ω^2
χ_3	1	ω^2	ω

$(\omega = e^{2\pi i/3})$

(3) Let $G = D_8 = \langle a, b: a^4 = b^2 = 1, b^{-1}ab = a^{-1}\rangle$. You found all the irreducible representations of G in Exercise 10.4. The conjugacy classes

of G are given by (12.12), and representatives are $1,\ a^2,\ a,\ b,\ ab$. Hence the character table of G is

	1	a^2	a	b	ab
χ_1	1	1	1	1	1
χ_2	1	1	1	-1	-1
χ_3	1	1	-1	1	-1
χ_4	1	1	-1	-1	1
χ_5	2	-2	0	0	0

The character tables of all dihedral groups will be found in Chapter 18.

The orthogonality relations

We have already seen many uses for the relations (14.13),

$$\langle \chi_r, \chi_s \rangle = \delta_{rs},$$

among the irreducible characters $\chi_1, \ldots, \chi_k$ of G. These relations can be expressed in terms of the rows of the character table, by writing them as

$$\sum_{i=1}^{k} \frac{\chi_r(g_i)\overline{\chi_s(g_i)}}{|C_G(g_i)|} = \delta_{rs}$$

(see Proposition 14.5(2)). Similar relations exist between the columns of the character table, and these are given by part (2) of our next result.

16.4 Theorem
Let $\chi_1, \ldots, \chi_k$ be the irreducible characters of G, and let $g_1, \ldots, g_k$ be representatives of the conjugacy classes of G. Then the following relations hold for any $r,\ s \in \{1, \ldots, k\}$.

(1) The row orthogonality relations:

$$\sum_{i=1}^{k} \frac{\chi_r(g_i)\overline{\chi_s(g_i)}}{|C_G(g_i)|} = \delta_{rs}.$$

(2) The column orthogonality relations:

$$\sum_{i=1}^{k} \chi_i(g_r)\overline{\chi_i(g_s)} = \delta_{rs}|C_G(g_r)|.$$

Proof The row orthogonality relations have already been proved. They are recorded here merely for comparison with the column relations.

For $1 \leqslant s \leqslant k$, let ψ_s be the class function which satisfies

$$\psi_s(g_r) = \delta_{rs} \quad (1 \leqslant r \leqslant k).$$

By Corollary 15.4, ψ_s is a linear combination of $\chi_1, \ldots, \chi_k$, say

$$\psi_s = \sum_{i=1}^{k} \lambda_i \chi_i \quad (\lambda_i \in \mathbb{C}).$$

We know that $\langle \chi_i, \chi_j \rangle = \delta_{ij}$, so

$$\lambda_i = \langle \psi_s, \chi_i \rangle = \frac{1}{|G|} \sum_{g \in G} \psi_s(g) \overline{\chi_i(g)}.$$

Now $\psi_s(g) = 1$ if g is conjugate to g_s, and $\psi_s(g) = 0$ otherwise; also there are $|G|/|C_G(g_s)|$ elements of G which are conjugate to g_s, by Theorem 12.8. Hence

$$\lambda_i = \frac{1}{|G|} \sum_{g \in g_s^G} \psi_s(g) \overline{\chi_i(g)} = \frac{\overline{\chi_i(g_s)}}{|C_G(g_s)|}.$$

Therefore

$$\delta_{rs} = \psi_s(g_r) = \sum_{i=1}^{k} \lambda_i \chi_i(g_r) = \sum_{i=1}^{k} \frac{\chi_i(g_r) \overline{\chi_i(g_s)}}{|C_G(g_s)|},$$

and the column orthogonality relations follow. ∎

16.5 Examples
We illustrate the column orthogonality relations.

(1) Let $G = D_6$. We copy the character table of G from Example 16.3(1), and this time we record the order of the centralizer $C_G(g_i)$ next to each conjugacy class representative g_i:

| g_i | 1 | a | b |
$\lvert C_G(g_i) \rvert$	6	3	2
χ_1	1	1	1
χ_2	1	1	-1
χ_3	2	-1	0

Consider the sums $\sum_{i=1}^{3} \chi_i(g_r)\overline{\chi_i(g_s)}$ for various cases:

$$r = 1, \quad s = 2: \quad 1\cdot 1 + 1\cdot 1 + 2\cdot(-1) = 0;$$

$$r = 2, \quad s = 2: \quad 1\cdot 1 + 1\cdot 1 + (-1)\cdot(-1) = 3;$$

$$r = 1, \quad s = 3: \quad 1\cdot 1 + 1\cdot(-1) + 2\cdot 0 = 0.$$

In each case, we read down columns r and s of the character table, taking the products of the numbers which appear. The sum of the products is 0 if $r \neq s$, and is the number at the top of the column (that is, the order of the centralizer of g_r) if $r = s$.

(2) Suppose we are given the following part of the character table of a group G of order 12 which has exactly four conjugacy classes:

g_i $\|C_G(g_i)\|$	g_1 12	g_2 4	g_3 3	g_4 3
χ_1	1	1	1	1
χ_2	1	1	ω	ω^2
χ_3	1	1	ω^2	ω
χ_4				

(where $\omega = e^{2\pi i/3}$). We shall use the column orthogonality relations to determine the last row of the character table.

The entries in the first column of the character table are the degrees of the irreducible characters, so they are positive integers. By the column orthogonality relations with $r = s = 1$, the sum of the squares of these numbers is 12 (this also follows from Theorem 11.12). Hence the last entry in the first column is 3.

Let x denote the number at the foot of the second column. The column orthogonality relation

$$\sum_{i=1}^{4} \chi_i(g_1)\overline{\chi_i(g_2)} = 0$$

gives

$$1\cdot 1 + 1\cdot 1 + 1\cdot 1 + 3\bar{x} = 0.$$

Therefore $x = -1$.

By considering the orthogonality relations between the first column and columns 3 and 4, we obtain the complete character table as follows:

g_i	g_1	g_2	g_3	g_4
$\|C_G(g_i)\|$	12	4	3	3
χ_1	1	1	1	1
χ_2	1	1	ω	ω^2
χ_3	1	1	ω^2	ω
χ_4	3	-1	0	0

Notice that the orthogonality relations hold between *all* pairs of columns, although our calculation has used only those relations which involve the first column. For example,

$$\sum_{i=1}^{4} \chi_i(g_2)\overline{\chi_i(g_2)} = 1 \cdot 1 + 1 \cdot 1 + 1 \cdot 1 + (-1) \cdot (-1) = 4,$$

$$\sum_{i=1}^{4} \chi_i(g_3)\overline{\chi_i(g_3)} = 1 \cdot 1 + \omega \cdot \overline{\omega} + \omega^2 \cdot \overline{\omega}^2 + 0 \cdot 0 = 3,$$

$$\sum_{i=1}^{4} \chi_i(g_3)\overline{\chi_i(g_4)} = 1 \cdot 1 + \omega \cdot \overline{\omega}^2 + \omega^2 \cdot \overline{\omega} + 0 \cdot 0 = 0.$$

We shall see later that the character table which we have constructed here is that of A_4.

Those column orthogonality relations which involve the first column of the character table were proved in Chapter 13, since Theorem 13.19 and Proposition 13.20 give

$$\sum_{i=1}^{k} d_i \chi_i(g) = \begin{cases} |G|, & \text{if } g = 1, \\ 0, & \text{if } g \neq 1, \end{cases}$$

where $d_i = \chi_i(1)$. By taking the complex conjugate of each side of this equation, we get

$$\sum_{i=1}^{k} \chi_i(1)\overline{\chi_i(g)} = \begin{cases} |G|, & \text{if } g = 1, \\ 0, & \text{if } g \neq 1, \end{cases}$$

and these are just the column orthogonality relations which involve the first column.

Rows versus columns

Notice that in Example 16.5(2), where we were given three of the four irreducible characters of G, we calculated the values of the last character one at a time using the column orthogonality relations. An alternative approach would have been to use the row orthogonality relations $\langle \chi_i, \chi_4 \rangle = \delta_{i4}$ to obtain four equations in the four unknown values $\chi_4(g_j)$ $(1 \leqslant j \leqslant 4)$. Although the calculation with the column orthogonality relations was easier to perform, it is a fact that the column orthogonality relations contain precisely the same information as the row orthogonality relations, as we shall now show.

The character table of G is a $k \times k$ matrix, and we adjust the entries $\chi_i(g_j)$ in this matrix to obtain another $k \times k$ matrix M, by letting the ij-entry of M be

$$\frac{\chi_i(g_j)}{|C_G(g_j)|^{1/2}} \, .$$

Let $\overline{M}^t$ denote the transpose of the complex conjugate of M.

Now the rs-entry in $M\overline{M}^t$ is

$$\sum_{i=1}^{k} \frac{\chi_r(g_i)\overline{\chi_s(g_i)}}{|C_G(g_i)|} = \delta_{rs},$$

by the row orthogonality relations, so $M\overline{M}^t = I$. Indeed, the equation $M\overline{M}^t = I$ is just another way of expressing the row orthogonality relations. On the other hand, the rs-entry in $\overline{M}^t M$ is

$$\frac{1}{|C_G(g_r)|^{1/2}|C_G(g_s)|^{1/2}} \sum_{i=1}^{k} \overline{\chi_i(g_r)}\chi_i(g_s) = \delta_{rs},$$

by the column orthogonality relations, so $\overline{M}^t M = I$.

Since the properties $\overline{M}^t M = I$ and $M\overline{M}^t = I$ of a square matrix M are equivalent to each other, we see that the row and column orthogonality relations are equivalent.

We could have used the above argument to deduce the column orthogonality relations from the row ones. More importantly, the row and column orthogonality relations encapsulate the same information, so when we are working with character tables, we can deduce exactly the same results using either set of relations.

Summary of Chapter 16

Let G be a finite group with irreducible characters $\chi_1, \ldots, \chi_k$ and conjugacy class representatives $g_1, \ldots, g_k$.

1. The character table of G is the $k \times k$ matrix with ij-entry $\chi_i(g_j)$.

2. The row orthogonality relations state that for all r, s,

$$\sum_{i=1}^{k} \frac{\chi_r(g_i)\overline{\chi_s(g_i)}}{|C_G(g_i)|} = \delta_{rs}.$$

3. The column orthogonality relations state that for all r, s,

$$\sum_{i=1}^{k} \chi_i(g_r)\overline{\chi_i(g_s)} = \delta_{rs}|C_G(g_r)|.$$

Exercises for Chapter 16

1. Write down the character table of $C_2 \times C_2$.

2. A certain group G of order 8 is known to have a total of five conjugacy classes, with representatives $g_1, \ldots, g_5$, and four linear characters $\chi_1, \ldots, \chi_4$ taking the following values:

g_i $|C_G(g_i)|$	g_1 8	g_2 8	g_3 4	g_4 4	g_5 4
χ_1	1	1	1	1	1
χ_2	1	1	1	-1	-1
χ_3	1	1	-1	1	-1
χ_4	1	1	-1	-1	1

Find the complete character table of G.

3. There exists a group G of order 10 which has precisely four conjugacy classes, with representatives $g_1, \ldots, g_4$, and has irreducible characters χ_1, χ_2 as follows:

g_i	g_1	g_2	g_3	g_4
$\|C_G(g_i)\|$	10	5	5	2
χ_1	1	1	1	1
χ_2	2	α	β	0

where $\alpha = (-1 + \sqrt{5})/2$ and $\beta = (-1 - \sqrt{5})/2$.

Find the complete character table of G.

(Hint: first find the values of the remaining irreducible characters on g_1, then on g_4 – use Corollary 13.10.)

4. A certain group G has two columns of its character table as follows:

g_i	g_1	g_2
$\|C_G(g_i)\|$	21	7
χ_1	1	1
χ_2	1	1
χ_3	1	1
χ_4	3	ζ
χ_5	3	$\bar{\zeta}$

where $g_1 = 1$ and $\zeta \in \mathbb{C}$.

(a) Find ζ.

(b) Find another column of the character table.

5. Let $\chi_1, \ldots, \chi_k$ be the irreducible characters of G. Show that

$$Z(G) = \left\{ g \in G: \sum_{i=1}^{k} \chi_i(g)\overline{\chi_i(g)} = |G| \right\}.$$

6. Let G be a finite group with conjugacy class representatives $g_1, \ldots, g_k$ and character table C. Show that $\det C$ is either real or purely imaginary, and that

$$|\det C|^2 = \prod_{i=1}^{k} |C_G(g_i)|.$$

Find $\pm(\det C)$ when $G = C_3$.

17

Normal subgroups and lifted characters

If N is a normal subgroup of the finite group G, and $N \neq \{1\}$, then the factor group G/N is smaller than G. The characters of G/N should therefore be easier to find than the characters of G. In fact, we can use the characters of G/N to get some of the characters of G, by a process which is known as lifting. Thus, normal subgroups help us to find characters of G. In the opposite direction, it is also true that the character table of G enables us to find the normal subgroups of G; in particular, it is easy to tell from the character table whether or not G is simple.

The linear characters of G (i.e. the characters of degree 1) are obtained by lifting the irreducible characters of G/N in the case where N is the derived subgroup of G. (The derived subgroup is defined below in Definition 17.7.) The linear characters, in turn, can be used to get new irreducible characters from a given irreducible character, in a way which we shall describe.

Lifted characters

We begin by constructing a character of G from a character of G/N.

17.1 Proposition
Assume that $N \lhd G$, and let $\tilde{\chi}$ be a character of G/N. Define $\chi: G \to \mathbb{C}$ by

$$\chi(g) = \tilde{\chi}(Ng) \quad (g \in G).$$

Then χ is a character of G, and χ and $\tilde{\chi}$ have the same degree.

Proof Let $\tilde{\rho}: G/N \to \mathrm{GL}\,(n, \mathbb{C})$ be a representation of G/N with character $\tilde{\chi}$. The function $\rho: G \to \mathrm{GL}(n, \mathbb{C})$ which is given by the composition

$$g \to Ng \to (Ng)\tilde{\rho} \quad (g \in G)$$

is a homomorphism from G to $\mathrm{GL}(n, \mathbb{C})$. Thus ρ is a representation of G. The character χ of ρ satisfies

$$\chi(g) = \mathrm{tr}(g\rho) = \mathrm{tr}((Ng)\tilde{\rho}) = \tilde{\chi}(Ng)$$

for all $g \in G$. Moreover, $\chi(1) = \tilde{\chi}(N)$, so χ and $\tilde{\chi}$ have the same degree. ∎

17.2 Definition
If $N \lhd G$ and $\tilde{\chi}$ is a character of G/N, then the character χ of G which is given by

$$\chi(g) = \tilde{\chi}(Ng) \quad (g \in G)$$

is called the *lift* of $\tilde{\chi}$ to G.

17.3 Theorem
Assume that $N \lhd G$. By associating each character of G/N with its lift to G, we obtain a bijective correspondence between the set of characters of G/N and the set of characters χ of G which satisfy $N \leqslant \mathrm{Ker}\,\chi$. Irreducible characters of G/N correspond to irreducible characters of G which have N in their kernel.

Proof If $\tilde{\chi}$ is a character of G/N, and χ is the lift of $\tilde{\chi}$ to G, then $\tilde{\chi}(N) = \chi(1)$. Also, if $k \in N$ then

$$\chi(k) = \tilde{\chi}(Nk) = \tilde{\chi}(N) = \chi(1),$$

so $N \leqslant \mathrm{Ker}\,\chi$.

Now let χ be a character of G with $N \leqslant \mathrm{Ker}\,\chi$. Suppose that $\rho: G \to \mathrm{GL}(n, \mathbb{C})$ is a representation of G with character χ. If g_1, $g_2 \in G$ and $Ng_1 = Ng_2$ then $g_1 g_2^{-1} \in N$, so $(g_1 g_2^{-1})\rho = I$, and hence $g_1\rho = g_2\rho$. We may therefore define a function $\tilde{\rho}: G/N \to \mathrm{GL}(n, \mathbb{C})$ by

$$(Ng)\tilde{\rho} = g\rho \quad (g \in G).$$

Then for all $g, h \in G$ we have

$$((Ng)(Nh))\tilde{\rho} = (Ngh)\tilde{\rho} = (gh)\rho = (g\rho)(h\rho)$$

$$= ((Ng)\tilde{\rho})((Nh)\tilde{\rho}),$$

so $\tilde{\rho}$ is a representation of G/N. If $\tilde{\chi}$ is the character of $\tilde{\rho}$ then

$$\tilde{\chi}(Ng) = \chi(g) \quad (g \in G).$$

Thus χ is the lift of $\tilde{\chi}$.

We have now established that the function which sends each character of G/N to its lift to G is a bijection between the set of characters of G/N and the set of characters of G which have N in their kernel. It remains to show that irreducible characters correspond to irreducible characters. To see this, let U be a subspace of $\mathbb{C}^n$, and note that

$$u(g\rho) \in U \text{ for all } u \in U \Leftrightarrow u(Ng)\tilde{\rho} \in U \text{ for all } u \in U.$$

Thus, U is a $\mathbb{C}G$-submodule of $\mathbb{C}^n$ if and only if U is a $\mathbb{C}(G/N)$-submodule of $\mathbb{C}^n$. The representation ρ is therefore irreducible if and only if the representation $\tilde{\rho}$ is irreducible. Hence χ is irreducible if and only if $\tilde{\chi}$ is irreducible. ∎

If we know the character table of G/N for some normal subgroup N of G, then Theorem 17.3 enables us to write down as many irreducible characters of G as there are irreducible characters of G/N.

17.4 Example
Let $G = S_4$ and

$$N = V_4 = \{1, (1\ 2)(3\ 4), (1\ 3)(2\ 4), (1\ 4)(2\ 3)\},$$

so that $N \triangleleft G$ (see Example 12.20). If we put $a = N(1\ 2\ 3)$ and $b = N(1\ 2)$ then

$$G/N = \langle a, b \rangle \text{ and } a^3 = b^2 = N, b^{-1}ab = a^{-1},$$

so $G/N \cong D_6$. We know from Example 16.3(1) that the character table of G/N is

	N	$N(1\ 2)$	$N(1\ 2\ 3)$
$\tilde{\chi}_1$	1	1	1
$\tilde{\chi}_2$	1	−1	1
$\tilde{\chi}_3$	2	0	−1

To calculate the lift χ of a character $\tilde{\chi}$ of G/N, we note that

$$\chi((1\ 2)(3\ 4)) = \tilde{\chi}(N) \quad \text{since } (1\ 2)(3\ 4) \in N,$$

$$\chi((1\ 2\ 3\ 4)) = \tilde{\chi}(N(1\ 3)) \quad \text{since } N(1\ 2\ 3\ 4) = N(1\ 3).$$

Hence the lifts of $\tilde{\chi}_1, \tilde{\chi}_2, \tilde{\chi}_3$ are χ_1, χ_2, χ_3, which are given by

	1	(1 2)	(1 2 3)	(1 2)(3 4)	(1 2 3 4)
χ_1	1	1	1	1	1
χ_2	1	−1	1	1	−1
χ_3	2	0	−1	2	0

Then χ_1, χ_2, χ_3 are irreducible characters of G, since $\tilde{\chi}_1, \tilde{\chi}_2, \tilde{\chi}_3$ are irreducible characters of G/N.

Finding normal subgroups

The character table contains accessible information about the structure of a group, as our next two propositions will demonstrate. First we shall show how to find all the normal subgroups of G, once the character table of G is known. Recall that we can easily locate the kernel of an irreducible character χ from the character table, since

$$\mathrm{Ker}\,\chi = \{g \in G : \chi(g) = \chi(1)\}$$

(see Definition 13.13). Also $\mathrm{Ker}\,\chi \lhd G$. Of course, any subgroup which is the intersection of the kernels of irreducible characters is a normal subgroup too. The following proposition shows that *every* normal subgroup arises in this way.

17.5 Proposition
If $N \lhd G$ then there exist irreducible characters $\chi_1, \ldots, \chi_s$ of G such that

$$N = \bigcap_{i=1}^{s} \mathrm{Ker}\,\chi_i.$$

Proof If g belongs to the kernel of each irreducible character of G, then $\chi(g) = \chi(1)$ for all characters χ, so $g = 1$ by Proposition 15.5. Hence the intersection of the kernels of all the irreducible characters of G is $\{1\}$.

Now let $\tilde{\chi}_1, \ldots, \tilde{\chi}_s$ be the irreducible characters of G/N. By the above observation,

$$\bigcap_{i=1}^{s} \operatorname{Ker} \tilde{\chi}_i = \{N\}.$$

For $1 \le i \le s$, let χ_i be the lift to G of $\tilde{\chi}_i$. If $g \in \operatorname{Ker} \chi_i$ then

$$\tilde{\chi}_i(N) = \chi_i(1) = \chi_i(g) = \tilde{\chi}_i(Ng),$$

and so $Ng \in \operatorname{Ker} \tilde{\chi}_i$. Therefore if $g \in \bigcap \operatorname{Ker} \chi_i$ then $Ng \in \bigcap \operatorname{Ker} \tilde{\chi}_i = \{N\}$, and so $g \in N$. Hence

$$N = \bigcap_{i=1}^{s} \operatorname{Ker} \chi_i. \qquad \blacksquare$$

It is particularly easy to tell from the character table of G whether or not G is simple:

17.6 Proposition
The group G is not simple if and only if

$$\chi(g) = \chi(1)$$

for some non-trivial irreducible character χ of G, and some non-identity element g of G.

Proof Suppose there is a non-trivial irreducible character χ such that $\chi(g) = \chi(1)$ for some non-identity element g. Then $g \in \operatorname{Ker} \chi$, so $\operatorname{Ker} \chi \neq \{1\}$. If ρ is a representation of G with character χ, then $\operatorname{Ker} \chi = \operatorname{Ker} \rho$ by Theorem 13.11(2). Since χ is non-trivial and irreducible, $\operatorname{Ker} \rho \neq G$; hence $\operatorname{Ker} \chi \neq G$. Thus $\operatorname{Ker} \chi$ is a normal subgroup of G which is not equal to $\{1\}$ of G, and so G is not simple.

Conversely, suppose that G is not simple, so that there is a normal subgroup N of G with $N \neq \{1\}$ and $N \neq G$. Then by Proposition 17.5, there is an irreducible character χ of G such that $\operatorname{Ker} \chi$ is not $\{1\}$ or G. As $\operatorname{Ker} \chi \neq G$, χ is non-trivial; and taking $1 \neq g \in \operatorname{Ker} \chi$, we have $\chi(g) = \chi(1)$. $\qquad \blacksquare$

Linear characters

Recall that a linear character of a group is a character of degree 1. We shall show how to find all linear characters of any group G, since the first move in constructing the character table of G is often to write down the linear characters. As a preliminary step, it is necessary to determine the derived subgroup of G, which is defined in the following way.

17.7 Definition
For a group G, let G' be the subgroup of G which is generated by all elements of the form

$$g^{-1}h^{-1}gh \quad (g, h \in G).$$

Then G' is called the *derived subgroup* of G.

We abbreviate $g^{-1}h^{-1}gh$ as $[g, h]$. Thus

$$G' = \langle [g, h]: g, h \in G \rangle.$$

17.8 Examples
(1) If G is abelian then $[g, h] = 1$ for all $g, h \in G$, so $G' = \{1\}$.

(2) Let $G = S_3$. Clearly $[g, h]$ is always an even permutation, so $G' \leqslant A_3$. If $g = (1\ 2)$ and $h = (2\ 3)$ then $[g, h] = (1\ 2\ 3)$. Hence $G' = \langle (1\ 2\ 3) \rangle = A_3$.

We are going to show that $G' \lhd G$ and that the linear characters of G are the lifts to G of the irreducible characters of G/G'. One step is provided by the following proposition.

17.9 Proposition
If χ is a linear character of G, then $G' \leqslant \operatorname{Ker}\chi$.

Proof Let χ be a linear character of G. Then χ is a homomorphism from G to the multiplicative group of non-zero complex numbers. Therefore, for all $g, h \in G$,

$$\chi(g^{-1}h^{-1}gh) = \chi(g)^{-1}\chi(h)^{-1}\chi(g)\chi(h) = 1.$$

Hence $G' \leqslant \operatorname{Ker}\chi$. ∎

Next, we explore some group-theoretic properties of the derived subgroup.

17.10 Proposition
Assume that $N \triangleleft G$.
  (1) $G' \triangleleft G$.
  (2) $G' \leqslant N$ *if and only if G/N is abelian. In particular, G/G' is abelian.*

Proof (1) Note that for all a, b, $x \in G$, we have

$$x^{-1}(ab)x = (x^{-1}ax)(x^{-1}bx), \text{ and}$$

$$x^{-1}a^{-1}x = (x^{-1}ax)^{-1}.$$

Now G' consists of products of elements of the form $[g, h]$ and their inverses. Therefore, to prove that $G' \triangleleft G$ it is sufficient by the first sentence to prove that $x^{-1}[g, h]x \in G'$ for all g, h, $x \in G$. But

$$x^{-1}[g, h]x = x^{-1}g^{-1}h^{-1}ghx$$

$$= (x^{-1}gx)^{-1}(x^{-1}hx)^{-1}(x^{-1}gx)(x^{-1}hx)$$

$$= [x^{-1}gx, x^{-1}hx].$$

Therefore $G' \triangleleft G$.
  (2) Let g, $h \in G$. We have

$$ghg^{-1}h^{-1} \in N \Leftrightarrow Ngh = Nhg \Leftrightarrow (Ng)(Nh) = (Nh)(Ng).$$

Hence $G' \leqslant N$ if and only if G/N is abelian. Since we have proved that $G' \triangleleft G$, we deduce that G/G' is abelian. ∎

It follows from Proposition 17.10 that G' is the smallest normal subgroup of G with abelian factor group.

Given the derived subgroup G', we can obtain the linear characters of G by applying the next theorem.

17.11 Theorem
The linear characters of G are precisely the lifts to G of the irreducible characters of G/G'. In particular, the number of distinct linear characters of G is equal to $|G/G'|$, and so divides $|G|$.

Proof Let $m = |G/G'|$. Since G/G' is abelian, Theorem 9.8 shows that G/G' has exactly m irreducible characters $\tilde{\chi}_1, \ldots, \tilde{\chi}_m$, all of degree 1. The lifts $\chi_1, \ldots, \chi_m$ of these characters to G also have degree 1, and by Theorem 17.3 they are precisely the irreducible characters χ of G

such that $G' \leqslant \text{Ker}\,\chi$. In view of Proposition 17.9, the characters $\chi_1, \ldots, \chi_m$ are therefore all the linear characters of G. ∎

17.12 Example

Let $G = S_n$. We shall show that $G' = A_n$. If $n = 1$ or 2 then S_n is abelian, so $G' = \{1\} = A_n$. We proved that $S_3' = A_3$ in Example 17.8(2), so we assume that $n \geqslant 4$.

As $S_n/A_n \cong C_2$, we have $G' \leqslant A_n$ by Proposition 17.10(2). If $g = (1\ 2)$, $h = (2\ 3)$ and $k = (1\ 2)(3\ 4)$, then

$$[g, h] = (1\ 2\ 3), [h, k] = (1\ 4)(2\ 3).$$

Since $G' \lhd G$, all the elements in $(1\ 2\ 3)^G$ and $(1\ 4)(2\ 3)^G$ belong to G'. Therefore, by Theorem 12.15, G' contains all 3-cycles and all elements of cycle-shape $(2, 2)$. But every product of two transpositions is equal to the identity, a 3-cycle or an element of cycle-shape $(2, 2)$; and A_n consists of permutations, each of which is the product of an even number of transpositions. Therefore $A_n \leqslant G'$. We have now proved that $G' = A_n$.

17.13 Example

We find the linear characters of S_n $(n \geqslant 2)$. From the last example, we know that $S_n' = A_n$. Since $S_n/S_n' = \{A_n, A_n(1\ 2)\} \cong C_2$, the group S_n/S_n' has two linear characters $\tilde{\chi}_1$ and $\tilde{\chi}_2$, where

$$\tilde{\chi}_1(A_n(1\ 2)) = 1,$$

$$\tilde{\chi}_2(A_n(1\ 2)) = -1.$$

Therefore by Theorem 17.11, S_n has exactly two linear characters χ_1, χ_2, which are given by

$$\chi_1 = 1_{S_n},$$

$$\chi_2(g) = \begin{cases} 1, & \text{if } g \in A_n, \\ -1, & \text{if } g \notin A_n. \end{cases}$$

Not only are the linear characters of G important in being irreducible characters, but they can also be used to construct new irreducible characters from old, as the next result shows.

17.14 Proposition

Suppose that χ is a character of G and λ is a linear character of G. Then the product $\chi\lambda$, defined by

$$\chi\lambda(g) = \chi(g)\lambda(g) \quad (g \in G)$$

is a character of G. Moreover, if χ is irreducible, then so is $\chi\lambda$.

Proof Let $\rho: G \to GL(n, \mathbb{C})$ be a representation with character χ. Define $\rho\lambda: G \to GL(n, \mathbb{C})$ by

$$g(\rho\lambda) = \lambda(g)(g\rho) \quad (g \in G).$$

Thus $g(\rho\lambda)$ is the matrix $g\rho$ multiplied by the complex number $\lambda(g)$. Since ρ and λ are homomorphisms it follows easily that $\rho\lambda$ is a homomorphism. The matrix $g(\rho\lambda)$ has trace $\lambda(g)\,\mathrm{tr}\,(g\rho)$, which is $\lambda(g)\chi(g)$. Hence $\rho\lambda$ is a representation of G with character $\chi\lambda$.

Now for all $g \in G$, the complex number $\lambda(g)$ is a root of unity, so $\lambda(g)\overline{\lambda(g)} = 1$. Therefore

$$\langle \chi\lambda, \chi\lambda \rangle = \frac{1}{|G|} \sum_{g \in G} \chi(g)\lambda(g)\overline{\chi(g)\lambda(g)}$$

$$= \frac{1}{|G|} \sum_{g \in G} \chi(g)\overline{\chi(g)} = \langle \chi, \chi \rangle.$$

By Theorem 14.20, it follows that $\chi\lambda$ is irreducible if and only if χ is irreducible. ∎

The general case of a product of two characters will be discussed in Chapter 19.

Summary of Chapter 17

1. Characters of G/N correspond to characters χ of G for which $N \leqslant \mathrm{Ker}\,\chi$. The character of G which corresponds to the character $\tilde{\chi}$ of G/N is the *lift* of $\tilde{\chi}$, and is given by $\chi(g) = \tilde{\chi}(Ng)$ $(g \in G)$.

2. The normal subgroups of G can be found from the character table of G.

3. The linear characters of G are precisely the lifts to G of the irreducible characters of G/G'.

Exercises for Chapter 17

1. Let $G = Q_8 = \langle a, b: a^4 = 1, b^2 = a^2, b^{-1}ab = a^{-1} \rangle$.
   (a) Find the five conjugacy classes of G.
   (b) Find G', and construct all the linear characters of G.
   (c) Complete the character table of G.
   Compare your table with the character table of D_8 (Example 16.3(3)).

2. Let a and b be the following permutations in S_7:

$$a = (1\ 2\ 3\ 4\ 5\ 6\ 7), \ b = (2\ 3\ 5)(4\ 7\ 6).$$

   Let $G = \langle a, b \rangle$. Check that

$$a^7 = b^3 = 1, \ b^{-1}ab = a^2.$$

   (a) Show that G has order 21.
   (b) Find the conjugacy classes of G.
   (c) Find the character table of G.

3. Show that every group of order 12 has 3, 4 or 12 linear characters, and hence cannot be simple.

4. A certain group G of order 12 has precisely six conjugacy classes, with representatives $g_1, \ldots, g_6$ (where $g_1 = 1$), and has irreducible characters χ, ϕ with values as follows:

	g_1	g_2	g_3	g_4	g_5	g_6
χ	1	$-i$	i	1	-1	-1
ϕ	2	0	0	-1	-1	2

   Use Proposition 17.14 to complete the character table of G. What are the sizes of the conjugacy classes of G?

5. The character table of D_8 is as shown (see Example 16.3(3)):

	1	a^2	a, a^3	b, a^2b	ab, a^3b
χ_1	1	1	1	1	1
χ_2	1	1	1	-1	-1
χ_3	1	1	-1	1	-1
χ_4	1	1	-1	-1	1
χ_5	2	-2	0	0	0

Express each normal subgroup of D_8 as an intersection of kernels of irreducible characters, as in Proposition 17.5.

6. You are given that the group

$$T_{4n} = \langle a, b: a^{2n} = 1, a^n = b^2, b^{-1}ab = a^{-1}\rangle$$

has order $4n$. (It is known as a *dicyclic group*.)
   (a) Show that if ε is any $(2n)$th root of unity in $\mathbb{C}$, then there is a representation of T_{4n} over $\mathbb{C}$ which sends

$$a \to \begin{pmatrix} \varepsilon & 0 \\ 0 & \varepsilon^{-1} \end{pmatrix}, b \to \begin{pmatrix} 0 & 1 \\ \varepsilon^n & 0 \end{pmatrix}.$$

   (b) Find all the irreducible representations of T_{4n}.

7. For $n \geqslant 1$, the group

$$U_{6n} = \langle a, b: a^{2n} = b^3 = 1, a^{-1}ba = b^{-1}\rangle$$

has order $6n$.
   (a) Let $\omega = e^{2\pi i/3}$. Show that if ε is any $(2n)$th root of unity in $\mathbb{C}$, then there is a representation of U_{6n} over $\mathbb{C}$ which sends

$$a \to \begin{pmatrix} 0 & \varepsilon \\ \varepsilon & 0 \end{pmatrix}, b \to \begin{pmatrix} \omega & 0 \\ 0 & \omega^2 \end{pmatrix}.$$

   (b) Find all the irreducible representations of U_{6n}.

8. Let n be an odd positive integer. The group

$$V_{8n} = \langle a, b: a^{2n} = b^4 = 1, ba = a^{-1}b^{-1}, b^{-1}a = a^{-1}b\rangle$$

has order $8n$.
   (a) Show that if ε is any nth root of unity in $\mathbb{C}$, then there is a representation of V_{8n} over $\mathbb{C}$ which sends

$$a \to \begin{pmatrix} \varepsilon & 0 \\ 0 & -\varepsilon^{-1} \end{pmatrix}, b \to \begin{pmatrix} 0 & 1 \\ -1 & 0 \end{pmatrix}.$$

   (b) Find all the irreducible representations of V_{8n}.

18
Some elementary character tables

We now illustrate the techniques we have presented so far by constructing the character tables of several groups, including the groups S_4 and A_4, and all dihedral groups.

18.1 The group S_4

In Example 17.4, we produced three irreducible characters χ_1, χ_2, χ_3 of S_4 by lifting characters of the factor group S_4/V_4. We shall now use Proposition 17.14, which deals with the product of a character with a linear character, to complete the character table of S_4.

Let χ_4 be the character

$$\chi_4(g) = |\text{fix}(g)| - 1 \quad (g \in S_4)$$

which is given in Proposition 13.24. By Proposition 17.14, the product $\chi_4\chi_2$ is also a character of S_4. The values of χ_2, χ_4 and $\chi_4\chi_2$ are as follows:

g_i	1	(1 2)	(1 2 3)	(1 2)(3 4)	(1 2 3 4)
$|C_G(g_i)|$	24	4	3	8	4
χ_2	1	-1	1	1	-1
χ_4	3	1	0	-1	-1
$\chi_4\chi_2$	3	-1	0	-1	1

Note that

$$\langle \chi_4, \chi_4 \rangle = \frac{9}{24} + \frac{1}{4} + \frac{1}{8} + \frac{1}{4} = 1,$$

179

so χ_4 is irreducible. The character $\chi_4\chi_2$ is also irreducible, either by using the same calculation or by quoting the result of Proposition 17.14. Let $\chi_5 = \chi_4\chi_2$. Since S_4 has five conjugacy classes, and we have produced five irreducible characters, we have now found the complete character table of S_4, as shown.

Character table of S_4

g_i $|C_G(g_i)|$	1 24	(1 2) 4	(1 2 3) 3	(1 2)(3 4) 8	(1 2 3 4) 4
χ_1	1	1	1	1	1
χ_2	1	−1	1	1	−1
χ_3	2	0	−1	2	0
χ_4	3	1	0	−1	−1
χ_5	3	−1	0	−1	1

18.2 The group A_4

Let $G = A_4$, the alternating group of degree 4. Then $|G| = 12$, and G has four conjugacy classes, with representatives

$$1, (1\,2)(3\,4), (1\,2\,3), (1\,3\,2)$$

(see Example 12.18(1)).

Let ν be the character of A_4 given by Proposition 13.24, so that $\nu(g) = |\mathrm{fix}\,(g)| - 1$ for all $g \in A_4$. The values of ν are as follows:

g_i $|C_G(g_i)|$	1 12	(1 2)(3 4) 4	(1 2 3) 3	(1 3 2) 3
ν	3	−1	0	0

Note that

$$\langle \nu, \nu \rangle = \frac{9}{12} + \frac{1}{4} = 1,$$

so ν is an irreducible character of G of degree 3.

Since G has four irreducible characters, and the sum of the squares of their degrees is 12, there must be exactly three linear characters of G. Thus $|G/G'| = 3$ by Theorem 17.11. It is not difficult to confirm this by showing that

$$G' = V_4 = \{1, (1\,2)(3\,4), (1\,3)(2\,4), (1\,4)(2\,3)\}.$$

Now $G/G' = \{G', G'(1\,2\,3), G'(1\,3\,2)\} \cong C_3$, and the character table of G/G' is

	G'	$G'(1\,2\,3)$	$G'(1\,3\,2)$
$\tilde{\chi}_1$	1	1	1
$\tilde{\chi}_2$	1	ω	ω^2
$\tilde{\chi}_3$	1	ω^2	ω

(where $\omega = e^{2\pi i/3}$). The lifts of $\tilde{\chi}_1$, $\tilde{\chi}_2$, $\tilde{\chi}_3$ to G, together with the character $\chi_4 = v$, give the complete character table of A_4:

Character table of A_4

g_i	1	$(1\,2)(3\,4)$	$(1\,2\,3)$	$(1\,3\,2)$
$|C_G(g_i)|$	12	4	3	3
χ_1	1	1	1	1
χ_2	1	1	ω	ω^2
χ_3	1	1	ω^2	ω
χ_4	3	-1	0	0

18.3 The dihedral groups

Let G be the dihedral group D_{2n} of order $2n$, with $n \geqslant 3$, so that

$$G = \langle a, b: a^n = b^2 = 1, b^{-1}ab = a^{-1} \rangle.$$

We shall derive the character table of G.

Write $\varepsilon = e^{2\pi i/n}$. For each integer j with $1 \leqslant j < n/2$, define

$$A_j = \begin{pmatrix} \varepsilon^j & 0 \\ 0 & \varepsilon^{-j} \end{pmatrix}, \quad B_j = \begin{pmatrix} 0 & 1 \\ 1 & 0 \end{pmatrix}.$$

Check that

$$A_j^n = B_j^2 = I, \quad B_j^{-1}A_jB_j = A_j^{-1}.$$

It follows that by defining $\rho_j: G \to GL(2, \mathbb{C})$ by

$$(a^r b^s)\rho_j = (A_j)^r (B_j)^s \quad (r, s \in \mathbb{Z}),$$

we obtain a representation ρ_j of G for each j with $1 \leqslant j < n/2$.

Each ρ_j is an irreducible representation, either by the proof of Example 5.5(2) or by applying the result of Exercise 8.4.

If i and j are distinct integers with $1 \le i < n/2$ and $1 \le j < n/2$, then $\varepsilon^i \ne \varepsilon^j$ and $\varepsilon^i \ne \varepsilon^{-j}$, so $a\rho_i$ and $a\rho_j$ have different eigenvalues. Therefore there is no matrix T with $a\rho_i = T^{-1}(a\rho_j)T$, and so ρ_i and ρ_j are not equivalent.

Let ψ_j be the character of ρ_j. We have now constructed distinct irreducible characters ψ_j of G, one for each j which satisfies $1 \le j < n/2$.

At this point it is convenient to consider separately the cases where n is odd and where n is even.

Case 1: *n* odd

By (12.11) the conjugacy classes of D_{2n} (n odd) are

$$\{1\}, \{a^r, a^{-r}\}(1 \le r \le (n-1)/2), \{a^s b: 0 \le s \le n-1\}.$$

Thus there are $(n+3)/2$ conjugacy classes.

The $(n-1)/2$ irreducible characters

$$\psi_1, \psi_2, \ldots, \psi_{(n-1)/2}$$

each have degree 2. As G has $(n+3)/2$ irreducible characters in all, there are two more to be found.

Since $\langle a \rangle \lhd G$ and $G/\langle a \rangle \cong C_2$, we obtain two linear characters χ_1, χ_2 of G by lifting the irreducible characters of $G/\langle a \rangle$ to G. These characters χ_1 and χ_2 are given by $\chi_1 = 1_G$ and

$$\chi_2(g) = \begin{cases} 1 & \text{if } g = a^r \text{ for some } r, \\ -1 & \text{if } g = a^r b \text{ for some } r. \end{cases}$$

We have now found all the irreducible characters of D_{2n} (n odd). (Incidentally, we have proved that $D'_{2n} = \langle a \rangle$ for n odd, in view of Theorem 17.11.)

The character table of D_{2n} (n odd) is therefore as follows (where $\varepsilon = e^{2\pi i/n}$):

g_i	1	$a^r \ (1 \le r \le (n-1)/2)$	b
$\|C_G(g_i)\|$	$2n$	n	2
χ_1	1	1	1
χ_2	1	1	-1
ψ_j $(1 \le j \le (n-1)/2)$	2	$\varepsilon^{jr} + \varepsilon^{-jr}$	0

Case 2: n even

If n is even, say $n = 2m$, then the conjugacy classes of D_{2n}, as supplied by (12.12), are

$$\{1\}, \{a^m\}, \{a^r, a^{-r}\}(1 \leqslant r \leqslant m - 1), \{a^s b: s \text{ even}\}, \{a^s b: s \text{ odd}\}.$$

Hence G has $m + 3$ irreducible characters, of which $m - 1$ are given by

$$\psi_1, \psi_2, \ldots, \psi_{m-1}.$$

To find the remaining four irreducible characters, we first note that $\langle a^2 \rangle = \{a^j: j \text{ even}\}$ is a normal subgroup of G and

$$G/\langle a^2 \rangle = \{\langle a^2 \rangle, \langle a^2 \rangle a, \langle a^2 \rangle b, \langle a^2 \rangle ab\}$$

$$\cong C_2 \times C_2.$$

Therefore G has four linear characters $\chi_1, \chi_2, \chi_3, \chi_4$ (and $G' = \langle a^2 \rangle$). Since these linear characters are the lifts of the irreducible characters of $G/\langle a^2 \rangle$, they are easy to calculate, and their values appear in the following complete character table of D_{2n} (n even, $n = 2m$, $\varepsilon = e^{2\pi i/n}$).

g_i	1	a^m	a^r $(1 \leqslant r \leqslant m - 1)$	b	ab
$\|C_G(g_i)\|$	$2n$	$2n$	n	4	4
χ_1	1	1	1	1	1
χ_2	1	1	1	-1	-1
χ_3	1	$(-1)^m$	$(-1)^r$	1	-1
χ_4	1	$(-1)^m$	$(-1)^r$	-1	1
ψ_j $(1 \leqslant j \leqslant m - 1)$	2	$2(-1)^j$	$\varepsilon^{jr} + \varepsilon^{-jr}$	0	0

18.4 Another group of order 12

We shall now describe a non-abelian group G of order 12 which is not isomorphic to either A_4 or D_{12}, and we shall construct the character table of G. It is in fact known that every non-abelian group of order

12 is isomorphic to A_4, D_{12} or G, but we shall not prove this result here.

Let a and b be the following permutations in S_{12}:

$$a = (1\ 2\ 3\ 4\ 5\ 6)(7\ 8\ 9\ 10\ 11\ 12),$$

$$b = (1\ 7\ 4\ 10)(2\ 12\ 5\ 9)(3\ 11\ 6\ 8),$$

and let $G = \langle a, b \rangle$, a subgroup of S_{12}. Since a has order 6 and $b \notin \langle a \rangle$, the group G has at least 12 elements, namely

$$a^r, a^r b \quad (0 \leqslant r \leqslant 5).$$

Check that a and b satisfy

$$a^6 = 1, \ a^3 = b^2, \ b^{-1}ab = a^{-1}.$$

It follows from these relations that every element of G has the form $a^r b^s$ with $0 \leqslant r \leqslant 5$, $0 \leqslant s \leqslant 1$ as given above, and so $|G| = 12$.

The relations further imply that

$$C_G(a) = \langle a \rangle, \ C_G(a^3) = G, \ C_G(b) = \{1, a^3, b, a^3 b\}.$$

These, and similar facts, help us to find the conjugacy classes of G, which are tabulated below:

Conjugacy class	Representative g_i	$|C_G(g_i)|$
$\{1\}$	1	12
$\{a^3\}$	a^3	12
$\{a, a^{-1}\}$	a	6
$\{a^2, a^{-2}\}$	a^2	6
$\{b, a^2 b, a^4 b\}$	b	4
$\{ab, a^3 b, a^5 b\}$	ab	4

Therefore G has six irreducible characters.

Observe that $\langle a^2 \rangle = \{1, a^2, a^4\} \triangleleft G$, and

$$G/\langle a^2 \rangle = \{\langle a^2 \rangle, \langle a^2 \rangle a, \langle a^2 \rangle b, \langle a^2 \rangle ab\}.$$

Since $\langle a^2 \rangle a = \langle a^2 \rangle b^2$, we have $G/\langle a^2 \rangle \cong C_4$. By lifting the irreducible characters of C_4 to G, we obtain the linear characters $\chi_1, \chi_2, \chi_3, \chi_4$ of G given below:

g_i	1	a^3	a	a^2	b	ab
$|C_G(g_i)|$	12	12	6	6	4	4
χ_1	1	1	1	1	1	1
χ_2	1	-1	-1	1	i	$-i$
χ_3	1	1	1	1	-1	-1
χ_4	1	-1	-1	1	$-i$	i
χ_5	α_1	α_2	α_3	α_4	α_5	α_6
χ_6	β_1	β_2	β_3	β_4	β_5	β_6

It remains to find the values α_r, β_r taken by the last two irreducible characters χ_5, χ_6. For this, we shall use the column orthogonality relations, Theorem 16.4(2).

Observe that α_1, β_1 are the degrees of χ_5, χ_6, so they are positive integers; also a^3 is an element of order 2, so α_2 and β_2 are integers by Corollary 13.10. By the column orthogonality relations applied to columns 1 and 2, we have

$$4 + \alpha_1^2 + \beta_1^2 = 12,$$

$$4 + \alpha_2^2 + \beta_2^2 = 12,$$

$$\alpha_1\alpha_2 + \beta_1\beta_2 = 0.$$

Since α_1, β_1 are positive integers, the first equation gives $\alpha_1 = \beta_1 = 2$. The other two equations then imply that $\alpha_2 = -\beta_2 = \pm 2$. Since we have not yet distinguished between χ_5 and χ_6, we may take $\alpha_2 = 2$ and $\beta_2 = -2$.

For $r > 2$, the column orthogonality relations

$$\sum_{i=1}^{6} \chi_i(g_r)\overline{\chi_i(g_1)} = 0 \quad \text{and} \quad \sum_{i=1}^{6} \chi_i(g_r)\overline{\chi_i(g_2)} = 0$$

now give us two equations involving $2\alpha_r + 2\beta_r$ and $2\alpha_r - 2\beta_r$, respectively, so we can solve them for α_r and β_r. Explicitly:

$$r = 3: \quad 2\alpha_3 + 2\beta_3 = 0, \quad 4 + 2\alpha_3 - 2\beta_3 = 0;$$

$$r = 4: \quad 4 + 2\alpha_4 + 2\beta_4 = 0, \quad 2\alpha_4 - 2\beta_4 = 0;$$

$$r = 5: \quad 2\alpha_5 + 2\beta_5 = 0, \quad 2\alpha_5 - 2\beta_5 = 0;$$

$$r = 6: \quad 2\alpha_6 + 2\beta_6 = 0, \quad 2\alpha_6 - 2\beta_6 = 0.$$

Hence

$$\alpha_3 = -1, \qquad \beta_3 = 1,$$
$$\alpha_4 = -1, \qquad \beta_4 = -1,$$
$$\alpha_5 = 0, \qquad \beta_5 = 0,$$
$$\alpha_6 = 0, \qquad \beta_6 = 0.$$

The complete character table of G is therefore as follows:

g_i $C_G(g_i)$	1 12	a^3 12	a 6	a^2 6	b 4	ab 4
χ_1	1	1	1	1	1	1
χ_2	1	-1	-1	1	i	$-i$
χ_3	1	1	1	1	-1	-1
χ_4	1	-1	-1	1	$-i$	i
χ_5	2	2	-1	-1	0	0
χ_6	2	-2	1	-1	0	0

We can deduce that G is not isomorphic to A_4 or D_{12} from the fact that the character table of G is different from those of A_4 and D_{12}.

It is instructive to note that we produced the last two irreducible characters of G by simply using the orthogonality relations, without constructing the corresponding $\mathbb{C}G$-modules. This is typical of more advanced calculations, and illustrates the fact that it is usually much easier to construct an irreducible character of a group than to obtain an irreducible representation. (In fact, it is not hard to construct the representations of the above group G with characters χ_5 and χ_6 – see Exercise 17.6.)

Summary of Chapter 18

In this chapter we gave the character tables of various groups, as follows.

1. Section 18.1: the group S_4.

2. Section 18.2: the group A_4.

3. Section 18.3: the dihedral groups.

Exercises for Chapter 18

1. Regard D_8 as a subgroup of S_4 permuting the four corners of a square, as in Example 1.1(3). Let π be the corresponding permutation character of D_8. Find the values of π on the elements of D_8, and express π as a sum of irreducible characters.

2. Write down explicitly the character table of D_{12}, and show that all its entries are integers.

   Use the character table to find seven distinct normal subgroups of D_{12}. (Hint: use Proposition 17.5.)

3. Let $G = T_{4n} = \langle a, b\colon a^{2n} = 1, a^n = b^2, b^{-1}ab = a^{-1} \rangle$, as in Exercise 17.6. Find the character table of G.

   (Hint: use the result of Exercise 17.6. It is a good idea to do the cases n odd and n even separately.)

4. Let $G = U_{6n} = \langle a, b\colon a^{2n} = b^3 = 1, \ a^{-1}ba = b^{-1} \rangle$, as in Exercise 17.7. Find the character table of G.

5. Let $G = V_{8n} = \langle a, b\colon a^{2n} = b^4 = 1, \ ba = a^{-1}b^{-1}, b^{-1}a = a^{-1}b \rangle$, with n odd, as in Exercise 17.8. Find the character table of G.

19

Tensor products

The idea of multiplying a character of a group G by a linear character of G was introduced at the end of Chapter 17, and it can be extended to include the product of any pair of characters χ and ψ. The value of the product $\chi\psi$ on an element g of G is simply $\chi(g)\psi(g)$. It is therefore straightforward to calculate the product, but a little ingenuity is required in order to justify the conclusion that the product $\chi\psi$ is a character of G. The plan is to take $\mathbb{C}G$-modules V and W with characters χ and ψ respectively, and to put them together to form a new $\mathbb{C}G$-module, called the tensor product of V and W, which has character $\chi\psi$.

An important special case of the product $\chi\psi$ occurs when $\chi = \psi$, so we consider the character χ^2, and more generally χ^3, χ^4, and so on. If χ is not linear, then the degrees of χ, χ^2, ... increase, and by taking successive powers of χ we obtain arbitrarily many new characters. Potentially, then, we have a chance of getting a large proportion of the character table of G from just one non-linear character χ of G; and indeed, products of characters provide a very good source of new characters from given ones. We shall illustrate this by constructing the character tables of S_5 and S_6.

At the end of the chapter, we apply tensor products in a different way, to find all the irreducible characters of a direct product $G \times H$, given those of G and H.

Tensor product spaces

Let V and W be vector spaces over $\mathbb{C}$ with bases $v_1, \ldots, v_m$ and $w_1, \ldots, w_n$, respectively. For each i, j with $1 \leqslant i \leqslant m$, $1 \leqslant j \leqslant n$, we introduce a symbol $v_i \otimes w_j$. The tensor product space $V \otimes W$ is defined to be the mn-dimensional vector space over $\mathbb{C}$ with a basis given by

$$\{v_i \otimes w_j \colon 1 \leqslant i \leqslant m, \, 1 \leqslant j \leqslant n\}.$$

Thus $V \otimes W$ consists of all expressions of the form

$$\sum_{i,j} \lambda_{ij}(v_i \otimes w_j) \quad (\lambda_{ij} \in \mathbb{C}).$$

For $v \in V$ and $w \in W$ with $v = \sum_{i=1}^{m} \lambda_i v_i$ and $w = \sum_{j=1}^{n} \mu_j w_j$ ($\lambda_i, \mu_j \in \mathbb{C}$), we define $v \otimes w \in V \otimes W$ by

$$v \otimes w = \sum_{i,j} \lambda_i \mu_j (v_i \otimes w_j).$$

For example,

$$(2v_1 - v_2) \otimes (w_1 + w_2)$$

$$= 2v_1 \otimes w_1 + 2v_1 \otimes w_2 - v_2 \otimes w_1 - v_2 \otimes w_2.$$

Do not be misled by the notation into believing that every element of $V \otimes W$ has the form $v \otimes w$, because this is not the case. For instance, it is impossible to express

$$v_1 \otimes w_1 + v_2 \otimes w_2$$

in the form $v \otimes w$.

19.1 Proposition
(1) *If* $v \in V$, $w \in W$ *and* $\lambda \in \mathbb{C}$, *then*

$$v \otimes (\lambda w) = (\lambda v) \otimes w = \lambda(v \otimes w).$$

(2) *If* $x_1, \ldots, x_a \in V$ *and* $y_1, \ldots, y_b \in W$, *then*

$$\left(\sum_{i=1}^{a} x_i\right) \otimes \left(\sum_{j=1}^{b} y_j\right) = \sum_{i,j} x_i \otimes y_j.$$

Proof (1) Let $v = \sum_{i=1}^{m} \lambda_i v_i$ and $w = \sum_{j=1}^{n} \mu_j w_j$. Then

$$v \otimes (\lambda w) = \left(\sum_i \lambda_i v_i\right) \otimes \left(\sum_j \lambda \mu_j w_j\right) = \sum_{i,j} \lambda \lambda_i \mu_j (v_i \otimes w_j),$$

$$(\lambda v) \otimes w = \left(\sum_i \lambda \lambda_i v_i\right) \otimes \left(\sum_j \mu_j w_j\right) = \sum_{i,j} \lambda \lambda_i \mu_j (v_i \otimes w_j),$$

$$\lambda(v \otimes w) = \lambda \sum_{i,j} \lambda_i \mu_j (v_i \otimes w_j) = \sum_{i,j} \lambda \lambda_i \mu_j (v_i \otimes w_j).$$

Therefore $v \otimes (\lambda w) = (\lambda v) \otimes w = \lambda(v \otimes w)$.

The proof of part (2) is equally straightforward, and we leave it as an exercise.   ∎

Our construction of $V \otimes W$ depended upon choosing a basis of V and a basis of W at the beginning; the next proposition shows that other bases of V and W work equally well.

19.2 Proposition
If $e_1, \ldots, e_m$ is a basis of V and $f_1, \ldots, f_n$ is a basis of W, then the elements in

$$\{e_i \otimes f_j \colon 1 \leqslant i \leqslant m, 1 \leqslant j \leqslant n\}$$

give a basis of $V \otimes W$.

Proof Write

$$v_i = \sum_{k=1}^{m} \lambda_{ik} e_k, \quad w_j = \sum_{l=1}^{n} \mu_{jl} f_l \quad (\lambda_{ik}, \mu_{jl} \in \mathbb{C}).$$

Then by Proposition 19.1, we have

$$v_i \otimes w_j = \sum_{k,l} \lambda_{ik} \mu_{jl} (e_k \otimes f_l).$$

Now the elements $v_i \otimes w_j$ $(1 \leqslant i \leqslant m, \ 1 \leqslant j \leqslant n)$ are a basis of $V \otimes W$, and hence the mn elements $e_k \otimes f_l$ $(1 \leqslant k \leqslant m, \ 1 \leqslant l \leqslant n)$ span $V \otimes W$. Since $V \otimes W$ has dimension mn, it follows that the elements $e_k \otimes f_l$ are also a basis of $V \otimes W$.

Tensor product modules

We have introduced the tensor product of two vector spaces, so we are now in a position to define the tensor product of two $\mathbb{C}G$-modules.

Let G be a finite group and let V and W be $\mathbb{C}G$-modules with bases $v_1, \ldots, v_m$ and $w_1, \ldots, w_n$, respectively. We know that the elements

$$v_i \otimes w_j \quad (1 \leqslant i \leqslant m, 1 \leqslant j \leqslant n)$$

give a basis of $V \otimes W$. The multiplication of $v_i \otimes w_j$ by an element of

G is defined in the following simple way, which is then extended linearly to a multiplication on the whole of $V \otimes W$.

19.3 Definition
Let $g \in G$. For all i, j, define

$$(v_i \otimes w_j)g = v_i g \otimes w_j g$$

and, more generally, let

$$\left(\sum_{i,j} \lambda_{ij}(v_i \otimes w_j)\right) g = \sum_{i,j} \lambda_{ij}(v_i g \otimes w_j g)$$

for arbitrary complex numbers λ_{ij}.

19.4 Proposition
For all $v \in V$, $w \in W$ and all $g \in G$, we have

$$(v \otimes w)g = vg \otimes wg.$$

Proof Let $v = \sum_{i=1}^m \lambda_i v_i$ and $w = \sum_{j=1}^n \mu_j w_j$. Then

$$(v \otimes w)g = \left(\sum_{i,j} \lambda_i \mu_j (v_i \otimes w_j)\right) g \quad \text{by Proposition 19.1}$$

$$= \sum_{i,j} \lambda_i \mu_j (v_i g \otimes w_j g)$$

$$= \left(\sum_i \lambda_i v_i g\right) \otimes \left(\sum_j \mu_j w_j g\right) \quad \text{by Proposition 19.1}$$

$$= vg \otimes wg. \qquad \blacksquare$$

You should be warned that $(v \otimes w)r \neq vr \otimes wr$ for most elements r in $\mathbb{C}G$. For example, consider what happens when r is a scalar multiple of g.

19.5 Proposition
The rule for multiplying an element of $V \otimes W$ by an element of G, given in Definition 19.3, makes the vector space $V \otimes W$ into a $\mathbb{C}G$-module.

Proof Let $1 \le i \le m$, $1 \le j \le n$, and $g, h \in G$. Then

$$(v_i \otimes w_j)g = v_i g \otimes w_j g \in V \otimes W,$$

$$(v_i \otimes w_j)(gh) = v_i(gh) \otimes w_j(gh)$$

$$= (v_i g)h \otimes (w_j g)h$$

$$= (v_i g \otimes w_j g)h \quad \text{by Proposition 19.4}$$

$$= ((v_i \otimes w_j)g)h,$$

$$(v_i \otimes w_j)1 = v_i \otimes w_j,$$

and

$$\left(\sum_{i,j} \lambda_{ij}(v_i \otimes w_j) \right) g = \sum_{i,j} \lambda_{ij}((v_i \otimes w_j)g).$$

Therefore all the conditions of Proposition 4.6 are fulfilled, and $V \otimes W$ is a $\mathbb{C}G$-module. ∎

We now calculate the character of $V \otimes W$.

19.6 Proposition
Let V and W be $\mathbb{C}G$-modules with characters χ and ψ, respectively.
Then the character of the $\mathbb{C}G$-module $V \otimes W$ is the product character $\chi\psi$, *where*

$$\chi\psi(g) = \chi(g)\psi(g) \quad \text{for all } g \in G.$$

Proof Let $g \in G$. By Proposition 9.11 we can choose a basis $e_1, \ldots, e_m$ of V and a basis $f_1, \ldots, f_n$ of W such that

$$e_i g = \lambda_i e_i \quad (1 \le i \le m) \text{ and } f_j g = \mu_j f_j \quad (1 \le j \le n)$$

for some complex numbers λ_i, μ_j. Then

$$\chi(g) = \sum_{i=1}^{m} \lambda_i, \ \psi(g) = \sum_{j=1}^{n} \mu_j.$$

Now for $1 \le i \le m$ and $1 \le j \le n$,

$$(e_i \otimes f_j)g = e_i g \otimes f_j g = \lambda_i \mu_j (e_i \otimes f_j),$$

and by Proposition 19.2, these vectors $e_i \otimes f_j$ form a basis of $V \otimes W$. Hence, if ϕ is the character of $V \otimes W$ then

$$\phi(g) = \sum_{i,j} \lambda_i \mu_j = \left(\sum_i \lambda_i\right)\left(\sum_j \mu_j\right) = \chi(g)\psi(g),$$

as required. ∎

19.7 Corollary
The product of two characters of G is again a character of G.

19.8 Example
The character table of S_4 was given in Section 18.1. We reproduce it here, and calculate $\chi_3\chi_4$ and $\chi_4\chi_4$.

g_i	1	(1 2)	(1 2 3)	(1 2)(3 4)	(1 2 3 4)
$|C_G(g_i)|$	24	4	3	8	4
χ_1	1	1	1	1	1
χ_2	1	−1	1	1	−1
χ_3	2	0	−1	2	0
χ_4	3	1	0	−1	−1
χ_5	3	−1	0	−1	1
$\chi_3\chi_4$	6	0	0	−2	0
$\chi_4\chi_4$	9	1	0	1	1

We see that

$$\chi_3\chi_4 = \chi_4 + \chi_5, \text{ and}$$

$$\chi_4\chi_4 = \chi_1 + \chi_3 + \chi_4 + \chi_5.$$

Powers of characters

Corollary 19.7 shows that if χ is a character of G then so is χ^2, where $\chi^2 = \chi\chi$, the product of χ with itself. More generally, for every non-negative integer n, we define χ^n by

$$\chi^n(g) = (\chi(g))^n \quad \text{for all } g \in G.$$

Thus $\chi^0 = 1_G$. An inductive proof using Corollary 19.7 shows that χ^n is a character of G. When χ is a faithful character (that is, Ker $= \{1\}$), the powers of χ carry a lot of information about the whole character table of G, as can be seen from Theorem 19.10 below.

In the course of the proof of Theorem 19.10, we shall need the following result concerning the so-called 'Vandermonde matrix'.

(19.9) *If* $\alpha_1, \ldots, \alpha_r$ *are distinct complex numbers, then the matrix*

$$A = \begin{pmatrix} 1 & \alpha_1 & \alpha_1^2 & \cdots & \alpha_1^{r-1} \\ 1 & \alpha_2 & \alpha_2^2 & \cdots & \alpha_2^{r-1} \\ \cdot & \cdot & \cdot & \cdots & \cdot \\ 1 & \alpha_r & \alpha_r^2 & \cdots & \alpha_r^{r-1} \end{pmatrix}$$

is invertible.

We first sketch a proof of this result.

Suppose that $x_1, \ldots, x_r$ are indeterminates, and consider

$$\Delta = \det \begin{pmatrix} 1 & x_1 & x_1^2 & \cdots & x_1^{r-1} \\ 1 & x_2 & x_2^2 & \cdots & x_2^{r-1} \\ \cdot & \cdot & \cdot & \cdots & \cdot \\ 1 & x_r & x_r^2 & \cdots & x_r^{r-1} \end{pmatrix}.$$

If $i \neq j$ and $x_i = x_j$ then two rows of the given matrix are equal, so $\Delta = 0$. It follows that Δ is divisible by

$$\prod_{i<j}(x_i - x_j) = (x_1 - x_2)(x_1 - x_3) \ldots (x_1 - x_r)$$

$$\times (x_2 - x_3) \ldots (x_2 - x_r)$$

$$\vdots$$

$$\times (x_{r-1} - x_r).$$

Now the coefficient of $x_1^{r-1}x_2^{r-2} \ldots x_{r-1}$ in this product is 1: for in the way we have displayed the product, the term x_1^{r-1} must come from all the factors in the first row, x_2^{r-2} must come from all the factors in the second row, and so on. On the other hand, to obtain $x_1^{r-1}x_2^{r-2} \ldots x_{r-1}$ in the expansion of the determinant Δ, we must take x_1^{r-1} from the first row of the matrix, x_2^{r-2} from the second row, and so on. Hence the coefficient of $x_1^{r-1}x_2^{r-2} \ldots x_{r-1}$ in Δ is ± 1. It follows that

$$\Delta = \pm \prod_{i<j}(x_i - x_j).$$

To obtain (19.9), we substitute α_i for x_i ($1 \leqslant i \leqslant r$), and deduce that the matrix A is invertible since its determinant is non-zero.

19.10 Theorem
Let χ be a faithful character of G, and suppose that $\chi(g)$ takes precisely r different values as g varies over all the elements of G. Then every irreducible character of G is a constituent of one of the powers $\chi^0, \chi^1, \ldots, \chi^{r-1}$.

Proof Let the r values taken by χ be $\alpha_1, \ldots, \alpha_r$, and for $1 \leqslant i \leqslant r$, define
$$G_i = \{g \in G : \chi(g) = \alpha_i\}.$$
Take $\alpha_1 = \chi(1)$, so that $G_1 = \mathrm{Ker}\,\chi$. As χ is faithful, $G_1 = \{1\}$.

Now let ψ be an irreducible character of G. We must show that $\langle \chi^j, \psi \rangle \neq 0$ for some j with $0 \leqslant j \leqslant r - 1$.

For $1 \leqslant i \leqslant r$, let
$$\beta_i = \sum_{g \in G_i} \overline{\psi(g)},$$
and note that $\beta_1 = \psi(1) \neq 0$. Then for all $j \geqslant 0$,
$$\langle \chi^j, \psi \rangle = \frac{1}{|G|} \sum_{g \in G} (\chi(g))^j \overline{\psi(g)} = \frac{1}{|G|} \sum_{i=1}^{r} (\alpha_i)^j \beta_i.$$

Let A be the $r \times r$ matrix with ij-entry $(\alpha_i)^{j-1}$, and let b be the row vector which is given by
$$b = (\beta_1, \ldots, \beta_r).$$
Now A is invertible by (19.9), and $b \neq 0$ since $\beta_1 \neq 0$; hence $bA \neq 0$. But the $(j + 1)$th entry in the row vector bA is equal to $|G|\langle \chi^j, \psi \rangle$, and thus $\langle \chi^j, \psi \rangle \neq 0$ for some j with $0 \leqslant j \leqslant r - 1$, as we wished to prove. ∎

19.11 Examples
(1) If $G \neq \{1\}$ and χ is the regular character of G, then $\chi(g)$ takes just two different values (see Proposition 13.20), so Theorem 19.10 says that every irreducible character of G is a constituent of 1_G or χ; we know this already, by Theorem 10.5.

(2) Let $G = S_4$, and refer to Example 19.8. Let $\chi = \chi_4$. Then $\chi(g)$ takes four different values. We have seen that
$$\chi^2 = \chi_1 + \chi_3 + \chi_4 + \chi_5$$

and we find that

$$\langle \chi^3, \chi_2 \rangle = 1.$$

Thus χ^0, χ^1, χ^2, χ^3 (indeed, in this case, just χ^2, χ^3) have between them as constituents all the irreducible characters $\chi_1, \ldots, \chi_5$ of G, illustrating Theorem 19.10.

Decomposing χ^2

In view of Theorem 19.10, it is of some importance to be able to decompose powers of a character χ into sums of irreducible characters. We are going to provide a method for decomposing χ^2, the square of χ. This special case is particularly useful in finding irreducible characters, as we shall see.

Let V be a $\mathbb{C}G$-module with character χ. By Proposition 19.6, the module $V \otimes V$ has character χ^2. Let $v_1, \ldots, v_n$ be a basis of V, and define a linear transformation $T: V \otimes V \rightarrow V \otimes V$ by

$$(v_i \otimes v_j)T = v_j \otimes v_i \quad \text{for all } i, j$$

and extending linearly – that is,

$$\left(\sum_{i,j} \lambda_{ij}(v_i \otimes v_j) \right) T = \sum_{i,j} \lambda_{ij}(v_j \otimes v_i).$$

Check that for all v, $w \in V$, we have

$$(v \otimes w)T = w \otimes v.$$

Hence T is independent of the choice of basis.

Now define subsets of $V \otimes V$ as follows:

$$S(V \otimes V) = \{x \in V \otimes V: xT = x\},$$

$$A(V \otimes V) = \{x \in V \otimes V: xT = -x\},$$

Since T is linear, it is easy to see that $S(V \otimes V)$ and $A(V \otimes V)$ are subspaces of $V \otimes V$ (indeed, they are eigenspaces of T). The subspace $S(V \otimes V)$ is called the *symmetric* part of $V \otimes V$, and the subspace $A(V \otimes V)$ is known as the *antisymmetric* part of $V \otimes V$.

19.12 Proposition
The subspaces $S(V \otimes V)$ and $A(V \otimes V)$ are $\mathbb{C}G$-submodules of $V \otimes V$.
Also,

$$V \otimes V = S(V \otimes V) \oplus A(V \otimes V).$$

Proof For all $\lambda_{ij} \in \mathbb{C}$ and $g \in G$,

$$\left(\sum_{i,j} \lambda_{ij}(v_i \otimes v_j) \right) Tg = \sum_{i,j} \lambda_{ij}(v_j g \otimes v_i g)$$

$$= \sum_{i,j} \lambda_{ij}(v_i g \otimes v_j g)T$$

$$= \left(\sum_{i,j} \lambda_{ij}(v_i \otimes v_j) \right) gT.$$

Therefore T is a $\mathbb{C}G$-homomorphism from $V \otimes V$ to itself. Hence, for $x \in S(V \otimes V)$, $y \in A(V \otimes V)$ and $g \in G$, we have

$$(xg)T = (xT)g = xg, \text{ and}$$

$$(yg)T = (yT)g = -yg,$$

so $xg \in S(V \otimes V)$ and $yg \in A(V \otimes V)$. Thus $S(V \otimes V)$ and $A(V \otimes V)$ are $\mathbb{C}G$-submodules of $V \otimes V$.

If $x \in S(V \otimes V) \cap A(V \otimes V)$ then $x = xT = -x$, so $x = 0$. Further, for all $x \in V$ we have

$$x = \tfrac{1}{2}(x + xT) + \tfrac{1}{2}(x - xT).$$

Since T^2 is the identity, $\tfrac{1}{2}(x + xT) \in S(V \otimes V)$ and $\tfrac{1}{2}(x - xT) \in A(V \otimes V)$. Therefore,

$$V \otimes V = S(V \otimes V) \oplus A(V \otimes V). \qquad \blacksquare$$

Note that the symmetric part of $V \otimes V$ contains all vectors which have the form $v \otimes w + w \otimes v$ with $v, w \in V$, while the antisymmetric part of $V \otimes V$ contains all vectors of the form $v \otimes w - w \otimes v$. We now present bases of the symmetric and antisymmetric parts of $V \otimes V$ which consist of elements like these.

19.13 Proposition
Let $v_1, \ldots, v_n$ be a basis of V.

(1) The vectors $v_i \otimes v_j + v_j \otimes v_i$ $(1 \leqslant i \leqslant j \leqslant n)$ form a basis of $S(V \otimes V)$. The dimension of $S(V \otimes V)$ is $n(n+1)/2$.

(2) The vectors $v_i \otimes v_j - v_j \otimes v_i$ $(1 \leqslant i < j \leqslant n)$ form a basis of $A(V \otimes V)$. The dimension of $A(V \otimes V)$ is $n(n-1)/2$.

Proof Clearly the vectors $v_i \otimes v_j + v_j \otimes v_i$ $(1 \le i \le j \le n)$ are linearly independent elements of $S(V \otimes V)$, and the vectors $v_i \otimes v_j - v_j \otimes v_i$ $(1 \le i < j \le n)$ are linearly independent elements of $A(V \otimes V)$. Hence

$$\dim S(V \otimes V) \ge n(n+1)/2, \quad \dim A(V \otimes V) \ge n(n-1)/2.$$

By Proposition 19.12,

$$\dim S(V \otimes V) + \dim A(V \otimes V) = \dim V \otimes V = n^2.$$

Hence the above inequalities are equalities, and the result follows. ∎

Define χ_S to be the character of the $\mathbb{C}G$-module $S(V \otimes V)$, and χ_A to be the character of the $\mathbb{C}G$-module $A(V \otimes V)$. By Proposition 19.12,

$$\chi^2 = \chi_S + \chi_A.$$

The next result gives the values of the characters χ_S and χ_A.

19.14 Proposition
For $g \in G$, we have

$$\chi_S(g) = \tfrac{1}{2}(\chi^2(g) + \chi(g^2)), \text{ and}$$

$$\chi_A(g) = \tfrac{1}{2}(\chi^2(g) - \chi(g^2)).$$

Proof By Proposition 9.11 we can choose a basis $e_1, \ldots, e_n$ of V such that $e_i g = \lambda_i e_i$ $(1 \le i \le n)$ for some complex numbers λ_i. Then

$$(e_i \otimes e_j - e_j \otimes e_i)g = \lambda_i \lambda_j (e_i \otimes e_j - e_j \otimes e_i),$$

and hence from Proposition 19.13(2),

$$\chi_A(g) = \sum_{i<j} \lambda_i \lambda_j.$$

Now $e_i g^2 = \lambda_i^2 e_i$, so $\chi(g) = \sum_i \lambda_i$ and $\chi(g^2) = \sum_i \lambda_i^2$. Therefore

$$\chi^2(g) = (\chi(g))^2 = \sum_i \lambda_i^2 + 2\sum_{i<j} \lambda_i \lambda_j = \chi(g^2) + 2\chi_A(g).$$

Hence

$$\chi_A(g) = \tfrac{1}{2}(\chi^2(g) - \chi(g^2)).$$

Also, $\chi^2 = \chi_S + \chi_A$, which implies that

$$\chi_S(g) = \chi^2(g) - \chi_A(g) = \tfrac{1}{2}(\chi^2(g) + \chi(g^2)).$$

∎

19.15 Example
Let $G = S_4$. The character table of G is given in Example 19.8. Let $\chi = \chi_4$. The values of χ, and the values of χ_S and χ_A, given by Proposition 19.14, appear below.

g_i $|C_G(g_i)|$	1 24	(1 2) 4	(1 2 3) 3	(1 2)(3 4) 8	(1 2 3 4) 4
χ	3	1	0	-1	-1
χ_S	6	2	0	2	0
χ_A	3	-1	0	-1	1

We find that $\chi_S = \chi_1 + \chi_3 + \chi_4$ and $\chi_A = \chi_5$.

The techniques which we have developed so far give a useful method for finding new irreducible characters of a group, given one or two irreducible characters to start with. The strategy is simple:

(1) Given a character χ, form χ_S and χ_A, and use inner products to analyse χ_S and χ_A for new irreducible characters.
(2) If ψ is a new character found in (1), then form ψ_S and ψ_A and repeat.

We illustrate this strategy with two examples.

19.16 Example   *The character table of S_5*
Let $G = S_5$, the symmetric group of degree 5. By Example 12.16(4), G has conjugacy class representatives g_i, conjugacy class sizes and centralizer orders $|C_G(g_i)|$ as follows:

g_i Class size $|C_G(g_i)|$	1 1 120	(1 2) 10 12	(1 2 3) 20 6	(1 2)(3 4) 15 8	(1 2 3 4) 30 4	(1 2 3)(4 5) 20 6	(1 2 3 4 5) 24 5

Thus G has exactly seven irreducible characters.

(a) *Linear characters* By Example 17.13, $G' = A_5$ and G has exactly two linear characters χ_1, χ_2, obtained by lifting the irreducible characters of G/G'. We have

$$\chi_1 = 1_G, \text{ and}$$

$$\chi_2(g) = \begin{cases} 1, & \text{if } g \text{ is an even permutation,} \\ -1, & \text{if } g \text{ is an odd permutation.} \end{cases}$$

(b) *Permutation character* Proposition 13.24 gives us a character χ_3 of G with values

$$\chi_3(g) = |\text{fix}(g)| - 1 \quad (g \in G).$$

Observe that

$$\langle \chi_3, \chi_3 \rangle = \frac{4^2}{120} + \frac{2^2}{12} + \frac{1^2}{6} + \frac{(-1)^2}{6} + \frac{(-1)^2}{5} = 1.$$

Hence χ_3 is irreducible, by Theorem 14.20. Next, Proposition 17.14 shows that $\chi_4 = \chi_3 \chi_2$ is also an irreducible character.

At this point we have the following portion of the character table of G:

g_i	1	(1 2)	(1 2 3)	(1 2)(3 4)	(1 2 3 4)	(1 2 3)(4 5)	(1 2 3 4 5)
$|C_G(g_i)|$	120	12	6	8	4	6	5
χ_1	1	1	1	1	1	1	1
χ_2	1	-1	1	1	-1	-1	1
χ_3	4	2	1	0	0	-1	-1
χ_4	4	-2	1	0	0	1	-1

(c) *Tensor products* We now use tensor products to construct the last three irreducible characters of G.

Write $\chi = \chi_3$. By Proposition 19.14 the values of the characters χ_S and χ_A are

g_i	1	(1 2)	(1 2 3)	(1 2)(3 4)	(1 2 3 4)	(1 2 3)(4 5)	(1 2 3 4 5)
$|C_G(g_i)|$	120	12	6	8	4	6	5
χ_S	10	4	1	2	0	1	0
χ_A	6	0	0	-2	0	0	1

Thus,

$$\langle \chi_A, \chi_A \rangle = \frac{36}{120} + \frac{4}{8} + \frac{1}{5} = 1,$$

and so χ_A is a new irreducible character, which we call χ_5.

Next,

$$\langle \chi_5, \chi_1 \rangle = \frac{10}{120} + \frac{4}{12} + \frac{1}{6} + \frac{2}{8} + \frac{1}{6} = 1,$$

$$\langle \chi_5, \chi_3 \rangle = \frac{40}{120} + \frac{8}{12} + \frac{1}{6} - \frac{1}{6} = 1, \text{ and}$$

$$\langle \chi_5, \chi_5 \rangle = \frac{100}{120} + \frac{16}{12} + \frac{1}{6} + \frac{4}{8} + \frac{1}{6} = 3,$$

Therefore,

$$\chi_5 = \chi_1 + \chi_3 + \psi,$$

where ψ is an irreducible character of degree 5. Let $\chi_6 = \psi$, so that $\chi_6 = \chi_5 - \chi_1 - \chi_3$.

Finally, $\chi_7 = \chi_6 \chi_2$ is a different irreducible character of degree 5.

We have now found all seven irreducible characters of S_5. The character table of S_5 is as shown.

Character table of S_5

g_i	1	(1 2)	(1 2 3)	(1 2)(3 4)	(1 2 3 4)	(1 2 3)(4 5)	(1 2 3 4 5)
$\lvert C_G(g_i) \rvert$	120	12	6	8	4	6	5
χ_1	1	1	1	1	1	1	1
χ_2	1	−1	1	1	−1	−1	1
χ_3	4	2	1	0	0	−1	−1
χ_4	4	−2	1	0	0	1	−1
χ_5	6	0	0	−2	0	0	1
χ_6	5	1	−1	1	−1	1	0
χ_7	5	−1	−1	1	1	−1	0

19.17 Example The character table of S_6

In this example, we use techniques similar to those of the previous example to find 8 of the 11 irreducible characters of the symmetric group S_6; we then find the last three irreducible characters by using the orthogonality relations.

Let $G = S_6$, of order 720. For ease of printing, it is convenient to label each conjugacy class by the cycle-shape of its elements. Using this notation, the conjugacy class sizes and centralizer orders are as follows (see Exercise 12.3):

Cycle-shape	(1)	(2)	(3)	(2,2)	(4)	(3,2)	(5)	(2,2,2)	(3,3)	(4,2)	(6)
Class size	1	15	40	45	90	120	144	15	40	90	120
$|C_G(g_i)|$	720	48	18	16	8	6	5	48	18	8	6

Since G has 11 conjugacy classes, it has 11 irreducible characters.

(a) *Linear characters* As with all symmetric groups S_n $(n \geqslant 2)$, the derived subgroup is A_n, and we get exactly two linear characters χ_1 and χ_2, where

$$\chi_1 = 1_G,$$

$$\chi_2(g) = \begin{cases} 1, & \text{if } g \text{ is even,} \\ -1, & \text{if } g \text{ is odd} \end{cases}$$

(see Example 17.13).

(b) *Permutation character and tensor products* The function χ_3 which is given by

$$\chi_3(g) = |\text{fix}(g)| - 1 \quad (g \in G)$$

is a character of G, by Proposition 13.24. Let $\chi = \chi_3$. The values of χ, χ_S and χ_A are as follows:

Class	(1)	(2)	(3)	(2,2)	(4)	(3,2)	(5)	(2,2,2)	(3,3)	(4,2)	(6)
$|C_G(g_i)|$	720	48	18	16	8	6	5	48	18	8	6
$\chi = \chi_3$	5	3	2	1	1	0	0	−1	−1	−1	−1
χ_S	15	7	3	3	1	1	0	3	0	1	0
χ_A	10	2	1	−2	0	−1	0	−2	1	0	1

We calculate that

$$\langle \chi_3, \chi_3 \rangle = 1, \quad \langle \chi_A, \chi_A \rangle = 1,$$

$$\langle \chi_S, \chi_1 \rangle = 1, \quad \langle \chi_S, \chi_3 \rangle = 1,$$

$$\langle \chi_S, \chi_S \rangle = 3.$$

Therefore χ_3 is irreducible; so is $\chi_4 = \chi_3\chi_2$. Also, $\chi_5 = \chi_A$ is irreducible, as is $\chi_6 = \chi_5\chi_2$. Further,

$$\chi_S = \chi_1 + \chi_3 + \chi_7,$$

where χ_7 is another irreducible character, of degree 9. Finally, $\chi_8 = \chi_7\chi_2$ is also irreducible.

The irreducible characters $\chi_1, \ldots, \chi_8$ are recorded in the following portion of the character table of G.

Class	(1)	(2)	(3)	(2,2)	(4)	(3,2)	(5)	(2,2,2)	(3,3)	(4,2)	(6)
$\|C_G(g_i)\|$	720	48	18	16	8	6	5	48	18	8	6
χ_1	1	1	1	1	1	1	1	1	1	1	1
χ_2	1	−1	1	1	−1	−1	1	−1	1	1	−1
χ_3	5	3	2	1	1	0	0	−1	−1	−1	−1
χ_4	5	−3	2	1	−1	0	0	1	−1	−1	1
χ_5	10	2	1	−2	0	−1	0	−2	1	0	1
χ_6	10	−2	1	−2	0	1	0	2	1	0	−1
χ_7	9	3	0	1	−1	0	−1	3	0	1	0
χ_8	9	−3	0	1	1	0	−1	−3	0	1	0

(c) *Orthogonality relations* We now use the column orthogonality relations to complete the character table of G. It will be shown later (Corollary 22.16) that all the entries in the character tables of all symmetric groups are integers, but for the moment we know for certain only that $\chi(g)$ is an integer if $g^2 = 1$ (see Corollary 13.10). It is therefore convenient first to concentrate on elements of order 2, so that we can guarantee that the solutions to the equations which we deal with are integers.

Let s denote the permutation $(1\,2)$ and t denote the permutation $(1\,2)(3\,4)$. Thus the conjugacy classes of s and t correspond to the second and fourth columns of the character table, respectively, in the ordering which we have adopted. From Corollary 13.10 we know that $\chi(s)$ and $\chi(t)$ are integers for all characters χ of G.

Ingeniously, we call the three irreducible characters of G which have yet to be found χ_9, χ_{10} and χ_{11}.

The column orthogonality relations give

$$\sum_{i=1}^{11} \chi_i(s)^2 = 48.$$

Hence

$$\chi_9(s)^2 + \chi_{10}(s)^2 + \chi_{11}(s)^2 = 2.$$

We can assume, without loss of generality, that

$$\chi_9(s)^2 = \chi_{10}(s)^2 = 1, \ \chi_{11}(s)^2 = 0.$$

Now $\chi_9\chi_2$ is an irreducible character, and is not equal to any of $\chi_1, \ldots, \chi_8$. Moreover, since $\chi_9\chi_2(s) = -\chi_9(s)$, we see that $\chi_9\chi_2$ is not χ_9 or χ_{11}. Therefore,

$$\chi_9\chi_2 = \chi_{10}.$$

Once more, we lose no generality in assuming that

$$\chi_9(s) = 1, \ \chi_{10}(s) = -1.$$

The plan is now to find $\chi_i(1)$ and $\chi_i(t)$ for $i = 9, 10, 11$. That is, we aim to evaluate the integers a, b, c, d, e, f in the following portion of the character table:

Element Class	1 (1)	s (2)	t (2,2)
χ_9	a	1	d
χ_{10}	b	-1	e
χ_{11}	c	0	f

The column orthogonality relations give

$$\sum_{i=1}^{11} \chi_i(1)\chi_i(s) = 0, \quad \sum_{i=1}^{11} \chi_i(s)\chi_i(t) = 0,$$

$$\sum_{i=1}^{11} \chi_i(t)\chi_i(t) = 16, \quad \sum_{i=1}^{11} \chi_i(1)\chi_i(t) = 0,$$

whence

$$a - b = 0, \qquad d - e = 0,$$

$$d^2 + e^2 + f^2 = 2, \qquad ad + be + cf = 10.$$

The only solution to these equations in integers with $a > 0$ and $b > 0$ is

$$d = e = 1, \ f = 0, \ a = b = 5.$$

Finally, we find that $c = 16$ by using the relation

$$\sum_{i=1}^{11} \chi_i(1)^2 = 720.$$

The above portion of the character table is therefore

Element Class	1 (1)	s (2)	t (2,2)
χ_9	5	1	1
χ_{10}	5	−1	1
χ_{11}	16	0	0

We can now determine the three unknown entries in each further column, since the column orthogonality relations will give three independent equations in these unknowns (as the above 3×3 matrix is invertible). Having done these calculations, we find that the complete character table of S_6 is as shown.

Character table of S_6

Class $|C_G(g_i)|$	(1) 720	(2) 48	(3) 18	(2,2) 16	(4) 8	(3,2) 6	(5) 5	(2,2,2) 48	(3,3) 18	(4,2) 8	(6) 6
χ_1	1	1	1	1	1	1	1	1	1	1	1
χ_2	1	−1	1	1	−1	−1	1	−1	1	1	−1
χ_3	5	3	2	1	1	0	0	−1	−1	−1	−1
χ_4	5	−3	2	1	−1	0	0	1	−1	−1	1
χ_5	10	2	1	−2	0	−1	0	−2	1	0	1
χ_6	10	−2	1	−2	0	1	0	2	1	0	−1
χ_7	9	3	0	1	−1	0	−1	3	0	1	0
χ_8	9	−3	0	1	1	0	−1	−3	0	1	0
χ_9	5	1	−1	1	−1	1	0	−3	2	−1	0
χ_{10}	5	−1	−1	1	1	−1	0	3	2	−1	0
χ_{11}	16	0	−2	0	0	0	1	0	−2	0	0

Direct products

We conclude the chapter by showing that tensor products can be used to determine the character table of a direct product $G \times H$, given the character tables of G and H.

Let V be a $\mathbb{C}G$-module, with basis $v_1, \ldots, v_m$, and let W be a

$\mathbb{C}H$-module, with basis $w_1, \ldots, w_n$. For all i, j with $1 \le i \le m$ and $1 \le j \le n$, and all $g \in G$, $h \in H$, define

$$(v_i \otimes w_j)(g, h) = v_i g \otimes w_j h$$

and extend this definition linearly to the whole of $V \otimes W$, that is, for $\lambda_{ij} \in \mathbb{C}$,

$$\left(\sum_{i,j} \lambda_{ij}(v_i \otimes w_j) \right)(g, h) = \sum_{i,j} \lambda_{ij}(v_i g \otimes w_j h).$$

As in Proposition 19.4, we find that

$$(v \otimes w)(g, h) = vg \otimes wh,$$

for all $v \in V$, $w \in W$. Then a proof similar to that of Proposition 19.5 shows that $V \otimes W$ is a $\mathbb{C}(G \times H)$-module.

Let χ be the character of V and ψ be the character of W. By the proof of Proposition 19.6, the character of $V \otimes W$ is $\chi \times \psi$, where

$$(\chi \times \psi)(g, h) = \chi(g)\psi(h) \quad (g \in G, h \in H).$$

19.18 Theorem
Let $\chi_1, \ldots, \chi_a$ be the distinct irreducible characters of G and let $\psi_1, \ldots, \psi_b$ be the distinct irreducible characters of H. Then $G \times H$ has precisely ab distinct irreducible characters, and these are

$$\chi_i \times \psi_j \ (1 \le i \le a, 1 \le j \le b).$$

Proof For all i, j, k, l,

$$\langle \chi_i \times \psi_j, \chi_k \times \chi_l \rangle_{G \times H} = \frac{1}{|G \times H|} \sum_{\substack{g \in G \\ h \in H}} \chi_i(g)\psi_j(h)\overline{\chi_k(g)\psi_l(h)}$$

$$= \left(\frac{1}{|G|} \sum_{g \in G} \chi_i(g)\overline{\chi_k(g)} \right) \left(\frac{1}{|H|} \sum_{h \in H} \psi_j(h)\overline{\psi_l(h)} \right)$$

$$= \langle \chi_i, \chi_k \rangle_G \langle \psi_j, \psi_l \rangle_H = \delta_{ik}\delta_{jl}.$$

(Here the subscripts $G \times H$, G and H indicate inner products of characters of $G \times H$, G and H, respectively.) Thus the ab characters $\chi_i \times \psi_j$ are distinct and irreducible.

Next, note that for all g, $x \in G$ and h, $y \in H$, we have

$$(x, y)^{-1}(g, h)(x, y) = (x^{-1}gx, y^{-1}hy).$$

Hence elements (g, h) and (g', h') of $G \times H$ are conjugate if and only if the elements g and g' are conjugate in G and the elements h and h' are conjugate in H. Consequently, if $g_1, \ldots, g_a$ are representatives of the conjugacy classes of G and $h_1, \ldots, h_b$ are representatives of the conjugacy classes of H, then the elements

$$(g_i, h_j) \quad (1 \le i \le a, 1 \le j \le b)$$

are representatives of the conjugacy classes of $G \times H$. In particular, $G \times H$ has precisely ab conjugacy classes.

By Theorem 15.3, $G \times H$ has exactly ab irreducible characters, so the irreducible characters $\chi_i \times \psi_j$ which we have found must be all the irreducible characters of $G \times H$. ∎

19.19 Example *The character table of $S_3 \times C_2$*
The character table of S_3 ($\cong D_6$) is given in Example 16.3(1). We reproduce it here, alongside the character table of C_2.

Character table of S_3

g_i	1	(1 2)	(1 2 3)
$|C_G(g_i)|$	6	2	3
χ_1	1	1	1
χ_2	1	-1	1
χ_3	2	0	-1

Character table of C_2

h_i	1	-1
$|C_H(h_i)|$	2	2
ψ_1	1	1
ψ_2	1	-1

The conjugacy classes of $S_3 \times C_2$ are represented by

$$(1, 1), ((1\ 2), 1), ((1\ 2\ 3), 1), (1, -1), ((1\ 2), -1), ((1\ 2\ 3), -1),$$

and by Theorem 19.18, the character table of $S_3 \times C_2$ is as shown.

Character table of $S_3 \times C_2$

(g_i, h_j)	(1, 1)	((1 2), 1)	((1 2 3), 1)	(1, -1)	((1 2), -1)	((1 2 3), -1)
$|C_{G \times H}(g_i, h_j)|$	12	4	6	12	4	6
$\chi_1 \times \psi_1$	1	1	1	1	1	1
$\chi_2 \times \psi_1$	1	-1	1	1	-1	1
$\chi_3 \times \psi_1$	2	0	-1	2	0	-1
$\chi_1 \times \psi_2$	1	1	1	-1	-1	-1
$\chi_2 \times \psi_2$	1	-1	1	-1	1	-1
$\chi_3 \times \psi_2$	2	0	-1	-2	0	1

Compare the solution to Exercise 18.2, where we give the character table of D_{12} (Exercise 1.5 shows that $D_{12} \cong S_3 \times C_2$).

Summary of Chapter 19

1. The product of any two characters of G is a character of G.

2. If χ is a character of G, then so are χ_S and χ_A, where

$$\chi_S(g) = \tfrac{1}{2}(\chi^2(g) + \chi(g^2)),$$

$$\chi_A(g) = \tfrac{1}{2}(\chi^2(g) - \chi(g^2))$$

   for all $g \in G$.

3. The irreducible characters of $G \times H$ are those characters $\chi \times \psi$, where χ is an irreducible character of G and ψ is an irreducible character of H. The values of $\chi \times \psi$ are given by

$$(\chi \times \psi)(g, h) = \chi(g)\psi(h)$$

   for all $g \in G$, $h \in H$.

Exercises for Chapter 19

1. Let χ, ψ and ϕ be characters of the group G. Show that

$$\langle \chi\psi, \phi \rangle = \langle \chi, \overline{\psi}\phi \rangle = \langle \psi, \overline{\chi}\phi \rangle.$$

2. Suppose that χ and ψ are irreducible characters of G. Prove that

$$\langle \chi\psi, 1_G \rangle = \begin{cases} 1, & \text{if } \chi = \overline{\psi}, \\ 0, & \text{if } \chi \neq \overline{\psi}. \end{cases}$$

3. Let χ be a character of G which is not faithful. Show that there is some irreducible character ψ of G such that $\langle \chi^n, \psi \rangle = 0$ for all integers n with $n \geq 0$.

   (This shows that the hypothesis that χ is faithful cannot be dropped from Theorem 19.10.)

4. In Example 20.13 of the next chapter we shall show that there exist irreducible characters χ and ϕ of A_5 which take the following values:

	1	(1 2 3)	(1 2)(3 4)	(1 2 3 4 5)	(1 3 4 5 2)
χ	5	-1	1	0	0
ϕ	3	0	-1	$(1 + \sqrt{5})/2$	$(1 - \sqrt{5})/2$

Calculate the values of χ_S, χ_A, ϕ_S and ϕ_A. Express these characters as linear combinations of the irreducible characters $\psi_1, \ldots, \psi_5$ of A_5 which are given in Example 20.13.

5. A certain group G or order 24 has precisely seven conjugacy classes with representatives $g_1, \ldots, g_7$; further, G has a character χ with values as follows:

g_i	g_1	g_2	g_3	g_4	g_5	g_6	g_7
$|C_G(g_i)|$	24	24	4	6	6	6	6
χ	2	-2	0	$-\omega^2$	$-\omega$	ω	ω^2

where $\omega = e^{2\pi i/3}$. Moreover, g_1^2, g_2^2, g_3^2, g_4^2, g_5^2, g_6^2, g_7^2 are conjugate to g_1, g_1, g_2, g_5, g_4, g_4, g_5, respectively.

Find χ_S and χ_A, and show that both are irreducible.

By forming products of the irreducible characters found so far, find the character table of G.

6. Write down the character table of $D_6 \times D_6$.

20

Restriction to a subgroup

In this chapter and the next, we are going to look at ways of relating the representations of a group to the representations of its subgroups. Here, we introduce the elementary idea of restricting a $\mathbb{C}G$-module to a subgroup H of G, and illustrate its use. The case where H is a normal subgroup of G is of particular interest, and Clifford's Theorem 20.8 gives important information in this case. We apply this result in the situation where H is of index 2 in G, which occurs, for example, when $G = S_n$ and $H = A_n$.

Restriction

Let H be a subgroup of the finite group G. Then $\mathbb{C}H$ is a subset of $\mathbb{C}G$. If V is a $\mathbb{C}G$-module, then V is also a $\mathbb{C}H$-module, since properties (1)–(5) of Definition 4.2 certainly hold for all $g, h \in H$ if they hold for all $g, h \in G$. This simple way of converting a $\mathbb{C}G$-module into a $\mathbb{C}H$-module is known as *restricting* from G to H. If V is a $\mathbb{C}G$-module then we write the corresponding $\mathbb{C}H$-module as $V \downarrow H$, and call it the *restriction* of V to H.

The character of $V \downarrow H$ is obtained from the character χ of V by evaluating χ on the elements of H only. We write this character of H as $\chi \downarrow H$, and refer to it as the *restriction* of χ to H. More generally, if $f : G \to \mathbb{C}$ is any function, then $f \downarrow H$ denotes the restriction of f to H (so that $(f \downarrow H)(h) = f(h)$ for all $h \in H$).

20.1 Example
Let $G = D_8 = \langle a, b : a^4 = b^2 = 1, b^{-1}ab = a^{-1} \rangle$. As in Example 4.5(1), let V be the $\mathbb{C}G$-module with basis v_1, v_2 for which

$$v_1 a = v_2, \qquad v_1 b = v_1,$$
$$v_2 a = -v_1, \qquad v_2 b = -v_2.$$

If H is the subgroup $\{1, a^2, b, a^2b\}$ of G, then $V \downarrow H$ is the $\mathbb{C}H$-module with basis v_1, v_2 for which

$$v_1a^2 = -v_1, \qquad v_1b = v_1,$$
$$v_2a^2 = -v_2, \qquad v_2b = -v_2.$$

The character χ of V is given by

g	1	a	a^2	a^3	b	ab	a^2b	a^3b
$\chi(g)$	2	0	-2	0	0	0	0	0

and the character $\chi \downarrow H$ of $V \downarrow H$ is given by

h	1	a^2	b	a^2b
$\chi(h)$	2	-2	0	0

If V is a $\mathbb{C}G$-module and H is a subgroup of G, then $\dim V = \dim(V \downarrow H)$. However, it might be the case that V is an irreducible $\mathbb{C}G$-module while $V \downarrow H$ is not an irreducible $\mathbb{C}H$-module; Example 20.1 illustrates this fact. On the other hand, if $V \downarrow H$ is an irreducible $\mathbb{C}H$-module then V is an irreducible $\mathbb{C}G$-module; for if U is a $\mathbb{C}G$-submodule of V, then $U \downarrow H$ is a $\mathbb{C}H$-submodule of $V \downarrow H$.

20.2 Example
Let $G = S_5$ and let H be the subgroup A_4 of G consisting of all even permutations of $\{1, 2, 3, 4\}$ fixing 5. By 18.2, the character table of H is

g_i $\lvert C_H(g_i)\rvert$	1 12	(1 2)(3 4) 4	(1 2 3) 3	(1 3 2) 3
ψ_1	1	1	1	1
ψ_2	1	1	ω	ω^2
ψ_3	1	1	ω^2	ω
ψ_4	3	-1	0	0

(where $\omega = e^{2\pi i/3}$). The character table of G is given in Example 19.16, with irreducible characters labelled $\chi_1, \ldots, \chi_7$.

For each i with $1 \le i \le 7$, we calculate the character $\chi_i \downarrow H$ as a sum of irreducible characters ψ_j. From Example 19.16 we see that

$$\chi_1 \downarrow H = \chi_2 \downarrow H, \chi_3 \downarrow H = \chi_4 \downarrow H \text{ and } \chi_6 \downarrow H = \chi_7 \downarrow H.$$

Therefore we need only consider $\chi_1 \downarrow H$, $\chi_3 \downarrow H$, $\chi_5 \downarrow H$ and $\chi_6 \downarrow H$. These have values

	1	(1 2)(3 4)	(1 2 3)	(1 3 2)
$\chi_1 \downarrow H$	1	1	1	1
$\chi_3 \downarrow H$	4	0	1	1
$\chi_5 \downarrow H$	6	−2	0	0
$\chi_6 \downarrow H$	5	1	−1	−1

It is easy to spot that

$$\chi_1 \downarrow H = \psi_1,$$

$$\chi_3 \downarrow H = \psi_1 + \psi_4,$$

$$\chi_5 \downarrow H = 2\psi_4,$$

$$\chi_6 \downarrow H = \psi_2 + \psi_3 + \psi_4.$$

Constituents of a restricted character

To help us discuss the way in which a restricted character $\chi \downarrow H$ can be expressed in terms of the irreducible characters of H, we introduce the following notation.

20.3 Definitions

The inner product $\langle , \rangle_G$ is the inner product on the vector space of functions from G to $\mathbb{C}$ which we have defined earlier, and $\langle , \rangle_H$ is the inner product on the vector space of functions from H to $\mathbb{C}$, defined similarly. Thus, if ϑ_1 and ϑ_2 are functions from G to $\mathbb{C}$, then

$$\langle \vartheta_1, \vartheta_2 \rangle_G = \frac{1}{|G|} \sum_{g \in G} \vartheta_1(g)\overline{\vartheta_2(g)},$$

and if ϕ_1 and ϕ_2 are functions from H to $\mathbb{C}$, then

$$\langle \phi_1, \phi_2 \rangle_H = \frac{1}{|H|} \sum_{h \in H} \phi_1(h) \overline{\phi_2(h)}.$$

If χ is a character of G and $\psi_1, \ldots, \psi_r$ are the irreducible characters of the subgroup H of G, then by Theorem 14.17,

$$\chi \downarrow H = d_1 \psi_1 + \ldots + d_r \psi_r$$

for some non-negative integers $d_1, \ldots, d_r$ which are given by

$$d_i = \langle \chi \downarrow H, \psi_i \rangle_H.$$

We say that ψ_i is a *constituent* of $\chi \downarrow H$ if the coefficient d_i in the above expression is non-zero.

The next proposition shows that every irreducible character of H is a constituent of the restriction of some irreducible character of G.

20.4 Proposition
Let H be a subgroup of G and let ψ be a non-zero character of H. Then there exists an irreducible character χ of G such that

$$\langle \chi \downarrow H, \psi \rangle_H \neq 0.$$

Proof Let $\chi_1, \ldots, \chi_k$ be the irreducible characters of G. Recall from Theorem 13.19 and Proposition 13.20 that the regular character χ_{reg} of G satisfies

$$\chi_{\text{reg}}(g) = \begin{cases} |G| & \text{if } g = 1, \\ 0 & \text{if } g \neq 1, \end{cases} \quad \text{and} \quad \chi_{\text{reg}} = \sum_{i=1}^{k} \chi_i(1) \chi_i.$$

Now

$$0 \neq \frac{|G|}{|H|} \psi(1) = \langle \chi_{\text{reg}} \downarrow H, \psi \rangle_H = \sum_{i=1}^{k} \chi_i(1) \langle \chi_i \downarrow H, \psi \rangle_H.$$

Therefore $\langle \chi_i \downarrow H, \psi \rangle_H \neq 0$ for some i. ∎

Suppose that we know the character table of G. In the light of Proposition 20.4, we could hope to find the character table of the subgroup H by restricting the irreducible characters χ of G to H. Unfortunately, it may be very difficult in practice to write down the restrictions $\chi \downarrow H$ in terms of irreducible characters of H. The best chance of doing this occurs when the index $|G: H|(= |G|/|H|)$ is small, since the restrictions $\chi \downarrow H$ then do not have many constituents, as the following result shows.

20.5 Proposition
Let H be a subgroup of G, let χ be an irreducible character of G, and let $\psi_1, \ldots, \psi_r$ be the irreducible characters of H. Then $\chi \downarrow H = d_1\psi_1 + \ldots + d_r\psi_r$, where the non-negative integers $d_1, \ldots, d_r$ satisfy

$$(20.6) \qquad \sum_{i=1}^{r} d_i^2 \leq |G: H|.$$

Moreover, we have equality in (20.6) if and only if $\chi(g) = 0$ for all elements g of G which lie outside H.

Proof By Theorem 14.17, we have

$$\sum_{i=1}^{r} d_i^2 = \langle \chi \downarrow H, \chi \downarrow H \rangle_H = \frac{1}{|H|} \sum_{h \in H} \chi(h)\overline{\chi(h)}.$$

Also, since χ is irreducible,

$$1 = \langle \chi, \chi \rangle_G = \frac{1}{|G|} \sum_{g \in G} \chi(g)\overline{\chi(g)}$$

$$= \frac{1}{|G|} \sum_{h \in H} \chi(h)\overline{\chi(h)} + K$$

$$= \frac{|H|}{|G|} \sum_{i=1}^{r} d_i^2 + K,$$

where $K = (1/|G|)\sum_{g \notin H}\chi(g)\overline{\chi(g)}$. Now $K \geq 0$, and $K = 0$ if and only if $\chi(g) = 0$ for all g with $g \notin H$. The conclusions of the proposition follow at once. ∎

We can say more about the constituents of $\chi \downarrow H$ in the case where H is a normal subgroup of G. For example, we will see that all the constituents of $\chi \downarrow H$ have the same degree.

The way to exploit the fact that $H \lhd G$ us revealed in the following proposition.

20.7 Proposition
Suppose that $H \lhd G$. Let V be an irreducible $\mathbb{C}G$-module and U be an irreducible $\mathbb{C}H$-submodule of $V \downarrow H$. For every $g \in G$ let $Ug = \{ug : u \in U\}$. Then

(1) The set Ug is an irreducible $\mathbb{C}H$-submodule of $V \downarrow H$ and $\dim Ug = \dim U$.
(2) As a $\mathbb{C}H$-module, V is a direct sum of some of the $\mathbb{C}H$-modules Ug.
(3) If g_1, g_2, $g \in G$ and Ug_1 and Ug_2 are isomorphic $\mathbb{C}H$-modules, then $Ug_1 g$ and $Ug_2 g$ are isomorphic $\mathbb{C}H$-modules.

Proof (1) Clearly, Ug is a subspace of V; and since $H \lhd G$ we have $ghg^{-1} \in H$ for all $h \in H$, so

$$(ug)h = u(ghg^{-1})g \in Ug \quad (u \in U),$$

proving that Ug is a $\mathbb{C}H$-submodule of $V \downarrow H$. Further, if W is a $\mathbb{C}H$-submodule of Ug, then Wg^{-1} is a $\mathbb{C}H$-submodule of U; since U is irreducible, $Wg^{-1} = \{0\}$ or U, whence $W = \{0\}$ or Ug. Therefore, Ug is an irreducible $\mathbb{C}H$-module, as claimed. Moreover, $u \to ug \ (u \in U)$ is an invertible linear transformation from U to Ug, so $\dim U = \dim Ug$.

(2) The sum of all the subspaces Ug with $g \in G$ is a $\mathbb{C}G$-submodule of V. Therefore, since V is irreducible, we have

$$V = \sum_{g \in G} Ug.$$

Then by Proposition 7.12, V is a direct sum of some of the $\mathbb{C}H$-modules Ug.

(3) Now let ϕ be a $\mathbb{C}H$-isomorphism from Ug_1 to Ug_2. Define $\theta : Ug_1 g \to Ug_2 g$ by

$$\theta : wg \to (w\phi)g \quad (w \in Ug_1).$$

Then θ is clearly an isomorphism of vector spaces. Suppose that $h \in H$. Then $gh = h'g$ for some $h' \in H$, and

$$(wgh)\theta = (wh'g)\theta = (wh'\phi)g = (w\phi)h'g$$

$$= (w\phi)gh = (wg\theta)h.$$

Therefore, θ is a $\mathbb{C}H$-isomorphism, and the proof of the proposition is complete. ∎

We now come the fundamental theorem on the restriction of a character to a normal subgroup.

20.8 Clifford's Theorem
Suppose that $H \lhd G$ and that χ is an irreducible character of G. Then

(1) *all the constituents of $\chi \downarrow H$ have the same degree; and*
(2) *if $\psi_1, \ldots, \psi_m$ are the constituents of $\chi \downarrow H$, then*

$$\chi \downarrow H = e(\psi_1 + \ldots + \psi_m)$$

for some positive integer e.

Proof Let V be a $\mathbb{C}G$-module with character χ. Then it follows from Proposition 20.7, parts (1) and (2), that all the constituents of $\chi \downarrow H$ have the same degree.

Let $e = \langle \chi \downarrow H, \psi_1 \rangle$. Then V contains a $\mathbb{C}H$-module X_1 whose character is $e\psi_1$, and which is therefore a direct sum of e isomorphic $\mathbb{C}H$-modules, each having character ψ_1; say

$$X_1 = U_1 \oplus \ldots \oplus U_e.$$

By Proposition 20.7(3), if $g \in G$ then $X_1 g$ is a direct sum of isomorphic $\mathbb{C}H$-modules. On the other hand, V is a sum of $\mathbb{C}H$-modules of the form $X_1 g$, by Proposition 20.7(2). Hence V has the form

$$V = X_1 \oplus \ldots \oplus X_m$$

where each X_i is a direct sum of e isomorphic $\mathbb{C}H$-modules, and $X_i \not\cong X_j$ if $i \neq j$. Therefore,

$$\chi \downarrow H = e(\psi_1 + \ldots + \psi_m). \qquad \blacksquare$$

Our main applications of Clifford's Theorem will concern the case where $|G: H| = 2$, but you might like to look at Corollary 22.14 in Chapter 22 to see an advanced use of the theorem.

Normal subgroups of index 2

We are shortly going to give more precise information about the constituents of $\chi \downarrow H$ when H is a normal subgroup of G of index 2 (that is, $|G: H| = 2$). Examples where this happens are $G = S_n$,

$H = A_n$, or $G = D_{2n} = \langle a, b: a^n = b^2 = 1, b^{-1}ab = a^{-1} \rangle$, $H = \langle a \rangle$. In fact, if H is a subgroup of index 2 in G, then H must be normal in G (see Exercise 1.10).

When H is a normal subgroup of index 2 in G, the character tables of G and H are closely related. We describe this relationship in (20.13) below, and we then illustrate the results by finding the character table of A_5 from that of S_5 (which we have already obtained in Example 19.16).

20.9 Proposition
Suppose that H is a normal subgroup of index 2 in G, and let χ be an irreducible character of G. Then either

(1) *$\chi \downarrow H$ is irreducible, or*
(2) *$\chi \downarrow H$ is the sum of two distinct irreducible characters of H of the same degree.*

Proof If $\psi_1, \ldots, \psi_r$ are the irreducible characters of H, then by Proposition 20.5,

$$\chi \downarrow H = d_1\psi_1 + \ldots + d_r\psi_r,$$

where $\sum_{i=1}^{r} d_i^2 \leqslant 2$. Since $d_1, \ldots, d_r$ are non-negative integers, we deduce that either $\chi \downarrow H = \psi_i$ for some i, or $\chi \downarrow H = \psi_i + \psi_j$ for some i, j with $i \neq j$. In the latter case, ψ_i and ψ_j have the same degree, by Clifford's Theorem 20.8 ∎

For practical purposes, it is often desirable to have more details about the two cases in Proposition 20.9, and we shall supply these next.

Since $G/H \cong C_2$, we may lift the non-trivial linear character of G/H to obtain a linear character λ of G which satisfies

$$\lambda(g) = \begin{cases} 1 & \text{if } g \in H, \\ -1 & \text{if } g \notin H. \end{cases}$$

Note that for all irreducible characters χ of G, χ and $\chi\lambda$ are irreducible characters of the same degree (see Proposition 17.14). Also, $\chi \downarrow H = \chi\lambda \downarrow H$, since $\lambda(h) = 1$ for all $h \in H$.

20.10 Proposition
Suppose that H is a normal subgroup of index 2 in G, and that χ is

an irreducible character of G. Then the following three conditions are equivalent:

(1) $\chi \downarrow H$ *is irreducible*;
(2) $\chi(g) \neq 0$ *for some* $g \in G$ *with* $g \notin H$;
(3) *the characters* χ *and* $\chi\lambda$ *of G are not equal*.

Proof We use Proposition 20.5; since $|G: H| = 2$, $\chi \downarrow H$ is irreducible if and only if the inequality in (20.6) is strict, and this happens if and only if $\chi(g) \neq 0$ for some $g \in G$ with $g \notin H$. Thus (1) is equivalent to (2).

To see that (2) is equivalent to (3), observe that

$$\chi\lambda(g) = \begin{cases} \chi(g) & \text{if } g \in H, \\ -\chi(g) & \text{if } g \notin H, \end{cases}$$

so $\chi(g) \neq 0$ for some g with $g \notin H$ if and only if $\chi\lambda \neq \chi$. ∎

According to Proposition 20.9, if H is a normal subgroup of G of index 2, and χ is an irreducible character of G, then $\chi \downarrow H$ is the sum of one or two irreducible characters of H. In the next proposition we consider the first possibility.

20.11 Proposition
Suppose that H is a normal subgroup of index 2 in G, and that χ is an irreducible character of G for which $\chi \downarrow H$ is irreducible. If ϕ is an irreducible character of G which satisfies

$$\phi \downarrow H = \chi \downarrow H,$$

then either $\phi = \chi$ or $\phi = \chi\lambda$.

Proof We have

$$(\chi + \chi\lambda)(g) = \begin{cases} 2\chi(g) & \text{if } g \in H, \\ 0 & \text{if } g \notin H. \end{cases}$$

Hence

$$\langle \chi + \chi\lambda, \phi \rangle_G = \frac{1}{|G|} \sum_{g \in H} 2\chi(g)\overline{\phi(g)}$$

$$= \frac{1}{|H|} \sum_{g \in H} \chi(g)\overline{\phi(g)}$$

$$= \langle \chi \downarrow H, \phi \downarrow H \rangle_H.$$

Now $\langle \chi \downarrow H, \phi \downarrow H \rangle_H = 1$, since $\chi \downarrow H$ is irreducible and $\phi \downarrow H = \chi \downarrow H$. Therefore $\langle \chi + \chi\lambda, \phi \rangle_G = 1$, and so either $\phi = \chi$ or $\phi = \chi\lambda$. ∎

Finally, we look at the case where $\chi \downarrow H$ is reducible.

20.12 Proposition
Suppose that H is a normal subgroup of index 2 in G, and that χ is an irreducible character of G for which $\chi \downarrow H$ is the sum of two irreducible characters of H, say $\chi \downarrow H = \psi_1 + \psi_2$. If ϕ is an irreducible character of G such that $\phi \downarrow H$ has ψ_1 or ψ_2 as a constituent, then $\phi = \chi$.

Proof In view of Proposition 20.10, $\chi(g) = 0$ for all g with $g \notin H$. Therefore,

$$\langle \phi, \chi \rangle_G = \frac{1}{|G|} \sum_{g \in G} \phi(g)\overline{\chi(g)} = \frac{1}{|G|} \sum_{g \in H} \phi(g)\overline{\chi(g)}$$

$$= \tfrac{1}{2}\langle \phi \downarrow H, \chi \downarrow H \rangle_H.$$

If $\phi \downarrow H$ has ψ_1 or ψ_2 as a constituent, then $\langle \phi \downarrow H, \chi \downarrow H \rangle_H \neq 0$, so $\langle \phi, \chi \rangle_G \neq 0$, and hence $\phi = \chi$. ∎

We summarize our results on subgroups of index 2, by explaining how to list the irreducible characters of H on the assumption that we know the character table of G.

(20.13) Let H be a normal subgroup of index 2 in the group G.
   (1) Each irreducible character χ of G which is non-zero somewhere outside H restricts to be an irreducible character of H. Such characters of G occur in pairs (χ and $\chi\lambda$) which have the same restriction to H (Propositions 20.10, 20.11).

(2) If χ is an irreducible character of G which is zero everywhere outside H, then χ restricts to be the sum of two distinct irreducible characters of H of the same degree. The two characters of H which we get from χ in this way come from no other irreducible character of G (Propositions 20.9, 20.10, 20.12).

(3) Every irreducible character of H appears among those obtained by restricting irreducible characters of G, as in parts (1) and (2) (Proposition 20.4).

In case (2) of (20.13), extra work is needed to calculate the values taken by the two constituents of $\chi \downarrow H$. Fortunately, in practice case (1) occurs more frequently than case (2).

20.14 Example   The character table of A_5

Write $H = A_5$, and note that H is a normal subgroup of index 2 in the group S_5. The conjugacy classes of H and their sizes are given in Example 12.18(2), and the irreducible characters $\chi_1, \ldots, \chi_7$ of S_5 can be found in Example 19.16.

Observe that χ_1, χ_3 and χ_6 are non-zero somewhere outside H, so by (20.13)(1), $\chi_1 \downarrow H$, $\chi_3 \downarrow H$ and $\chi_6 \downarrow H$ are irreducible characters of H. Call them ψ_1, ψ_2 and ψ_3, respectively. Also, $\chi_5(g) = 0$ for all $g \notin H$, so by (20.13)(2), $\chi_5 \downarrow H = \psi_4 + \psi_5$ where ψ_4 and ψ_5 are distinct irreducible characters of H of degree 3. Note that $\chi_2 \downarrow H = \chi_1 \downarrow H$, $\chi_4 \downarrow H = \chi_3 \downarrow H$ and $\chi_7 \downarrow H = \chi_6 \downarrow H$, and hence $\psi_1, \ldots, \psi_5$ are the distinct irreducible characters of H by (20.13)(3).

The results we have obtained so far have been deduced from our summary (20.13) of facts about characters of subgroups of index 2. They can also be verified by calculating inner products. We have established that the character table of H is

g_i	1	(1 2 3)	(1 2)(3 4)	(1 2 3 4 5)	(1 3 4 5 2)
$|C_G(g_i)|$	60	3	4	5	5
ψ_1	1	1	1	1	1
ψ_2	4	1	0	-1	-1
ψ_3	5	-1	1	0	0
ψ_4	3	α_2	α_3	α_4	α_5
ψ_5	3	β_2	β_3	β_4	β_5

We use the column orthogonality relations to calculate the unknowns α_i and β_i. The values of $\alpha_i + \beta_i$ for $2 \leqslant i \leqslant 5$ are given by noting that $\psi_4 + \psi_5 = \chi_5 \downarrow H$ (or by using the column orthogonality relations for column 1 and column i). We get

$$\alpha_2 + \beta_2 = 0, \ \alpha_3 + \beta_3 = -2, \ \alpha_4 + \beta_4 = \alpha_5 + \beta_5 = 1.$$

Using Proposition 12.13, we see that each element of A_5 is conjugate to its inverse. Hence by Proposition 13.9(4), all the numbers in the character table are real. By the orthogonality relation for column i with itself $(2 \leqslant i \leqslant 5)$, we obtain

$$3 = 3 + \alpha_2^2 + \beta_2^2,$$
$$4 = 2 + \alpha_3^2 + \beta_3^2,$$
$$5 = 2 + \alpha_4^2 + \beta_4^2 = 2 + \alpha_5^2 + \beta_5^2.$$

Hence

$$\alpha_2 = \beta_2 = 0, \ \alpha_3 = \beta_3 = -1,$$

and we find that α_4 and β_4 are the solutions of the quadratic $x^2 - x - 1 = 0$. Since we have not yet distinguished between ψ_4 and ψ_5, we may take

$$\alpha_4 = \tfrac{1}{2}(1 + \sqrt{5}), \ \beta_4 = \tfrac{1}{2}(1 - \sqrt{5}).$$

Similarly, α_5 and β_5 are $\tfrac{1}{2}(1 \pm \sqrt{5})$. Since $\psi_4 \neq \psi_5$, we have

$$\alpha_5 = \tfrac{1}{2}(1 - \sqrt{5}), \ \beta_5 = \tfrac{1}{2}(1 + \sqrt{5}).$$

Thus the character table of A_5 is as shown.

Character table of A_5

g_i	1	(1 2 3)	(1 2)(3 4)	(1 2 3 4 5)	(1 3 4 5 2)
$|C_G(g_i)|$	60	3	4	5	5
ψ_1	1	1	1	1	1
ψ_2	4	1	0	-1	-1
ψ_3	5	-1	1	0	0
ψ_4	3	0	-1	α	β
ψ_5	3	0	-1	β	α

where $\alpha = \tfrac{1}{2}(1 + \sqrt{5})$, $\beta = \tfrac{1}{2}(1 - \sqrt{5})$.

Proposition 17.6 allows us to deduce from the character table that A_5 is a simple group.

Summary of Chapter 20

Assume throughout that H is a subgroup of G.

1. If χ is a character of G, then $\chi \downarrow H$ is a character of H. The values of $\chi \downarrow H$ are given by

$$(\chi \downarrow H)(h) = \chi(h)$$

for all $h \in H$. In particular, χ and $\chi \downarrow H$ have the same degree.

2. The number of irreducible constituents of $\chi \downarrow H$ is bounded above by $|G : H|$. Indeed, if $\psi_1, \ldots, \psi_r$ are the irreducible characters of H, and $\chi \downarrow H = d_1\psi_1 + \ldots + d_r\psi_r$, then

$$\sum_{i=1}^{r} d_i^2 \leqslant |G : H|.$$

3. If $H \lhd G$ and χ is an irreducible character of G, then all the constituents of $\chi \downarrow H$ have the same degree.

Exercises for Chapter 20

1. Let $G = S_4$ and let H be the subgroup $\langle (1\ 2\ 3\ 4), (1\ 3) \rangle$ of G.
   (a) Show that $H \cong D_8$.
   (b) For each irreducible character χ of G (given in Section 18.1), express $\chi \downarrow H$ as a sum of irreducible characters of H.

2. Use the restrictions of the irreducible characters of S_6, given in Example 19.17, to find the character table of A_6.
   (The seven conjugacy classes of A_6 can be found by consulting the solutions to Exercises 12.3 and 12.4.)

3. Let G be a group with an abelian subgroup H of index n. Prove that $\chi(1) \leqslant n$ for every irreducible character χ of G.

4. Suppose that G is a group with a subgroup H of index 3, and let χ be an irreducible character of G. Prove that

$$\langle \chi \downarrow H, \chi \downarrow H \rangle_H = 1, 2 \text{ or } 3.$$

Give examples to show that each possibility can occur.

5. It is known that the complete list of degrees of the irreducible characters of S_7 is

$$1, 1, 6, 6, 14, 14, 14, 14, 15, 15, 20, 21, 21, 35, 35.$$

Also, A_7 has nine conjugacy classes. Find the complete list of degrees of the irreducible characters of A_7.

21
Induced modules and characters

Throughout this chapter we assume that H is a subgroup of the finite group G. We saw in the last chapter that restriction gives a simple way of converting a $\mathbb{C}G$-module into a $\mathbb{C}H$-module. Much more subtle than this is the process of induction, which constructs a $\mathbb{C}G$-module from a given $\mathbb{C}H$-module, and induction is the main concern of this chapter. As H is smaller than G, it is usually the case that $\mathbb{C}H$-modules are easier to understand and construct than $\mathbb{C}G$-modules, so induction can often give us an important handle on the representations of a group if we know some representations of its subgroups. We shall see many applications of this method in later chapters.

Before describing the process of induction, we require some results which connect $\mathbb{C}H$-homomorphisms with $\mathbb{C}G$-homomorphisms.

$\mathbb{C}H$-homomorphisms and $\mathbb{C}G$-homomorphisms

Let U be a $\mathbb{C}H$-submodule of the regular $\mathbb{C}H$-module $\mathbb{C}H$. If $r \in \mathbb{C}G$, then

$$\vartheta: u \to ru \quad (u \in U)$$

defines a $\mathbb{C}H$-homomorphism from U to $\mathbb{C}G$, since for all $s \in \mathbb{C}H$, $(us)\vartheta = rus = (u\vartheta)s$. We now prove the striking fact that *every* $\mathbb{C}H$-homomorphism from U to $\mathbb{C}G$ has this simple form.

21.1 Proposition
Assume that $H \leqslant G$, and let U be a $\mathbb{C}H$-submodule of $\mathbb{C}H$. If ϑ is a $\mathbb{C}H$-homomorphism from U to $\mathbb{C}G$, then there exists $r \in \mathbb{C}G$ such that

$$u\vartheta = ru \quad \text{for all } u \in U.$$

Proof By Maschke's Theorem 8.1, there is a $\mathbb{C}H$-submodule W of $\mathbb{C}H$ such that $\mathbb{C}H = U \oplus W$. Define $\phi: \mathbb{C}H \to \mathbb{C}G$ by

$$\phi: u + w \to u\vartheta \quad (u \in U, w \in W).$$

Then ϕ is easily seen to be a $\mathbb{C}H$-homomorphism. Let $r = 1\phi$. For $u \in U$,

$$u\vartheta = u\phi = (1u)\phi = (1\phi)u = ru,$$

and so ϑ is of the required form. ∎

We give two corollaries of Proposition 21.1, the first of which is just the case $H = G$ of the proposition.

21.2 Corollary
Let U be a $\mathbb{C}G$-submodule of $\mathbb{C}G$. Then every $\mathbb{C}G$-homomorphism from U to $\mathbb{C}G$ has the form

$$u \to ru \quad (u \in U)$$

for some $r \in \mathbb{C}G$.

21.3 Corollary
Let U and V be $\mathbb{C}G$-submodules of $\mathbb{C}G$. Then the following two statements are equivalent:

(1) $U \cap V = \{0\}$;
(2) *there exists $r \in \mathbb{C}G$ such that for all $u \in U, v \in V$,*

$$ru = u \text{ and } rv = 0.$$

Proof Assume that $U \cap V = \{0\}$. Then the sum $U + V$ is a direct sum, so

$$u + v \to u \quad (u \in U, v \in V)$$

is a function; moreover, it is a $\mathbb{C}G$-homomorphism from $U \oplus V$ to $\mathbb{C}G$ (see Proposition 7.11). Therefore by Corollary 21.2, there exists $r \in \mathbb{C}G$ such that for all $u \in U, v \in V$,

$$r(u + v) = u.$$

Then $ru = u$ if $u \in U$, and $rv = 0$ if $v \in V$.

Conversely, assume that for some $r \in \mathbb{C}G$ we have $ru = u$ and $rv = 0$ for all $u \in U, v \in V$. If $x \in U \cap V$ then $rx = x$ and $rx = 0$, and so $x = 0$. Consequently $U \cap V = \{0\}$. ∎

Induction from H to G

For any subset X of $\mathbb{C}G$, we write $X(\mathbb{C}G)$ for the subspace of $\mathbb{C}G$ which is spanned by all the elements xg with $x \in X$, $g \in G$. That is,

$$X(\mathbb{C}G) = \text{sp}\{xg: x \in X, g \in G\}.$$

Clearly, $X(\mathbb{C}G)$ is then a $\mathbb{C}G$-submodule of $\mathbb{C}G$.

Remember that H is a subgroup of G, so $\mathbb{C}H$ is a subset of $\mathbb{C}G$.

21.4 Definition

Assume that H is a subgroup of G. Let U be a $\mathbb{C}H$-submodule of $\mathbb{C}H$, and let $U \uparrow G$ denote the $\mathbb{C}G$-module $U(\mathbb{C}G)$. Then $U \uparrow G$ is called the $\mathbb{C}G$-module *induced* from U.

21.5 Example

Let $G = D_6 = \langle a, b: a^3 = b^2 = 1, b^{-1}ab = a^{-1} \rangle$, and let $H = \langle a \rangle$, a cyclic subgroup of G of order 3. Let $\omega = e^{2\pi i/3}$, and define

$$W_0 = \text{sp}(1 + a + a^2),$$
$$W_1 = \text{sp}(1 + \omega^2 a + \omega a^2),$$
$$W_2 = \text{sp}(1 + \omega a + \omega^2 a^2).$$

These are $\mathbb{C}H$-submodules of $\mathbb{C}H$ (see Example 10.8(1)). Clearly,

$$W_0 \uparrow G = \text{sp}(1 + a + a^2, b + ab + a^2b),$$
$$W_1 \uparrow G = \text{sp}(1 + \omega^2 a + \omega a^2, b + \omega^2 ab + \omega a^2 b),$$
$$W_2 \uparrow G = \text{sp}(1 + \omega a + \omega^2 a^2, b + \omega ab + \omega^2 a^2 b).$$

Recall from Example 10.8(2) that

$$\mathbb{C}G = U_1 \oplus U_2 \oplus U_3 \oplus U_4,$$

a direct sum of irreducible $\mathbb{C}G$-modules U_i, where

$$U_1 = \text{sp}(1 + a + a^2 + b + ab + a^2b),$$
$$U_2 = \text{sp}(1 + a + a^2 - b - ab - a^2b),$$
$$U_3 = \text{sp}(1 + \omega^2 a + \omega a^2, b + \omega^2 ab + \omega a^2 b),$$
$$U_4 = \text{sp}(1 + \omega a + \omega^2 a^2, b + \omega ab + \omega^2 a^2 b).$$

Thus

$$W_0 \uparrow G = U_1 \oplus U_2, \quad W_1 \uparrow G = U_3, \quad W_2 \uparrow G = U_4.$$

In particular, $W_0 \uparrow G$ is reducible, while $W_1 \uparrow G$ and $W_2 \uparrow G$ are irreducible.

We now show that isomorphic $\mathbb{C}H$-modules give isomorphic induced $\mathbb{C}G$-modules.

21.6 Proposition
Assume that $H \leqslant G$. Suppose that U and V are $\mathbb{C}H$-submodules of $\mathbb{C}H$ and that U is $\mathbb{C}H$-isomorphic to V. Then $U \uparrow G$ is $\mathbb{C}G$-isomorphic to $V \uparrow G$.

Proof Let $\vartheta: U \to V$ be a $\mathbb{C}H$-isomorphism. By Proposition 21.1, there exists $r \in \mathbb{C}G$ such that $u\vartheta = ru$ for all $u \in U$, and also there exists $s \in \mathbb{C}G$ such that $v\vartheta^{-1} = sv$ for all $v \in V$. Consequently

$$sru = u \text{ and } rsv = v \quad \text{for all } u \in U, v \in V.$$

If $a \in U \uparrow G$ then a is a linear combination of elements ug ($u \in U, g \in G$), so ra is a linear combination of elements rug, and hence $ra \in V \uparrow G$. Therefore

$$\phi: a \to ra \quad (a \in U \uparrow G)$$

is a function from $U \uparrow G$ to $V \uparrow G$. Moreover, ϕ is a $\mathbb{C}G$-homomorphism, as $(a\phi)g = rag = (ag)\phi$ ($a \in U \uparrow G, g \in G$). Since $sru = u$ and $rsv = v$ for all $u \in U, v \in V$, we have

$$sra = a, rsb = b \quad \text{for all } a \in U \uparrow G, b \in V \uparrow G.$$

Hence the function

$$b \to sb \quad (b \in V \uparrow G)$$

is the inverse of ϕ. Therefore ϕ is a $\mathbb{C}G$-isomorphism, proving that $U \uparrow G$ is $\mathbb{C}G$-isomorphic to $V \uparrow G$. ∎

The next proposition and its corollary enable us to define the induced module $U \uparrow G$ for an arbitrary $\mathbb{C}H$-module U.

21.7 Proposition
Assume that U and V are $\mathbb{C}H$-submodules of $\mathbb{C}H$ with $U \cap V = \{0\}$. Then $(U \uparrow G) \cap (V \uparrow G) = \{0\}$.

Proof By Corollary 21.3, there exists $r \in \mathbb{C}G$ such that $ru = u$ and $rv = 0$ for all $u \in U$, $v \in V$. Then for all $u \in U$, $v \in V$ and all $g \in G$,

$$ru g = ug \text{ and } rvg = 0.$$

Since $U \uparrow G$ is spanned by elements of the form ug $(u \in U, g \in G)$, this implies that

$$ru' = u' \quad \text{for all } u' \in U \uparrow G,$$

and similarly,

$$rv' = 0 \quad \text{for all } v' \in V \uparrow G.$$

Therefore $(U \uparrow G) \cap (V \uparrow G) = \{0\}$ by Corollary 21.3. ∎

21.8 Corollary
Let U be a $\mathbb{C}H$-submodule of $\mathbb{C}H$, and suppose that

$$U = U_1 \oplus \ldots \oplus U_m,$$

a direct sum of $\mathbb{C}H$-submodules U_i. Then

$$U \uparrow G = (U_1 \uparrow G) \oplus \ldots \oplus (U_m \uparrow G).$$

Proof We prove this by induction on m. It is trivial for $m = 1$. Now $U = U_1 \oplus V$, where $V = U_2 \oplus \ldots \oplus U_m$. The definition of $U \uparrow G$ implies that

$$U \uparrow G = (U_1 \uparrow G) + (V \uparrow G).$$

Therefore by Proposition 21.7,

$$U \uparrow G = (U_1 \uparrow G) \oplus (V \uparrow G).$$

By induction, $V \uparrow G = (U_2 \uparrow G) \oplus \ldots \oplus (U_m \uparrow G)$, and hence, using (2.10), we obtain

$$U \uparrow G = (U_1 \uparrow G) \oplus \ldots \oplus (U_m \uparrow G),$$

as required. ∎

We can now define the induced module $U \uparrow G$ for an arbitrary $\mathbb{C}H$-module U (where U is not necessarily a $\mathbb{C}H$-submodule of $\mathbb{C}H$).

21.9 Definition
Let U be a $\mathbb{C}H$-module. Then (by Theorems 8.7 and 10.5),

$$U \cong U_1 \oplus \ldots \oplus U_m$$

for certain irreducible $\mathbb{C}H$-submodules U_i of $\mathbb{C}H$. Define $U \uparrow G$ to be the following (external) direct sum:

$$U \uparrow G = (U_1 \uparrow G) \oplus \ldots \oplus (U_m \uparrow G).$$

Proposition 21.6 and Corollary 21.8 ensure that this definition is consistent with Definition 21.4.

We emphasize that the definition of the induced module $U \uparrow G$ in the case where U is a $\mathbb{C}H$-submodule of $\mathbb{C}H$ is a natural one:

$$U \uparrow G = U(\mathbb{C}G).$$

We shall always prove results for general induced modules $U \uparrow G$ by first dealing with the special case where U is a $\mathbb{C}H$-submodule of $\mathbb{C}H$, and then applying the fact (which is immediate from Definition 21.9) that

(21.10) $\qquad (U_1 \oplus \ldots \oplus U_m) \uparrow G \cong (U_1 \uparrow G) \oplus \ldots \oplus (U_m \uparrow G).$

Our first major result on general induced modules is known as 'the transitivity of induction'.

21.11 Theorem
Suppose that H and K are subgroups of G such that $H \leqslant K \leqslant G$. If U is a $\mathbb{C}H$-module, then

$$(U \uparrow K) \uparrow G \cong U \uparrow G.$$

Proof Assume first that U is a $\mathbb{C}H$-submodule of $\mathbb{C}H$. Then $U(\mathbb{C}K)$ is spanned by elements of the form

$$uk \quad (u \in U, k \in K).$$

Therefore, $(U(\mathbb{C}K))(\mathbb{C}G)$ is spanned by elements of the form

$$ukg \quad (u \in U, k \in K, g \in G).$$

Since $K \leqslant G$, it follows that $(U(\mathbb{C}K))(\mathbb{C}G) = U(\mathbb{C}G)$. That is,

(21.12) $\qquad\qquad (U \uparrow K) \uparrow G = U \uparrow G.$

Now let U be an arbitrary $\mathbb{C}H$-module. Then

$$U \cong U_1 \oplus \ldots \oplus U_m$$

for certain irreducible $\mathbb{C}H$-submodules U_i of $\mathbb{C}H$. By (21.10),

$$U \uparrow K \cong (U_1 \uparrow K) \oplus \ldots \oplus (U_m \uparrow K).$$

Therefore

$$(U \uparrow K) \uparrow G \cong (U_1 \uparrow K) \uparrow G \oplus \ldots \oplus (U_m \uparrow K) \uparrow G$$

by (21.10)

$$= (U_1 \uparrow G) \oplus \ldots \oplus (U_m \uparrow G) \quad \text{by (21.12)}$$

$$\cong U \uparrow G \quad \text{by Definition 21.9.} \qquad \blacksquare$$

Induced characters

21.13 Definition

If ψ is the character of a $\mathbb{C}H$-module U, then the character of the induced $\mathbb{C}G$-module $U \uparrow G$ is denoted by $\psi \uparrow G$, and is called the character *induced* from ψ.

The next example illustrates an important relationship between induced characters and restrictions of characters.

21.14 Example

Let $G = S_5$ and let H be the subgroup A_4 of G, as in Example 20.2. We showed in that example that if $\chi_1, \ldots, \chi_7$ are the irreducible characters of G (given in Example 19.16), and $\psi_1, \ldots, \psi_4$ are the irreducible characters of H (given in 18.2) then

$$\chi_1 \downarrow H = \psi_1,$$

$$\chi_2 \downarrow H = \psi_1,$$

$$\chi_3 \downarrow H = \psi_1 + \psi_4,$$

$$\chi_4 \downarrow H = \psi_1 + \psi_4,$$

$$\chi_5 \downarrow H = 2\psi_4,$$

$$\chi_6 \downarrow H = \psi_2 + \psi_3 + \psi_4,$$

$$\chi_7 \downarrow H = \psi_2 + \psi_3 + \psi_4.$$

By Theorem 14.17, the coefficients which appear here are the values of $\langle \chi_i \downarrow H, \psi_j \rangle_H$ for appropriate i, j. We record these coefficients in a matrix whose ij-entry is $\langle \chi_i \downarrow H, \psi_j \rangle_H$:

$$
\begin{array}{c}
\\
\chi_1 \\
\chi_2 \\
\chi_3 \\
\chi_4 \\
\chi_5 \\
\chi_6 \\
\chi_7
\end{array}
\begin{array}{cccc}
\psi_1 & \psi_2 & \psi_3 & \psi_4 \\
\left(\begin{array}{cccc}
1 & 0 & 0 & 0 \\
1 & 0 & 0 & 0 \\
1 & 0 & 0 & 1 \\
1 & 0 & 0 & 1 \\
0 & 0 & 0 & 2 \\
0 & 1 & 1 & 1 \\
0 & 1 & 1 & 1
\end{array}\right)
\end{array}
$$

The rows of this matrix tell us how the irreducible characters of G restrict to H. For example, row 3 gives

$$\chi_3 \downarrow H = 1 \cdot \psi_1 + 0 \cdot \psi_2 + 0 \cdot \psi_3 + 1 \cdot \psi_4.$$

Remarkably, it turns out that the columns of the matrix tell us how the irreducible characters of H induce to G. To be precise, the seven integers in column 1 give

$$\psi_1 \uparrow G = 1 \cdot \chi_1 + 1 \cdot \chi_2 + 1 \cdot \chi_3 + 1 \cdot \chi_4 + 0 \cdot \chi_5 + 0 \cdot \chi_6 + 0 \cdot \chi_7.$$

Similarly,

$$\psi_2 \uparrow G = \psi_3 \uparrow G = \chi_6 + \chi_7, \text{ and}$$

$$\psi_4 \uparrow G = \chi_3 + \chi_4 + 2\chi_5 + \chi_6 + \chi_7.$$

Thus the ij-entry in our matrix, which we already know to be equal to $\langle \chi_i \downarrow H, \psi_j \rangle_H$, is also equal to $\langle \chi_i, \psi_j \uparrow G \rangle_G$.

In fact, it is true that

$$\langle \chi, \psi \uparrow G \rangle_G = \langle \chi \downarrow H, \psi \rangle_H$$

for all characters χ of G and ψ of H, and we devote the next section to a proof of this result, which is known as the Frobenius Reciprocity Theorem.

The Frobenius Reciprocity Theorem

Before proving the theorem, we need the following preliminary result.

21.15 Proposition
Assume that $H \leqslant G$. Let U be a $\mathbb{C}H$-submodule of $\mathbb{C}H$, and let V be a $\mathbb{C}G$-submodule of $\mathbb{C}G$. Then the vector spaces

$$Hom_{\mathbb{C}G}(U \uparrow G, V) \text{ and } Hom_{\mathbb{C}H}(U, V \downarrow H)$$

have equal dimensions.

Proof Suppose that $\vartheta \in \mathrm{Hom}_{\mathbb{C}G}(U \uparrow G, V)$. Then by Corollary 21.2, there is an element $r \in \mathbb{C}G$ such that

$$s\vartheta = rs \quad \text{for all } s \in U \uparrow G.$$

Define $\bar{\vartheta}: U \to \mathbb{C}G$ to be the restriction of ϑ to U; that is,

$$u\bar{\vartheta} = ru \quad \text{for all } u \in U.$$

Then $\bar{\vartheta} \in \mathrm{Hom}_{\mathbb{C}H}(U, V \downarrow H)$. Thus the function

$$\vartheta \to \bar{\vartheta}$$

is a linear transformation from $\mathrm{Hom}_{\mathbb{C}G}(U \uparrow G, V)$ to $\mathrm{Hom}_{\mathbb{C}H}(U, V \downarrow H)$. We shall show that this linear transformation is invertible.

Let $\phi \in \mathrm{Hom}_{\mathbb{C}H}(U, V \downarrow H)$. Then by Proposition 21.1, there exists $r \in \mathbb{C}G$ such that $u\phi = ru$ for all $u \in U$. Define ϑ from $U \uparrow G$ to $\mathbb{C}G$ by

$$s\vartheta = rs \quad (s \in U \uparrow G).$$

Then $\vartheta \in \mathrm{Hom}_{\mathbb{C}G}(U \uparrow G, V)$. Moreover, $\phi = \bar{\vartheta}$. Therefore the function $\vartheta \to \bar{\vartheta}$ is surjective.

Finally, note that if $r_1, r_2 \in \mathbb{C}G$ and $r_1 u = r_2 u$ for all $u \in U$, then $r_1 s = r_2 s$ for all $s \in U \uparrow G$, since s is a linear combination of elements ug with $u \in U$, $g \in G$. Hence the function $\vartheta \to \bar{\vartheta}$ is injective, and so it is an invertible linear transformation from $\mathrm{Hom}_{\mathbb{C}G}(U \uparrow G, V)$ to $\mathrm{Hom}_{\mathbb{C}H}(U, V \downarrow H)$. These two vector spaces therefore have the same dimension, as required. ∎

21.16 The Frobenius Reciprocity Theorem
Assume that $H \leqslant G$. Let χ be a character of G and let ψ be a character of H. Then

$$\langle \psi \uparrow G, \chi \rangle_G = \langle \psi, \chi \downarrow H \rangle_H.$$

Proof First assume that the characters χ and ψ are irreducible. Then there is a $\mathbb{C}H$-submodule U of $\mathbb{C}H$ which has character ψ, and there is a $\mathbb{C}G$-submodule V of $\mathbb{C}G$ which has character χ. By Theorem 14.24, we have

$$\langle \psi \uparrow G, \chi \rangle_G = \dim (\mathrm{Hom}_{\mathbb{C}G}(U \uparrow G, V)), \text{ and}$$

$$\langle \psi, \chi \downarrow H \rangle_H = \dim (\mathrm{Hom}_{\mathbb{C}H}(U, V \downarrow H)).$$

From Proposition 21.15, we therefore deduce that

$$(21.17) \qquad \langle \psi \uparrow G, \chi \rangle_G = \langle \psi, \chi \downarrow H \rangle_H$$

in the special case we are considering, namely when χ and ψ are irreducible.

For the general case, let $\chi_1, \ldots, \chi_k$ be the irreducible characters of G and let $\psi_1, \ldots, \psi_m$ be the irreducible characters of H. Then for some integers d_i, e_j we have

$$\chi = \sum_{i=1}^{k} d_i \chi_i \text{ and } \psi = \sum_{j=1}^{m} e_j \psi_j.$$

Therefore

$$\langle \psi \uparrow G, \chi \rangle_G = \left\langle \sum_{j=1}^{m} e_j \psi_j \uparrow G, \sum_{i=1}^{k} d_i \chi_i \right\rangle_G$$

$$= \sum_{j=1}^{m} \sum_{i=1}^{k} e_j d_i \langle \psi_j \uparrow G, \chi_i \rangle_G$$

$$= \sum_{j=1}^{m} \sum_{i=1}^{k} e_j d_i \langle \psi_j, \chi_i \downarrow H \rangle_H \quad \text{by (21.17)}$$

$$= \left\langle \sum_{j=1}^{m} e_j \psi_j, \sum_{i=1}^{k} d_i \chi_i \downarrow H \right\rangle_H$$

$$= \langle \psi, \chi \downarrow H \rangle_H.$$

This completes the proof of the Frobenius Reciprocity Theorem. ∎

21.18 Corollary
If f is a class function on G, and ψ is a character of H, then

$$\langle \psi \uparrow G, f \rangle_G = \langle \psi, f \downarrow H \rangle_H.$$

Proof This follows at once from the Frobenius Reciprocity Theorem, since by Corollary 15.4, the characters of G span the vector space of class functions on G. ∎

The values of induced characters

We now show how to evaluate induced characters. Let ψ be a character of the subgroup H of G, and for convenience of notation, define the function $\psi\colon G \to \mathbb{C}$ by

$$\dot\psi(g) = \begin{cases} \psi(g) & \text{if } g \in H, \\ 0 & \text{if } g \notin H. \end{cases}$$

21.19 Proposition
The values of the induced character $\psi \uparrow G$ are given by

$$(\psi \uparrow G)(g) = \frac{1}{|H|} \sum_{y \in G} \dot\psi(y^{-1}gy)$$

for all $g \in G$.

Proof Let $f\colon G \to \mathbb{C}$ be the function given by

$$f(g) = \frac{1}{|H|} \sum_{y \in G} \dot\psi(y^{-1}gy) \quad (g \in G).$$

We aim to prove that $f = \psi \uparrow G$.
　　If $w \in G$ then

$$f(w^{-1}gw) = \frac{1}{|H|} \sum_{y \in G} \dot\psi(y^{-1}w^{-1}gwy) = f(g)$$

since wy runs through G as y runs through G. Therefore f is a class function, and so by Corollary 15.4, it is sufficient to show that $\langle f, \chi \rangle_G = \langle \psi \uparrow G, \chi \rangle_G$ for all irreducible characters χ of G.
　　Let χ be an irreducible character of G. Then

$$\langle f, \chi \rangle_G = \frac{1}{|G|} \sum_{g \in G} f(g)\overline{\chi(g)}$$

$$= \frac{1}{|G|}\frac{1}{|H|} \sum_{g \in G} \sum_{y \in G} \dot\psi(y^{-1}gy)\overline{\chi(g)}.$$

Put $x = y^{-1}gy$. Then

$$\langle f, \chi \rangle_G = \frac{1}{|G|}\frac{1}{|H|} \sum_{x \in G} \sum_{y \in G} \dot\psi(x)\overline{\chi(yxy^{-1})}.$$

$$= \frac{1}{|H|} \sum_{x \in H} \psi(x)\overline{\chi(x)}$$

since $\dot\psi(x) = 0$ if $x \notin H$, and $\chi(yxy^{-1}) = \chi(x)$ for all $y \in G$. Therefore

$$\langle f, \chi \rangle_G = \langle \psi, \chi \downarrow H \rangle_H$$

and so by the Frobenius Reciprocity Theorem,

$$\langle f, \chi \rangle_G = \langle \psi \uparrow G, \chi \rangle_G.$$

This was the equation required to show that $f = \psi \uparrow G$, so the proof is complete. ∎

21.20 Corollary
If ψ is a character of the subgroup H of G, then the degree of $\psi \uparrow G$ is given by

$$(\psi \uparrow G)(1) = \frac{|G|}{|H|} \psi(1).$$

Proof This follows immediately by evaluating $(\psi \uparrow G)(1)$ using Proposition 21.19. Alternatively, the stated degree of $\psi \uparrow G$ can be found using just the definition of induced modules (see Exercise 21.3). ∎

For practical purposes, a formula for the values of induced characters different from that given in Proposition 21.19 is more useful, and we shall derive this next (it is given in Proposition 21.23 below).

For $x \in G$, define the class function f_x^G on G by

$$f_x^G(y) = \begin{cases} 1 & \text{if } y \in x^G \\ 0 & \text{if } y \notin x^G \end{cases} \quad (y \in G).$$

Thus f is the characteristic function of the conjugacy class x^G.

21.21 Proposition
If χ is a character of G and $x \in G$, then

$$\langle \chi, f_x^G \rangle_G = \frac{\chi(x)}{|C_G(x)|}.$$

Proof We have

$$\langle \chi, f_x^G \rangle_G = \frac{1}{|G|} \sum_{g \in G} \chi(g) f_x^G(g)$$

$$= \frac{1}{|G|} \sum_{g \in x^G} \chi(g) = \frac{|x^G|}{|G|} \chi(x)$$

$$= \frac{\chi(x)}{|C_G(x)|} \quad \text{by Theorem 12.8.} \qquad \blacksquare$$

Note that a result similar to Proposition 21.21 was used in the proof of Theorem 16.4.

If $H \leqslant G$ and $h \in H$ then $h^H \subseteq h^G$; but if $g \in G$ then g^G may contain 0, 1, 2 or more conjugacy classes of H. To put this another way, we have:

(21.22) *Suppose that $x \in G$.*
    (1) *If no element of x^G lies in H, then $f_x^G \downarrow H = 0$.*
    (2) *If some element of x^G lies in H, then there are elements*
        $x_1, \ldots, x_m \in H$ *such that*

$$f_x^G \downarrow H = f_{x_1}^H + \ldots + f_{x_m}^H.$$

Statement (2) just says that $H \cap x^G$ breaks up into m conjugacy classes of H, with representatives $x_1, \ldots, x_m$.

21.23 Proposition
Let ψ be a character of the subgroup H of G, and suppose that $x \in G$.

(1) *If no element of x^G lies in H, then $(\psi \uparrow G)(x) = 0$.*
(2) *If some element of x^G lies in H, then*

$$(\psi \uparrow G)(x) = |C_G(x)| \left(\frac{\psi(x_1)}{|C_H(x_1)|} + \ldots + \frac{\psi(x_m)}{|C_H(x_m)|} \right),$$

where $x_1, \ldots, x_m \in H$ and $f_x^G \downarrow H = f_{x_1}^H + \ldots + f_{x_m}^H$ (as in (21.22)).

Proof By Proposition 21.21 and Corollary 21.18, we have

$$\frac{(\psi \uparrow G)(x)}{|C_G(x)|} = \langle \psi \uparrow G, f_x^G \rangle_G = \langle \psi, f_x^G \downarrow H \rangle_H.$$

If no element of x^G lies in H, then $f_x^G \downarrow H = 0$, and hence $(\psi \uparrow G)(x) = 0$. And if some element of x^G lies in H, and $f_x^G \downarrow H = f_{x_1}^H + \ldots + f_{x_m}^H$ as in (21.22)(2), then

$$\frac{(\psi \uparrow G)(x)}{|C_G(x)|} = \langle \psi, f_{x_1}^H + \ldots + f_{x_m}^H \rangle_H$$

$$= \langle \psi, f_{x_1}^H \rangle_H + \ldots + \langle \psi, f_{x_m}^H \rangle_H$$

$$= \frac{\psi(x_1)}{|C_H(x_1)|} + \ldots + \frac{\psi(x_m)}{|C_H(x_m)|} \quad \text{by Proposition 21.21.}$$

The result follows. ∎

21.24 Example
Let $G = S_4$ and let $H = \langle a, b \rangle$, where

$$a = (1\ 2\ 3\ 4),\ b = (1\ 3).$$

Then $H \cong D_8$, since $a^4 = b^2 = 1$ and $b^{-1}ab = a^{-1}$. By (12.12), the conjugacy classes of H are

$$\{1\},$$

$$\{a^2 = (1\ 3)(2\ 4)\},$$

$$\{a = (1\ 2\ 3\ 4),\ a^3 = (1\ 4\ 3\ 2)\},$$

$$\{b = (1\ 3),\ a^2b = (2\ 4)\},$$

We have

$$f_1^G \downarrow H = f_1^H, \quad f_{(1\ 3)}^G \downarrow H = f_{(1\ 3)}^H, \quad f_{(1\ 2\ 3)}^G \downarrow H = 0,$$

$$f_{(1\ 2)(3\ 4)}^G \downarrow H = f_{(1\ 3)(2\ 4)}^H + f_{(1\ 2)(3\ 4)}^H, \text{ and}$$

$$f_{(1\ 2\ 3\ 4)}^G \downarrow H = f_{(1\ 2\ 3\ 4)}^H.$$

For example, the statement

$$f_{(1\ 2)(3\ 4)}^G \downarrow H = f_{(1\ 3)(2\ 4)}^H + f_{(1\ 2)(3\ 4)}^H$$

records the fact that the G-conjugacy class $(1\ 2)(3\ 4)^G$ contains exactly two H-conjugacy classes, with representatives $(1\ 3)(2\ 4)$ and $(1\ 2)(3\ 4)$.

The orders of the centralizers of the elements of H are as follows:

h	1	(1 3)(2 4)	(1 2 3 4)	(1 3)	(1 2)(3 4)
$\|C_G(h)\|$	24	8	4	4	8
$\|C_H(h)\|$	8	8	4	4	4

Suppose that ψ is a character of H. Then according to Proposition 21.23, we have

$$(\psi \uparrow G)(1) = 24\frac{\psi(1)}{8},$$

$$(\psi \uparrow G)((1\ 3)) = 4\frac{\psi((1\ 3))}{4},$$

$$(\psi \uparrow G)((1\ 2\ 3)) = 0,$$

$$(\psi \uparrow G)((1\ 2)(3\ 4)) = 8\left(\frac{\psi((1\ 3)(2\ 4))}{8} + \frac{\psi((1\ 2)(3\ 4))}{4}\right),$$

$$(\psi \uparrow G)((1\ 2\ 3\ 4)) = 4\frac{\psi((1\ 2\ 3\ 4))}{4}.$$

Referring to Example 16.3(3) for the irreducible characters $\chi_1, \ldots, \chi_5$ of $H \cong D_8$, we therefore have

	1	(1 2)	(1 2 3)	(1 2)(3 4)	(1 2 3 4)
$\chi_1 \uparrow G$	3	1	0	3	1
$\chi_2 \uparrow G$	3	−1	0	−1	1
$\chi_3 \uparrow G$	3	1	0	−1	−1
$\chi_4 \uparrow G$	3	−1	0	3	−1
$\chi_5 \uparrow G$	6	0	0	−2	0

In the next example, we use induced characters to find the character table of a group of order 21.

21.25 Example (cf. Exercise 17.2)
Define permutations a, b in S_7 by

$$a = (1\ 2\ 3\ 4\ 5\ 6\ 7),\ b = (2\ 3\ 5)(4\ 7\ 6)$$

and let G be the subgroup $\langle a, b \rangle$ of S_7. Check that

$$a^7 = b^3 = 1, \; b^{-1}ab = a^2.$$

It follows from these relations that the elements of G are all of the form $a^i b^j$ with $0 \leqslant i \leqslant 6$, $0 \leqslant j \leqslant 2$. Also, G has order 21.

We aim to find the character table of G. First we find the conjugacy classes. Since $\langle a \rangle \leqslant C_G(a)$, 7 divides $|C_G(a)|$; and since $b \notin C_G(a)$, $|C_G(a)| < 21$. Hence $|C_G(a)| = 7$, and similarly $|C_G(b)| = 3$. Using this, we see that the conjugacy classes of G are

$$\{1\},$$

$$\{a, a^2, a^4\},$$

$$\{a^3, a^5, a^6\},$$

$$\{a^i b : 0 \leqslant i \leqslant 6\},$$

$$\{a^i b^2 : 0 \leqslant i \leqslant 6\}.$$

We take 1, a, a^3, b and b^2 to be representatives of the conjugacy classes. Notice that G has exactly five irreducible characters.

Since $\langle a \rangle \lhd G$ and $G/\langle a \rangle \cong C_3$, we obtain three linear characters χ_1, χ_2, χ_3 of G as the lifts of the linear characters of $G/\langle a \rangle$. Their values are shown below:

g $|C_G(g)|$	1 21	a 7	a^3 7	b 3	b^2 3
χ_1	1	1	1	1	1
χ_2	1	1	1	ω	ω^2
χ_3	1	1	1	ω^2	ω

where $\omega = e^{2\pi i/3}$.

Let $H = \langle a \rangle$. We shall obtain the last two irreducible characters of G by inducing linear characters of H. Let $\eta = e^{2\pi i/7}$. For $1 \leqslant k \leqslant 6$, there is a character ψ_k of H given by

$$\psi_k(a^j) = \eta^{jk} \quad (0 \leqslant j \leqslant 6).$$

To use the formula in Proposition 21.23 for calculating $\psi_k \uparrow G$, note that

$$f_a^G \downarrow H = f_a^H + f_{a^2}^H + f_{a^4}^H$$

since no two of the elements a, a^2, a^4 are conjugate in H. Hence by Proposition 21.23,

$$(\psi_1 \uparrow G)(a) = \eta + \eta^2 + \eta^4$$

and similarly

$$(\psi_1 \uparrow G)(a^3) = \eta^3 + \eta^5 + \eta^6, \quad (\psi_1 \uparrow G)(1) = 3.$$

And since no G-conjugate of b or b^2 lies in H,

$$(\psi_1 \uparrow G)(b) = (\psi_1 \uparrow G)(b^2) = 0.$$

Similarly

$$(\psi_3 \uparrow G)(1) = 3, \quad (\psi_3 \uparrow G)(a) = \eta^3 + \eta^5 + \eta^6,$$

$(\psi_3 \uparrow G)(a^3) = \eta + \eta^2 + \eta^4$, and $(\psi_3 \uparrow G)(b) = (\psi_3 \uparrow G)(b^2) = 0$.

Thus if we write $\chi_4 = \psi_1 \uparrow G$ and $\chi_5 = \psi_3 \uparrow G$, then the values of χ_4 and χ_5 are

	1	a	a^3	b	b^2
χ_4	3	$\eta + \eta^2 + \eta^4$	$\eta^3 + \eta^5 + \eta^6$	0	0
χ_5	3	$\eta^3 + \eta^5 + \eta^6$	$\eta + \eta^2 + \eta^4$	0	0

Now $\chi_4 \downarrow H = \psi_1 + \psi_2 + \psi_4$ and $\chi_5 \downarrow H = \psi_3 + \psi_5 + \psi_6$. Therefore $\chi_4 \ne \chi_5$, since $\psi_1, \ldots, \psi_6$ are linearly independent.

We now calculate that

$$\langle \chi_4, \chi_4 \rangle_G = \frac{9}{21} + \frac{2}{7} + \frac{2}{7} + \frac{0}{3} + \frac{0}{3} = 1,$$

and similarly $\langle \chi_5, \chi_5 \rangle_G = 1$. Thus χ_4 and χ_5 are our last two irreducible characters, and the character table of G is as shown.

Character table of $\langle a, b: a^7 = b^3 = 1, b^{-1}ab = a^2 \rangle$

g $|C_G(g_i)|$	1 21	a 7	a^3 7	b 3	b^2 3
χ_1	1	1	1	1	1
χ_2	1	1	1	ω	ω^2
χ_3	1	1	1	ω^2	ω
χ_4	3	$\eta + \eta^2 + \eta^4$	$\eta^3 + \eta^5 + \eta^6$	0	0
χ_5	3	$\eta^3 + \eta^5 + \eta^6$	$\eta + \eta^2 + \eta^4$	0	0

Summary of Chapter 21

Assume that H is a subgroup of G.

1. For each $\mathbb{C}H$-module U, an induced $\mathbb{C}G$-module $U \uparrow G$ can be defined. If U is a $\mathbb{C}H$-module of $\mathbb{C}H$, then $U \uparrow G$ is simply $U(\mathbb{C}G)$.

2. If ψ is a character of H then the induced character $\psi \uparrow G$ is given by

$$(\psi \uparrow G)(g) = \frac{1}{|H|} \sum_{y \in G} \dot{\psi}(y^{-1}gy).$$

   In particular, the degree of $\psi{\uparrow}G$ is $|G:H|\psi(1)$.

3. If no element of g^G lies in H, then

$$(\psi \uparrow G)(g) = 0.$$

   If some element of g^G lies in H, then

$$(\psi \uparrow G)(g) = |C_G(g)| \left(\frac{\psi(x_1)}{|C_H(x_1)|} + \ldots + \frac{\psi(x_m)}{|C_H(x_m)|} \right)$$

   where $f_g^G \downarrow H = f_{x_1}^H + \ldots + f_{x_m}^H$.

4. The Frobenius Reciprocity Theorem states that

$$\langle \psi \uparrow G, \chi \rangle_G = \langle \psi, \chi \downarrow H \rangle_H,$$

   where ψ is a character of H and χ is a character of G.

Exercises for Chapter 21

1. Let $G = D_8 = \langle a, b: a^4 = b^2 = 1, b^{-1}ab = a^{-1} \rangle$ and let H be the subgroup $\langle a^2, b \rangle$. Define U to be the 1-dimensional subspace of $\mathbb{C}H$ spanned by

$$1 - a^2 + b - a^2 b.$$

   (a) Check that U is a $\mathbb{C}H$-submodule of $\mathbb{C}H$.
   (b) Find a basis of the induced $\mathbb{C}G$-module $U \uparrow G$.
   (c) Write down the character of the $\mathbb{C}H$-module U and the character of the $\mathbb{C}G$-module $U \uparrow G$. Is $U \uparrow G$ irreducible?

2. Let $G = S_4$ and let H be the subgroup $\langle (1\,2\,3) \rangle \cong C_3$.
   (a) If $\chi_1, \ldots, \chi_5$ are the irreducible characters of G, as given in

Section 18.1, work out the restrictions $\chi_i \downarrow H$ $(1 \leqslant i \leqslant 5)$ as sums of the irreducible characters ψ_1, ψ_2, ψ_3 of C_3.

(b) Calculate the induced characters $\psi_j \uparrow G$ $(1 \leqslant j \leqslant 3)$ as sums of the irreducible characters χ_i of G.

3. Show direct from the definition that if $H \leqslant G$ and ψ is a character of H, then

$$(\psi \uparrow G)(1) = \frac{|G|}{|H|} \psi(1).$$

4. Let H be a subgroup of G, let ψ be a character of H, and let χ be a character of G. Prove that

$$(\psi(\chi \downarrow H)) \uparrow G = (\psi \uparrow G)\chi.$$

(Hint: consider the inner product of each side with an arbitrary irreducible character of G, and use the Frobenius Reciprocity Theorem.)

5. Let $G = S_7$ and let $H = \langle a, b \rangle$, where

$$a = (1\ 2\ 3\ 4\ 5\ 6\ 7),\ b = (2\ 3\ 5)(4\ 7\ 6),$$

as in Example 21.25. Let ϕ and ψ be the irreducible characters of H which are given by

g_i	1	a	a^3	b	b^2
$|C_H(g_i)|$	21	7	7	3	3
ϕ	1	1	1	1	1
ψ	3	$\eta + \eta^2 + \eta^4$	$\eta^3 + \eta^5 + \eta^6$	0	0

where $\eta = e^{2\pi i/7}$ (see Example 21.25).

You are given that $|C_G(a)| = 7$ and $|C_G(b)| = 18$. Calculate the values of the induced characters $\phi \uparrow G$ and $\psi \uparrow G$.

6. Suppose that H is a subgroup of G, and let $\chi_1, \ldots, \chi_k$ be the irreducible characters of G. Let ψ be an irreducible character of H. Show that the integers $d_1, \ldots, d_k$, which are given by $\psi \uparrow G = d_1\chi_1 + \ldots d_k\chi_k$, satisfy

$$\sum_{i=1}^{k} d_i^2 \leqslant |G : H|.$$

(Compare Proposition 20.5.)

7. Suppose that H is a normal subgroup of index 2 in G, and let ψ be an irreducible character of H. Discover and prove results for $\psi \uparrow G$ which are similar to those presented in Chapter 20 for the restriction of irreducible characters of G to H.

22

Algebraic integers

Among the properties of characters which may be regarded as fundamental, perhaps the most opaque is that which states that the degree of
an irreducible character of a finite group G must divide the order of
G. This is one of several results which we shall prove in this chapter,
using algebraic integers. Most of the results concern arithmetic properties of character values. We discuss properties of a group element
$g \in G$ which ensure that $\chi(g)$ is an integer for all characters χ of G.
And we prove some useful congruence properties; for example, if p is
a prime number and $g \in G$ is an element of order p^r for some r, then
$\chi(g) \equiv \chi(1) \bmod p$ for any character χ of G for which $\chi(g)$ is an
integer.

Algebraic integers

22.1 Definition
A complex number λ is an *algebraic integer* if and only if λ is an
eigenvalue of some matrix, all of whose entries are integers.

Thus, for λ to be an algebraic integer, we require that

$$\det(A - \lambda I) = 0$$

for some square matrix A with integer entries. Equivalently, for the
same matrix A, we have

$$uA = \lambda u$$

for some non-zero row vector u.

We remark that λ is an algebraic integer if and only if λ is a root of
a polynomial of the form

$$x^n + a_{n-1}x^{n-1} + \ldots + a_1 x + a_0$$

where $a_0, \ldots, a_{n-1}$ are integers (see Exercise 22.7). In fact, algebraic integers are usually defined in this way.

22.2 Examples
(1) Every integer n is an algebraic integer, since n is an eigenvalue of the 1×1 matrix (n).

(2) $\sqrt{2}$ is an algebraic integer, since it is an eigenvalue of the matrix $\begin{pmatrix} 0 & 1 \\ 2 & 0 \end{pmatrix}$.

(3) If λ is an algebraic integer, then so are $-\lambda$ and the complex conjugate $\bar{\lambda}$ of λ. To see this, note that if A is an integer matrix and u is a row vector with $uA = \lambda u$, then

$$u(-A) = (-\lambda)u \text{ and } \bar{u}A = \bar{\lambda}\bar{u},$$

where $\bar{u}$ is the row vector obtained from u by replacing each entry by its complex conjugate.

(4) Let A be the $n \times n$ matrix given by

$$A = \begin{pmatrix} 0 & 0 & 0 & \ldots & 0 & 0 & 1 \\ 1 & 0 & 0 & \ldots & 0 & 0 & 0 \\ 0 & 1 & 0 & \ldots & 0 & 0 & 0 \\ & & & \ldots & & & \\ 0 & 0 & 0 & \ldots & 1 & 0 & 0 \\ 0 & 0 & 0 & \ldots & 0 & 1 & 0 \end{pmatrix}.$$

Suppose that ω is an nth root of unity, and let u be the row vector $(1, \omega, \omega^2, \ldots, \omega^{n-1})$. Then

$$uA = (\omega, \omega^2, \ldots, \omega^{n-1}, 1) = \omega u.$$

This shows that every nth root of unity is an algebraic integer.

22.3 Theorem
If λ and μ are algebraic integers, then $\lambda\mu$ and $\lambda + \mu$ are also algebraic integers.

Proof There exist square matrices A and B, all of whose entries are integers, and non-zero row vectors u and v, such that
$$uA = \lambda u, \, vB = \mu v.$$
Suppose that A is an $m \times m$ matrix and B is an $n \times n$ matrix.

Let $e_1, \ldots, e_m$ be a basis of $\mathbb{C}^m$ and $f_1, \ldots, f_n$ be a basis of $\mathbb{C}^n$. Then the vectors $e_i \otimes f_j$ $(1 \leq i \leq m, 1 \leq j \leq n)$ form a basis of the tensor product space $V = \mathbb{C}^m \otimes \mathbb{C}^n$. Define an endomorphism $A \otimes B$ of V by

$$(e_i \otimes f_j)(A \otimes B) = e_i A \otimes f_j B \quad (1 \leq i \leq m, 1 \leq j \leq n),$$

extending linearly (that is, $(\sum \lambda_{ij}(e_i \otimes f_j))(A \otimes B) = \sum \lambda_{ij}(e_i A \otimes f_j B)$). It is easy to check as in the proof of Proposition 19.4 that for all vectors $x \in \mathbb{C}^m$, $y \in \mathbb{C}^n$, we have

$$(x \otimes y)(A \otimes B) = xA \otimes yB.$$

Hence

$$(u \otimes v)(A \otimes B) = uA \otimes vB = \lambda u \otimes \mu v = \lambda \mu (u \otimes v).$$

Therefore $\lambda \mu$ is an eigenvalue of $A \otimes B$. Since the matrix of $A \otimes B$ relative to the basis $e_i \otimes f_j$ $(1 \leq i \leq m, 1 \leq j \leq n)$ has integer entries, it follows that $\lambda \mu$ is an algebraic integer.

Let I_m and I_n denote the identity $m \times m$ and $n \times n$ matrices, respectively. Then

$$(u \otimes v)(A \otimes I_n + I_m \otimes B) = uA \otimes vI_n + uI_m \otimes vB$$

$$= \lambda u \otimes v + u \otimes \mu v$$

$$= (\lambda + \mu)(u \otimes v),$$

and we deduce as above that $\lambda + \mu$ is an algebraic integer. ∎

Theorem 22.3 shows that the set of all algebraic integers forms a subring of $\mathbb{C}$. The next result provides a link between algebraic integers and characters.

22.4 Corollary
If χ is a character of G and $g \in G$, then $\chi(g)$ is an algebraic integer.

Proof By Proposition 13.9, $\chi(g)$ is a sum of roots of unity. Each root of unity is an algebraic integer, by Example 22.2(4), so any sum of roots of unity is an algebraic integer by Theorem 22.3. Hence $\chi(g)$ is an algebraic integer. ∎

22.5 Proposition
If λ is both a rational number and an algebraic integer, then λ is an integer.

Proof Suppose that λ is a rational number which is not an integer. We shall show that λ is not an algebraic integer, which is enough to establish the proposition.

Write $\lambda = r/s$, where r and s are coprime integers and $s \neq \pm 1$. Let p be a prime number which divides s. For every $n \times n$ matrix A of integers, the entries of $sA - rI$ which are not on the diagonal are divisible by s, and hence also by p. Therefore

$$\det(sA - rI) = (-r)^n + mp$$

for some integer m. As p does not divide r (since r and s are coprime), we deduce that $\det(sA - rI) \neq 0$. Thus

$$\det(A - \lambda I) = \left(\frac{1}{s}\right)^n \det(sA - rI) \neq 0,$$

and hence λ is not an algebraic integer. ∎

The next result is an immediate consequence of Corollary 22.4 and Proposition 22.5.

22.6 Corollary
Let χ be a character of G and let $g \in G$. If $\chi(g)$ is a rational number, then $\chi(g)$ is an integer.

In passing, note that we have, as a special case of Proposition 22.5, the well known result that $\sqrt{2}$ is irrational. (Example 22.2(2) shows that $\sqrt{2}$ is an algebraic integer.)

The degree of every irreducible character divides $|G|$

To prepare for the proof that $|G|$ is divisible by the degree of each irreducible character of G, we establish two preliminary lemmas. Recall from Definition 12.21 that if C is a conjugacy class of G, then

$$\bar{C} = \sum_{x \in C} x \in \mathbb{C}G.$$

22.7 Lemma
Suppose that $g \in G$ and that C is the conjugacy class of G which contains g. Let U be an irreducible $\mathbb{C}G$-module, with character χ. Then

$$u\bar{C} = \lambda u \quad \text{for all } u \in U,$$

where

$$\lambda = \frac{|G|}{|C_G(g)|} \frac{\chi(g)}{\chi(1)}.$$

Proof Since $\bar{C}$ lies in the centre of $\mathbb{C}G$ (see Proposition 12.22), we know by Proposition 9.14 that there exists $\lambda \in \mathbb{C}$ such that $u\bar{C} = \lambda u$ for all $u \in U$; that is,

$$u \sum_{x \in C} x = \lambda u \quad \text{for all } u \in U.$$

Consequently if $\mathscr{B}$ is a basis of U, then

$$\sum_{x \in C} [x]_{\mathscr{B}} = \lambda I.$$

Taking the traces of both sides of this equation, we obtain

$$\sum_{x \in C} \chi(x) = \lambda \chi(1),$$

and since χ is constant on the conjugacy class C, this yields

$$|C|\chi(g) = \lambda \chi(1).$$

Thus $\lambda = |C|\chi(g)/\chi(1)$. As $|C| = |G:C_G(g)|$ by Theorem 12.8, the result follows. ∎

22.8 Lemma

Let $r = \sum_{g \in G} \alpha_g g \in \mathbb{C}G$, where each α_g is an integer. Suppose that u is a non-zero element of $\mathbb{C}G$ such that

$$ur = \lambda u,$$

where $\lambda \in \mathbb{C}$. Then λ is an algebraic integer.

Proof Let $g_1, \ldots, g_n$ be the elements of G. Then for $1 \leqslant i \leqslant n$, we have

$$g_i r = \sum_{j=1}^{n} a_{ij} g_j$$

for certain integers a_{ij}. (In fact, $a_{ij} = \alpha_g$ where $g = g_i^{-1}g_j$.) The statement that $ur = \lambda u$ (with $u \neq 0$) says that λ is an eigenvalue of the integer matrix $A = (a_{ij})$. Therefore λ is an algebraic integer. ∎

22.9 Example

Let $G = C_n = \langle x : x^n = 1 \rangle$, and define

$$u = 1 + \omega x^{-1} + \omega^2 x^{-2} + \ldots + \omega^{n-1} x \in \mathbb{C}G,$$

where ω is an nth root of unity. Then

$$ux = \omega u$$

and so Lemma 22.8 confirms that ω is an algebraic integer.

Notice that this example is just a reworking of Example 22.2(4).

22.10 Corollary

If χ is an irreducible character of G and $g \in G$, then

$$\lambda = \frac{|G|}{|C_G(g)|} \frac{\chi(g)}{\chi(1)}$$

is an algebraic integer.

Proof Let U be an irreducible $\mathbb{C}G$-submodule of $\mathbb{C}G$ with character χ, and let $\bar{C}$ be the sum of the elements in the conjugacy class of G which contains g. Then $u\bar{C} = \lambda u$ for all $u \in U$, by Lemma 22.7. Therefore λ is an algebraic integer, by Lemma 22.8. ∎

22.11 Theorem

If χ is an irreducible character of G, then $\chi(1)$ divides $|G|$.

Proof Let $g_1, \ldots, g_k$ be representatives of the conjugacy classes of G. Then for all i, both

$$\frac{|G|}{|C_G(g_i)|} \frac{\chi(g_i)}{\chi(1)} \quad \text{and} \quad \overline{\chi(g_i)}$$

are algebraic integers, by Corollaries 22.10 and 22.4. Hence by Theorem 22.3,

$$\sum_{i=1}^{k} \frac{|G|}{|C_G(g_i)|} \frac{\chi(g_i)\overline{\chi(g_i)}}{\chi(1)}$$

is an algebraic integer. This algebraic integer is equal to $|G|/\chi(1)$, by the row orthogonality relations, Theorem 16.4(1). As $|G|/\chi(1)$ is a rational number. Proposition 22.5 now implies that $|G|/\chi(1)$ is an integer. That is, $\chi(1)$ divides $|G|$.   ∎

22.12 Examples
(1) If p is a prime number and G is a group of order p^n for some n, then $\chi(1)$ is a power of p for all irreducible characters χ of G.

In particular, if $|G| = p^2$ then $\chi(1) = 1$ for all irreducible characters χ. (Note that $\chi(1) < p$, as the sum of the squares of the degrees of the irreducible characters is equal to $|G|$.) Hence, using Proposition 9.18, we recover the well known result that groups of order p^2 are abelian.

(2) Let G be a group of order $2p$, where p is prime. By Theorem 22.11, the degree of each irreducible character of G is 1 or 2 (it cannot be p for the reason noted in (1) above). By Theorem 17.11, the number of linear characters of G divides $|G|$. Hence, either the degrees of the irreducible characters of G are all 1, or they are

$$1, 1, 2, \ldots, 2 \quad (\text{with } (p-1)/2 \text{ degrees } 2).$$

(3) If $G = S_n$ then every prime p which divides the degree of an irreducible character of G also divides $n!$, and hence satisfies $p \leq n$.

Theorem 22.11 also has the following interesting consequences concerning irreducible characters of simple groups. (Recall that a group G is simple if it has no normal subgroups apart from $\{1\}$ and G.)

22.13 Corollary
No simple group has an irreducible character of degree 2.

Proof Suppose that G is a simple group which has an irreducible character χ of degree 2. Let $\rho: G \to \mathrm{GL}(2, \mathbb{C})$ be a representation of G with character χ. Since $\mathrm{Ker}\,\rho \lhd G$ and G is simple, we have $\mathrm{Ker}\,\rho = \{1\}$, and so ρ is injective.

First, observe that G is non-abelian, by Proposition 9.5. Hence $G' \neq 1$, and so $G' = G$ as G is simple. Therefore by Theorem 17.11, G has no non-trivial linear characters. But $g \to \det(g\rho)$ is a linear character of G (see Exercise 13.7(a)), and this implies that

$$\det(g\rho) = 1 \quad \text{for all } g \in G.$$

Now G has even order, by Theorem 22.11. Therefore G contains an element x of order 2 (see Exercise 1.8).

Consider the 2×2 matrix $x\rho$. As ρ is injective, $x\rho$ has order 2; and by Proposition 9.11, there is a 2×2 matrix T such that $T^{-1}(x\rho)T$ is a diagonal matrix with diagonal entries ± 1. Since $\det(x\rho) = 1$, we conclude that

$$T^{-1}(x\rho)T = \begin{pmatrix} -1 & 0 \\ 0 & -1 \end{pmatrix}.$$

Thus

$$x\rho = T(-I)T^{-1} = -I.$$

Consequently $(x\rho)(g\rho) = (g\rho)(x\rho)$ for all $g \in G$. As ρ is injective, this means that $xg = gx$ for all $g \in G$, and hence

$$\langle x \rangle \lhd G.$$

This is a contradiction, as G is simple. ∎

Our next result again shows that information about character degrees can sometimes be used to learn about the structure of a group. This time, we assume that every irreducible character of G has degree a power of a prime p, and we deduce that G has an abelian normal p-complement N; that is, N is an abelian normal subgroup of G, and $|N|$ is coprime to p, while $|G : N|$ is a power of p.

22.14 Corollary
Suppose that p is a prime and the degree of every irreducible character of G is a power of p. Then G has an abelian normal p-complement. In particular, G is not simple unless G has prime order.

Proof The result is correct if G is abelian (see Theorem 9.6), so we assume that G is non-abelian.

Theorems 11.12 and 17.11 give us the equation

$$|G| = |G/G'| + \sum \chi(1)^2,$$

the sum being over the irreducible characters χ of G for which $\chi(1) > 1$.

Since G is non-abelian, $\chi(1) > 1$ for some irreducible character χ of G. Then $\chi(1)$ is divisible by p, by our hypothesis, so p divides $|G|$ by

Theorem 22.11; and we deduce from the equation above that p divides the order of the abelian group G/G'. Since every finite abelian group is isomorphic to a direct product of cyclic groups, it follows that G/G' has a subgroup of index p. Hence G has a normal subgroup H of index p.

Let ψ be an irreducible character of H. Then $\langle \chi \downarrow H, \psi \rangle \neq 0$ for some irreducible character ψ of H, by Proposition 20.4. Next, Clifford's Theorem 20.8 shows that $\psi(1)$ divides $\chi(1)$, so $\psi(1)$ is a power of p. We may now apply induction on $|G|$ to deduce that H has an abelian normal p-complement N.

We have that $|N|$ is coprime to p and $|G : N|$ is a power of p, so it remains to prove that $N \lhd G$.

Suppose that $g \in G$ and the order of g is coprime to p. Then $g \in H$, since otherwise p divides the order of gH which in turn divides the order of g; a similar argument shows that $g \in N$. Hence N consists of those elements of G whose order is coprime to p, and from this fact it follows easily that $N \lhd G$.

Finally, assume that G is simple, so either $N = \{1\}$ or $N = G$. If $N = \{1\}$ then G is a p-group so $Z(G) \neq \{1\}$ (see Exercise 12.7); because G is simple, we have $Z(G) = G$, so G is abelian. On the other hand, if $N = G$ then G is again abelian. But an abelian simple group has prime order, by Exercise 1.1. Therefore, G has prime order. ∎

A condition which ensures that $\chi(g)$ is an integer

In Theorem 22.16 below we give a group-theoretic condition on an element g of G which implies that $\chi(g)$ is an integer for every character χ of G. This result implies, for example, that for all n, every entry in the character table of S_n is an integer (see Corollary 22.17). Bearing in mind the difficulties we encountered in constructing the character tables of S_n for small values of n (we reached $n = 6$ in Example 19.17), Theorem 22.16 is evidently a useful result.

Before proving Theorem 22.16, we require a preliminary lemma concerning roots of unity. If a and b are positive integers, then we denote their highest common factor by (a, b). Also, for integers d and n, we write $d|n$ to denote the fact that d divides n.

22.15 Lemma
If ω is an nth root of unity, then

$$\sum_{\substack{1 \leqslant i \leqslant n, \\ (i,n)=1}} \omega^i$$

is an integer.

Proof We prove the result by induction on n. It is trivial for $n = 1$. Also, if $\omega = 1$ then the result is immediate. So suppose that ω is an nth root of unity and $\omega \neq 1$. Then ω is a root of the polynomial $(x^n - 1)/(x - 1) = x^{n-1} + \ldots + x + 1$. Therefore $\sum_{i=1}^{n} \omega^i = 0$.

Now we partition the sum $\sum_{i=1}^{n} \omega^i$ according to the highest common factor d of i and n:

$$0 = \sum_{i=1}^{n} \omega^i = \sum_{d \mid n} \sum_{\substack{1 \leqslant i \leqslant n \\ (i,n)=d}} \omega^i = \sum_{d \mid n} \sum_{\substack{1 \leqslant j \leqslant n/d, \\ (j,n/d)=1}} \omega^{dj}.$$

If $d \mid n$ then ω^d is an (n/d)th root of unity, and if in addition $d > 1$, then by our induction hypothesis,

$$\sum_{\substack{1 \leqslant j \leqslant n/d, \\ (j,n/d)=1}} \omega^{dj} \in \mathbb{Z}.$$

It follows that

$$\sum_{\substack{1 \leqslant i \leqslant n, \\ (i,n)=1}} \omega^i = \sum_{i=1}^{n} \omega^i - \sum_{\substack{d \mid n, \\ d > 1}} \sum_{\substack{1 \leqslant j \leqslant n/d, \\ (j,n/d)=1}} \omega^{dj} \in \mathbb{Z},$$

as required. ∎

22.16 Theorem
Let g be an element of order n in G. Suppose that g is conjugate to g^i for all i with $1 \leqslant i \leqslant n$ and $(i, n) = 1$. Then $\chi(g)$ is an integer for all characters χ of G.

Proof Let V be a $\mathbb{C}G$-module with character χ of degree m. By Proposition 9.11, there is a basis $\mathcal{B}$ of V such that

$$[g]_{\mathcal{B}} = \begin{pmatrix} \omega_1 & & 0 \\ & \ddots & \\ 0 & & \omega_m \end{pmatrix}$$

where $\omega_1, \ldots, \omega_m$ are nth roots of unity. For $1 \leqslant i \leqslant n$, the matrix $[g^i]_{\mathscr{B}}$ has diagonal entries $\omega_1^i, \ldots, \omega_m^i$, and so

$$\chi(g^i) = \omega_1^i + \ldots + \omega_m^i.$$

Therefore by Lemma 22.15,

$$\sum_{\substack{1 \leqslant i \leqslant n, \\ (i,n)=1}} \chi(g^i) \in \mathbb{Z}.$$

As g is conjugate to g^i if $1 \leqslant i \leqslant n$ and $(i, n) = 1$, we have $\chi(g^i) = \chi(g)$ for such i, and hence

$$s\chi(g) \in \mathbb{Z},$$

where s is the number of integers i with $1 \leqslant i \leqslant n$ and $(i, n) = 1$. Consequently $\chi(g)$ is a rational number, and so $\chi(g)$ is an integer by Corollary 22.6. ∎

We remark that using Galois theory it is possible to prove the converse of Theorem 22.16, namely that if $\chi(g^i) \in \mathbb{Z}$ for all characters χ of G, then g is conjugate to g^i whenever i is coprime to n.

22.17 Corollary
All the character values of symmetric groups are integers.

Proof If $g \in S_n$ and i is coprime to the order of g, then the permutations g and g^i have the same cycle-shape, and hence are conjugate by Theorem 12.15. The result now follows from Theorem 22.16. ∎

The p'-part of a group element

The rest of the chapter is devoted to some important congruence properties of character values. For example, one particularly useful consequence of our results is that if p is a prime number, g is an element of G of order p^r for some r, and χ is a character of G such that $\chi(g) \in \mathbb{Z}$, then $\chi(g) \equiv \chi(1) \bmod p$.

Before going into the character theory, we need to define the p'-part of a group element. The definition will emerge from the following lemma.

22.18 Lemma

Let p be a prime number and let $g \in G$. Then there exist $x, y \in G$ such that

(1) $g = xy = yx$,
(2) *the order of x is a power of p, and*
(3) *the order of y is coprime to p.*

Moreover, the elements x and y of G which satisfy conditions (1)–(3) are unique.

Proof Let the order of g be up^v, where $u, v \in \mathbb{Z}$ and $(u, p) = 1$. Then there exist integers a, b such that

$$au + bp^v = 1.$$

Put $x = g^{au}$ and $y = g^{bp^v}$. *Then*

$$xy = yx = g^{au+bp^v} = g,$$
$$x^{p^v} = g^{aup^v} = 1,$$
$$y^u = g^{bup^v} = 1.$$

Hence the order of x is a power of p and the order of y divides u, so is coprime to p. Therefore x and y satisfy conditions (1)–(3).

Now suppose that $x', y' \in G$ also satisfy (1)–(3); that is, $g = x'y' = y'x'$, the order of x' is a power of p and the order of y' is coprime to p. We must show that $x = x'$ and $y = y'$.

We have

$$x'g = x'y'x' = gx',$$

so x' commutes with g, hence also with $g^{au} = x$. Since both x and x' have order a power of p, it follows that $x^{-1}x'$ has order a power of p. Similarly, y' commutes with y and $y(y')^{-1}$ has order coprime to p. Finally, $xy = g = x'y'$, so

$$x^{-1}x' = y(y')^{-1}.$$

If $z = x^{-1}x' = y(y')^{-1}$, then we have shown that the order of z is both a power of p and coprime to p. Therefore $z = 1$, and so $x = x'$ and $y = y'$, as required. ∎

22.19 Definition
We call the element y which appears in Lemma 22.18 the p'-part of g.

We extract the following statement from the proof of Lemma 22.18.

(22.20) *Let the order of g be up^v, where $u, v \in \mathbb{Z}$ and $(u, p) = 1$, and choose integers a, b with $au + bp^v = 1$. Then the p'-part of g is g^{bp^v}.*

For example, if $p = 2$ and g has order 6, then the p'-part of g is g^{-2}; the expression $g = xy$ in Lemma 22.18 has $x = g^3$, $y = g^{-2}$.

A little ring theory

To prepare for our main result on congruence properties of character values, we need a few basic facts about a certain subring of $\mathbb{C}$ in which all our character values will lie.

Let n be a positive integer and let $\zeta = e^{2\pi i/n}$. Define $\mathbb{Z}[\zeta]$ to be the subring of $\mathbb{C}$ generated by $\mathbb{Z}$ and ζ; that is,

$$\mathbb{Z}[\zeta] = \{f(\zeta): f(x) \in \mathbb{Z}[x]\}.$$

Clearly, every element of $\mathbb{Z}[\zeta]$ is an integer combination of the powers $1, \zeta, \zeta^2, \ldots, \zeta^{n-1}$, so in fact

$$\mathbb{Z}[\zeta] = \{f(\zeta): f(x) \in \mathbb{Z}[x] \text{ of degree } \leq n - 1\}.$$

Now let p be a prime number and let

$$p\mathbb{Z}[\zeta] = \{pr: r \in \mathbb{Z}[\zeta]\},$$

a principal ideal of $\mathbb{Z}[\zeta]$.

22.21 Proposition
There are only finitely many ideals I of $\mathbb{Z}[\zeta]$ which contain $p\mathbb{Z}[\zeta]$.

Proof Consider the factor ring $\mathbb{Z}[\zeta]/p\mathbb{Z}[\zeta]$. By definition, this has as its elements all the cosets $p\mathbb{Z}[\zeta] + r$ with $r \in \mathbb{Z}[\zeta]$. Every such coset contains an element of the form

$$a_0 + a_1\zeta + \ldots + a_{n-1}\zeta^{n-1}, \text{ with } a_i \in \mathbb{Z}, 0 \leq a_i \leq p - 1 \text{ for all } i.$$

As there are only finitely many such elements, we conclude that $\mathbb{Z}[\zeta]/p\mathbb{Z}[\zeta]$ is finite. The ideals of $\mathbb{Z}[\zeta]$ which contain $p\mathbb{Z}[\zeta]$ are in bijective correspondence with the ideals of $\mathbb{Z}[\zeta]/p\mathbb{Z}[\zeta]$ (the correspondence being $I \to I/p\mathbb{Z}[\zeta]$). Therefore there are only finitely many such ideals, and the proof is complete. ∎

We deduce from Proposition 22.21 that there is a *maximal* ideal of P of $\mathbb{Z}[\zeta]$ which contains $p\mathbb{Z}[\zeta]$; that is, P is a proper ideal which is contained in no larger proper ideal of $\mathbb{Z}[\zeta]$. (A *proper* ideal of $\mathbb{Z}[\zeta]$ is an ideal which is not equal to $\mathbb{Z}[\zeta]$.)

We now prove two easy results about the maximal ideal P.

22.22 Proposition
If $r, s \in \mathbb{Z}[\zeta]$ and $rs \in P$, then either $r \in P$ or $s \in P$. In particular, if $r^n \in P$ for some positive integer n, then $r \in P$.

Proof Assume that $rs \in P$ and $r \notin P$. We must show that $s \in P$.

Since $r \notin P$, the ideal $r\mathbb{Z}[\zeta] + P$ of $\mathbb{Z}[\zeta]$ strictly contains P. As P is maximal, we therefore have

$$r\mathbb{Z}[\zeta] + P = \mathbb{Z}[\zeta].$$

Consequently, there exist $a \in \mathbb{Z}[\zeta]$, $b \in P$ such that

$$1 = ra + b.$$

Then

$$s = rsa + sb.$$

As $rs \in P$ and $b \in P$, it follows that $s \in P$, as required.

For the last statement of the proposition, assume that $r^n \in P$. Since $r^n = r r^{n-1}$, this implies that either $r \in P$ or $r^{n-1} \in P$. Repeating this argument, we conclude that $r \in P$. ∎

22.23 Proposition
We have $P \cap \mathbb{Z} = p\mathbb{Z}$.

Proof Let $m \in P \cap \mathbb{Z}$. If $p \nmid m$ then there are integers a, b with $am + bp = 1$; but this implies that $1 \in P$, which is false, since $P \neq \mathbb{Z}[\zeta]$. Thus $p \mid m$, which establishes that $P \cap \mathbb{Z} \subseteq p\mathbb{Z}$. Since $p \in P$, we also have $p\mathbb{Z} \subseteq P \cap \mathbb{Z}$. ∎

Congruences

At last we are ready to prove our results on congruences of character values. Let G be a group of order n and let $\zeta = e^{2\pi i/n}$. The ring $\mathbb{Z}[\zeta]$ is of interest because all the character values of G lie in $\mathbb{Z}[\zeta]$ (see Proposition 9.11). As in the previous section, let p be a prime number and let P be a maximal ideal of $\mathbb{Z}[\zeta]$ containing $p\mathbb{Z}[\zeta]$.

22.24 Theorem
Let $g \in G$ and let y be the p'-part of g. If χ is any character of G, then

$$\chi(g) - \chi(y) \in P.$$

Proof Suppose that g has order $m = up^v$, where $u, v \in \mathbb{Z}$ and $(u, p) = 1$. Choose integers a, b with $au + bp^v = 1$. Then $y = g^{bp^v}$ (see (22.20)).

The orders of g and of y divide $n = |G|$, so $\chi(g)$ and $\chi(y)$ are both sums of nth roots of unity, and hence lie in $\mathbb{Z}[\zeta]$.

Now let ω be an mth root of unity (so $\omega \in \mathbb{Z}[\zeta]$ as $m|n$). Then $\omega = \omega^{au+bp^v}$, and so

$$\omega^{p^v} = \omega^{aup^v} \cdot \omega^{bp^{2v}} = \omega^{bp^{2v}}.$$

Consider the number $(\omega - \omega^{bp^v})^{p^v}$. By the Binomial Theorem,

$$(\omega - \omega^{bp^v})^{p^v} = \omega^{p^v} - p^v \omega^{p^v - 1} \omega^{bp^v} + \ldots \pm \binom{p^v}{r} \omega^{p^v - r} \omega^{rbp^v}$$

$$+ \ldots + (-1)^{p^v} \omega^{bp^{2v}}.$$

For $0 < r < p^v$, the binomial coefficient $\binom{p^v}{r}$ is divisible by p. Hence

$$(\omega - \omega^{bp^v})^{p^v} = \omega^{p^v} + (-1)^{p^v} \omega^{bp^{2v}} + p\alpha,$$

where $\alpha \in \mathbb{Z}[\zeta]$. Moreover, since $\omega^{p^v} = \omega^{bp^{2v}}$, we have

$$\omega^{p^v} + (-1)^{p^v} \omega^{bp^{2v}} = \begin{cases} 0, & \text{if } p \neq 2, \\ 2\omega^{p^v}, & \text{if } p = 2, \end{cases}$$

so it follows that

$$(\omega - \omega^{bp^v})^{p^v} \in p\mathbb{Z}[\zeta].$$

Thus $(\omega - \omega^{bp^v})^{p^v}$ lies in the maximal ideal P. Application of Proposition 22.22 now forces

(22.25) $$(\omega - \omega^{bp^v}) \in P.$$

By Proposition 9.11, there are mth roots of unity $\omega_1, \ldots, \omega_d$ such that

$$\chi(g) = \omega_1 + \ldots + \omega_d \text{ and } \chi(y) = \omega_1^{bp^v} + \ldots + \omega_d^{bp^v}.$$

Then

$$\chi(g) - \chi(y) = (\omega_1 - \omega_1^{bp^v}) + \ldots + (\omega_d - \omega_d^{bp^v}),$$

which, by (22.25), lies in P. ∎

22.26 Corollary

Let p be a prime number. Suppose that $g \in G$ and that y is the p'-part of g. If χ is a character of G such that $\chi(g)$ and $\chi(y)$ are both integers, then

$$\chi(g) \equiv \chi(y) \bmod p.$$

Proof As $\chi(g)$ and $\chi(y)$ are both integers, Theorem 22.24 and Proposition 22.23 give

$$\chi(g) - \chi(y) \in P \cap \mathbb{Z} = p\mathbb{Z}.$$

Therefore $\chi(g) \equiv \chi(y) \bmod p$. ∎

22.27 Corollary

Let p be a prime number. Suppose that $g \in G$ and the order of g is a power of p. If χ is a character of G such that $\chi(g) \in \mathbb{Z}$, then

$$\chi(g) \equiv \chi(1) \bmod p.$$

Proof As the order of g is a power of p, the p'-part of g is 1, so the result is immediate from Corollary 22.26. ∎

Notice that Corollary 13.10 is the special case of Corollary 22.27 in which g has order 2.

We shall use the congruence results 22.24–22.27 extensively in our character calculations in Chapters 25–7. For the moment, we just illustrate the results with reference to some character tables which we already know.

22.28 Example

Recall from Example 20.14 that the character table of A_5 is as shown.
where $\alpha = (1 + \sqrt{5})/2$, $\beta = (1 - \sqrt{5})/2$.

If $g = (1\ 2\ 3)$ then Corollary 22.26 implies that $\chi(g) \equiv \chi(1) \bmod 3$ whenever $\chi(g) \in \mathbb{Z}$. Thus the corresponding entries in columns 1 and 2

Character table of A_5

	1	(1 2 3)	(1 2)(3 4)	(1 2 3 4 5)	(1 3 4 5 2)
χ_1	1	1	1	1	1
χ_2	4	1	0	-1	-1
χ_3	5	-1	1	0	0
χ_4	3	0	-1	α	β
χ_5	3	0	-1	β	α

of the character table are congruent modulo 3, as can be seen by inspecting the table. Similarly the entries in columns 1 and 3 are congruent modulo 2. Also

$$\chi_i((1\ 2\ 3\ 4\ 5)) \equiv \chi_i(1) \bmod 5 \quad \text{for } i = 1, 2, 3.$$

However, $\chi_4((1\ 2\ 3\ 4\ 5)) = \alpha \notin \mathbb{Z}$. We illustrate Theorem 22.24 for this value. If we take $p = 5$ and $g = (1\ 2\ 3\ 4\ 5)$, then the p'-part of g is 1, and

$$\chi_4(g) - \chi_4(1) = \alpha - 3 = \tfrac{1}{2}(1 + \sqrt{5} - 6)$$

$$= \sqrt{5} \cdot \tfrac{1}{2}(1 - \sqrt{5}) = \beta\sqrt{5}.$$

Put $\zeta = e^{2\pi i/60}$, and let P be a maximal ideal of $\mathbb{Z}[\zeta]$ containing $5\mathbb{Z}[\zeta]$. Then $(\sqrt{5})^2 \in P$, so $\sqrt{5} \in P$ by Proposition 22.22. Since $\beta \in \mathbb{Z}[\zeta]$ (see Proposition 9.11), we have $\beta\sqrt{5} \in P$. That is,

$$\chi_4(g) - \chi_4(1) \in P.$$

This illustrates Theorem 22.24.

Summary of Chapter 22

1. Character values are algebraic integers.

2. The degree of every irreducible character of G divides $|G|$.

3. If g is conjugate to g^i for all integers i which are coprime to the order of g, then $\chi(g)$ is an integer, for all characters χ.

4. Let p be a prime number. If $g \in G$ and y is the p'-part of g, then $\chi(g) \equiv \chi(y) \bmod p$ for all characters χ of G such that $\chi(g)$ and $\chi(y)$ are integers.

Exercises for Chapter 22

1. Let G be a group of order 15. Use Theorems 11.12, 17.11 and 22.11 to show that every irreducible character of G has degree 1. Deduce that G is abelian.

2. Prove that the number of conjugacy classes in a group of order 16 is 7, 10 or 16.

3. Let p and q be prime numbers with $p > q$, and let G be a non-abelian group of order pq.
   (a) Find the degrees of all the irreducible characters of G.
   (b) Show that $|G'| = p$.
   (c) Show that q divides $p - 1$ and that G has $q + ((p-1)/q)$ conjugacy classes.

4. Let G be a group and let ϕ be a character of G such that $\phi(g) = \phi(h)$ for all non-identity elements g and h of G.
   (a) Show that $\phi = a1_G + b\chi_{\text{reg}}$ for some $a, b \in \mathbb{C}$.
   (b) Show that $a + b$ and $a + b|G|$ are integers.
   (c) Show that if χ is a non-trivial irreducible character of G, then $b\chi(1)$ is an integer.
   (d) Deduce that both a and b are integers.

5. Suppose that G is a group of odd order. This exercise shows that the only irreducible character χ of G such that $\chi = \bar{\chi}$ is the trivial character.
   (a) Prove that if $g \in G$ and $g = g^{-1}$, then $g = 1$.
   (b) Now let χ be an irreducible character of G with $\chi = \bar{\chi}$. Prove that

$$\langle \chi, 1_G \rangle = \frac{1}{|G|}(\chi(1) + 2\alpha),$$

   where α is an algebraic integer.
   (c) Deduce that $\chi = 1_G$.

6. It is often possible to calculate the character table from limited arithmetic information about the group G. This exercise illustrates this point with the group $G = S_5$.

   A certain group G of order 120 has exactly seven conjugacy classes, and contains an element g of order 5 such that $|C_G(g)| = 5$. Moreover, g, g^2, g^3 and g^4 are conjugate in G.
   (a) Show that for every irreducible character χ of G, $\chi(g)$ is 0, 1 or -1.
   (b) Use Corollary 22.27 to deduce that G has two irreducible characters of degree 5.
   (c) Find $\chi(1)$ and $\chi(g)$ for all irreducible characters χ of G.
   (d) You are given that all entries in the character table of G are integers, and that the conjugacy classes of G have representatives $g_1, \ldots, g_7$ with orders and centralizer orders as follows:

g_i	g_1	g_2	g_3	g_4	g_5	g_6	g_7
Order of g_i	1	2	2	3	4	6	5
$|C_G(g_i)|$	120	12	8	6	4	6	5

   Using Corollary 22.26 and the column orthogonality relations, find the character table of G.

7. Prove that a complex number λ is an algebraic integer if and only if λ is a root of a polynomial of the form

$$x^n + a_{n-1}x^{n-1} + \ldots + a_1 x + a_0,$$

   where each a_r $(0 \leq r \leq n - 1)$ is an integer.

23
Real representations

Since Chapter 9, we have always taken our representations to be over the field $\mathbb{C}$ of complex numbers. However, some results in representation theory work equally well for the field $\mathbb{R}$ of real numbers. There is a subtle interplay between representations over $\mathbb{C}$ and representations over $\mathbb{R}$, which we shall explore in this chapter.

Often, characters of $\mathbb{C}G$-modules are real-valued, and the first main result of the chapter describes the number of real-valued irreducible characters of G.

Let ρ be a representation of G. If all the matrices $g\rho$ $(g \in G)$ have real entries, then of course the character of ρ is real-valued. However, the converse is not true: it can happen that the character of ρ is real-valued, while there is no representation σ equivalent to ρ with all the matrices $g\sigma$ having real entries. Various criteria for whether or not a character corresponds to a representation over $\mathbb{R}$ lead us to the remarkable Frobenius–Schur Count of Involutions. This is used in the last section to prove a famous result of Brauer and Fowler concerning centralizers of involutions in finite simple groups.

The material in this chapter is perhaps at a slightly more advanced level than that in the rest of the book, and is not used in the ensuing chapters, which consist largely of the calculation of character tables and applications of character theory. Nevertheless, the subject of real representations not only is elegant and interesting, but also gives delicate information about characters which often comes into play in more difficult calculations.

Real characters

An element g of the finite group G is said to be *real* if g is conjugate to g^{-1}; and if g is real, then the conjugacy class g^G is also said to be

263

real. Notice that if a conjugacy class is real, then it contains the inverse of each of its elements, since $(g^{-1})^G = \{x^{-1}: x \in g^G\}$.

On the other hand, a character χ of G is *real* if $\chi(g)$ is real for all $g \in G$. Thus for example, the conjugacy class $\{1\}$ of G is real, and the trivial character of G is real.

23.1 Theorem
The number of real irreducible characters of G is equal to the number of real conjugacy classes of G.

Proof Let X denote the character table of G, and let $\bar{X}$ denote the complex conjugate of the matrix X.

For every irreducible character χ of G, the complex conjugate $\bar{\chi}$ is also an irreducible character (see Proposition 13.15), so $\bar{X}$ can be obtained from X by permuting the rows. Hence there is a permutation matrix P such that

$$PX = \bar{X}$$

(see Exercise 4.4).

For every conjugacy class g^G of G, the entries in the column of X which corresponds to g^G are the complex conjugates of the entries in the column of X which corresponds to $(g^{-1})^G$. Therefore $\bar{X}$ can be obtained from X be permuting the columns, and so there is a permutation matrix Q such that

$$XQ = \bar{X}$$

By Proposition 16.2, X is invertible. Therefore

$$Q = X^{-1}\bar{X} = X^{-1}PX.$$

Consequently P and Q have the same trace, by Proposition 13.2. Since the trace of a permutation matrix is equal to the number of points fixed by the corresponding permutation, we have

> the number of real irreducible characters of G is $\operatorname{tr}(P)$,

and

> the number of real conjugacy classes of G is $\operatorname{tr}(Q)$.

As these numbers are equal, the result is proved.   ∎

Part of the following corollary was obtained by a different method in Exercise 22.5.

23.2 Corollary
The group G has a non-trivial real irreducible character if and only if the order of G is even.

Proof If G has odd order, then no non-identity element of G is real (see the solution to Exercise 23.1). Therefore by Theorem 23.1, the only real character of G is the trivial character.

If G has even order, then by Exercise 1.8, G has an element g of order 2. Hence G has at least two real conjugacy classes, $\{1\}$ and g^G, and so G has at least two real irreducible characters by Theorem 23.1. ∎

Characters which can be realized over $\mathbb{R}$

Let χ be a character of the group G. We say that χ *can be realized over* $\mathbb{R}$ if there is a representation $\rho: G \to \mathrm{GL}\,(n, \mathbb{C})$ with character χ, such that all the entries in each matrix $g\rho$ $(g \in G)$ are real. This is the same as saying that there is some $\mathbb{C}G$-module V with character χ, and there is a basis $v_1, \ldots, v_n$ of V, such that for all $g \in G$ and $1 \leqslant i \leqslant n$, $v_i g$ is a linear combination of $v_1, \ldots, v_n$ with *real* coefficients.

23.3 Examples
(1) Let $G = D_8 = \langle a, b: a^4 = b^2 = 1, b^{-1}ab = a^{-1} \rangle$, and let χ be the irreducible character of G of degree 2 (see Example 16.3(3)). Then χ can be realized over $\mathbb{R}$, since

$$a\rho = \begin{pmatrix} 0 & 1 \\ -1 & 0 \end{pmatrix}, \quad b\rho = \begin{pmatrix} -1 & 0 \\ 0 & 1 \end{pmatrix}$$

provides a representation ρ of G with character χ such that all the matrices $g\rho$ $(g \in G)$ have real entries.

(2) Let $G = Q_8 = \langle a, b: a^4 = 1, b^2 = a^2, b^{-1}ab = a^{-1} \rangle$, and let χ be the irreducible character of G of degree 2 (see Exercise 17.1). The values of χ are as follows:

	1	a^2	a	b	ab
χ	2	-2	0	0	0

Thus χ is real. In fact, χ cannot be realized over $\mathbb{R}$, but it is at the moment unclear how to prove this. (We shall eventually establish this in Example 23.18(3) below.)

Although every character which can be realized over $\mathbb{R}$ is perforce a real character, Example 23.3(2) tells us that the converse is false.

$\mathbb{R}G$-modules

Recall that in Chapter 4 we defined an FG-module, where F is $\mathbb{R}$ or $\mathbb{C}$. Thus an $\mathbb{R}G$-module is a vector space over $\mathbb{R}$, with a multiplication by elements of G satisfying the conditions of Definition 4.2. In this section we shall study the relationship between $\mathbb{R}G$-modules and $\mathbb{C}G$-modules.

23.4 Examples
Let V be a 2-dimensional vector space over $\mathbb{R}$, with basis v_1, v_2.
  (1) V becomes an $\mathbb{R}D_8$-module if we define

$$v_1 a = v_2, \qquad v_1 b = -v_1,$$
$$v_2 a = -v_1, \qquad v_2 b = v_2$$

(compare Example 23.3(1)).
  (2) V becomes an $\mathbb{R}C_3$-module, where $C_3 = \langle x: x^3 = 1 \rangle$, if we define

$$v_1 x = v_2,$$
$$v_2 x = -v_1 - v_2.$$

(This gives the representation $x \to \begin{pmatrix} 0 & 1 \\ -1 & -1 \end{pmatrix}$ of Exercise 3.2.)

Every $\mathbb{R}G$-module can easily be converted into a $\mathbb{C}G$-module. Simply take a basis $v_1, \ldots, v_n$ of the $\mathbb{R}G$-module, and consider the vector space over $\mathbb{C}$ with basis $v_1, \ldots, v_n$. This new vector space is clearly a $\mathbb{C}G$-module (with $v_i g$ defined as before). The construction is even easier to understand in terms of representations: if ρ: $G \to \mathrm{GL}\,(n, \mathbb{R})$ is a representation then for each $g \in G$, the matrix $g\rho$ has its entries in $\mathbb{R}$, and hence also in $\mathbb{C}$. Therefore we obtain a representation ρ: $G \to \mathrm{GL}\,(n, \mathbb{C})$. Notice that a character χ of G can be realized over $\mathbb{R}$ if and only if there exists an $\mathbb{R}G$-module with character χ.

Rather more subtle than this is the construction of an $\mathbb{R}G$-module from a given $\mathbb{C}G$-module. Let V be a $\mathbb{C}G$-module with basis $v_1, \ldots, v_n$, and let $g \in G$. There exist complex numbers z_{jk} such that

$$v_j g = \sum_{k=1}^{n} z_{jk} v_k \quad (1 \leqslant j \leqslant n).$$

Now let $V_{\mathbb{R}}$ be the vector space over $\mathbb{R}$ with basis

$$v_1, \ldots, v_n, iv_1, \ldots, iv_n.$$

Write $z_{jk} = x_{jk} + iy_{jk}$ with $x_{jk}, y_{jk} \in \mathbb{R}$. We define a multiplication of $V_{\mathbb{R}}$ by g by putting

$$(23.5) \qquad v_j g = \sum_{k=1}^{n} (x_{jk} v_k + y_{jk}(iv_k)), \text{ and}$$

$$(iv_j)g = \sum_{k=1}^{n} (-y_{jk} v_k + x_{jk}(iv_k)) \quad (1 \leqslant j \leqslant n),$$

and extending linearly to define vg for all $v \in V_{\mathbb{R}}$. In this way we define vg for all $v \in V_{\mathbb{R}}$ and all $g \in G$. Regarding v_j as an element of the $\mathbb{C}G$-module V, we have

$$(v_j g)h = v_j(gh) \quad \text{for all } g, h \in G, 1 \leqslant j \leqslant n.$$

It follows easily that, regarding v_j and iv_j as elements of $V_{\mathbb{R}}$, we have

$$(v_j g)h = v_j(gh) \text{ and } ((iv_j)g)h = (iv_j)(gh).$$

Hence using Proposition 4.6, we see that (23.5) makes $V_{\mathbb{R}}$ into an $\mathbb{R}G$-module.

If χ is the character of V, then

$$\chi(g) = \sum_{k=1}^{n} z_{kk}.$$

The character of $V_{\mathbb{R}}$, evaluated at g, is

$$2 \sum_{k=1}^{n} x_{kk} = \chi(g) + \overline{\chi(g)}.$$

Hence the character of $V_{\mathbb{R}}$ is $\chi + \bar{\chi}$.

We summarize the basic properties of $V_{\mathbb{R}}$ in the next proposition.

23.6 Proposition
Let V be a $\mathbb{C}G$-module with character χ.
(1) *The* $\mathbb{R}G$*-module* $V_{\mathbb{R}}$ *has character* $\chi + \bar{\chi}$; *in particular,* $\dim V_{\mathbb{R}}$
$= 2 \dim V$.
(2) *If* V *is an irreducible* $\mathbb{C}G$*-module and* $V_{\mathbb{R}}$ *is a reducible* $\mathbb{R}G$-
module, then χ *can be realized over* $\mathbb{R}$.

Proof We have already proved part (1).
For part (2), suppose that V is an irreducible $\mathbb{C}G$-module and $V_{\mathbb{R}}$ is
a reducible $\mathbb{R}G$-module. Then by part (1), $V_{\mathbb{R}} = U \oplus W$ where U is an
$\mathbb{R}G$-module with character χ and W is an $\mathbb{R}G$-module with character $\bar{\chi}$.
Thus there is an $\mathbb{R}G$-module, namely U, with character χ, and so χ can
be realized over $\mathbb{R}$.   ∎

23.7 Examples
(1) Let $G = C_3 = \langle x : x^3 = 1 \rangle$, and let V be the 1-dimensional $\mathbb{C}G$-
module with basis v_1 such that

$$v_1 x = \tfrac{1}{2}(-1 + i\sqrt{3})v_1$$

(note that $\tfrac{1}{2}(-1 + i\sqrt{3}) = e^{2\pi i/3}$). Then $V_{\mathbb{R}}$ has basis v_1, iv_1, and with
respect to this basis, x is represented by the matrix

$$\begin{pmatrix} -1/2 & \sqrt{3}/2 \\ -\sqrt{3}/2 & -1/2 \end{pmatrix}.$$

(2) Let $G = D_8 = \langle a, b : a^4 = b^2 = 1, b^{-1}ab = a^{-1} \rangle$, and let V be the 2-
dimensional $\mathbb{C}G$-module with basis v_1, v_2 such that

$$v_1 a = iv_1, \qquad v_1 b = v_2,$$
$$v_2 a = -iv_2, \qquad v_2 b = v_1.$$

Then $V_{\mathbb{R}}$ has basis v_1, v_2, v_3, v_4, where $v_3 = iv_1$, $v_4 = iv_2$. With
respect to this basis, we obtain the representation ρ, where

$$a\rho = \begin{pmatrix} 0 & 0 & 1 & 0 \\ 0 & 0 & 0 & -1 \\ -1 & 0 & 0 & 0 \\ 0 & 1 & 0 & 0 \end{pmatrix}, \quad b\rho = \begin{pmatrix} 0 & 1 & 0 & 0 \\ 1 & 0 & 0 & 0 \\ 0 & 0 & 0 & 1 \\ 0 & 0 & 1 & 0 \end{pmatrix}.$$

The subspace of $V_{\mathbb{R}}$ which is spanned by $v_1 + v_4$ and $v_2 + v_3$ is an
$\mathbb{R}G$-submodule. Therefore the character of V can be realized over $\mathbb{R}$,

by Proposition 23.6(2). In fact, we already know this from Example 23.3(1).

Bilinear forms

The question of whether or not a given character can be realized over $\mathbb{R}$ turns out to be related to the existence of a certain bilinear form on the corresponding $\mathbb{C}G$-module.

Let V be a vector space over F, where F is $\mathbb{R}$ or $\mathbb{C}$. A *bilinear form* β on V is a function which associates with each ordered pair (u, v) of vectors in V an element $\beta(u, v)$ of F, and which has the following properties:

$$\beta(\lambda_1 u_1 + \lambda_2 u_2, v) = \lambda_1 \beta(u_1, v) + \lambda_2 \beta(u_2, v),$$

$$\beta(u, \lambda_1 v_1 + \lambda_2 v_2) = \lambda_1 \beta(u, v_1) + \lambda_2 \beta(u, v_2),$$

for all u, v, u_1, u_2, v_1, $v_2 \in V$ and λ_1, $\lambda_2 \in F$. (Thus for fixed u, v, the functions $x \to \beta(x, v)$ and $y \to \beta(u, y)$ are both linear – hence the term *bilinear*.)

The bilinear form β is *symmetric* if

$$\beta(u, v) = \beta(v, u) \quad \text{for all } u, v \in V.$$

And the bilinear form β is *skew-symmetric* if

$$\beta(u, v) = -\beta(v, u) \quad \text{for all } u, v \in V.$$

If V is an FG-module, then a bilinear form β on V is said to be G-*invariant* if

$$\beta(ug, vg) = \beta(u, v) \quad \text{for all } u, v \in V \text{ and } g \in G.$$

Our next result shows that every $\mathbb{R}G$-module has a G-invariant symmetric bilinear form with a strong positivity property. A similar result for $\mathbb{C}G$-modules was given in Exercise 8.6.

23.8 Theorem

If V is an $\mathbb{R}G$-module, then there exists a G-invariant symmetric bilinear form β on V such that

$$\beta(v, v) > 0 \quad \text{for all non-zero } v \in V.$$

Proof Let $v_1, \ldots, v_n$ be a basis of V. For $u = \sum_{j=1}^{n} \lambda_j v_j$, $v = \sum_{j=1}^{n} \mu_j v_j \in V$ with $\lambda_j, \mu_j \in \mathbb{R}$, define

$$\gamma(u, v) = \sum_{j=1}^{n} \lambda_j \mu_j.$$

Then γ is a symmetric bilinear form on V. Moreover, for non-zero $v \in V$,

$$\gamma(v, v) = \sum_{j=1}^{n} \mu_j^2 > 0.$$

Now let

$$\beta(u, v) = \sum_{x \in G} \gamma(ux, vx) \quad (u, v \in V).$$

Again, β is a symmetric bilinear form on V, and $\beta(v, v) > 0$ for all non-zero $v \in V$.

If $g \in G$, then gx runs through G as x runs through G, and hence

$$\beta(ug, vg) = \sum_{x \in G} \gamma(ugx, vgx) = \beta(u, v).$$

Therefore β is G-invariant and the theorem is proved.   ∎

23.9 Proposition
Let V be an $\mathbb{R}G$-module and let β be a G-invariant bilinear form on V. If U is an $\mathbb{R}G$-submodule of V, then so is

$$W = \{w \in V \colon \beta(u, w) = 0 \text{ for all } u \in U\}.$$

Proof It is easy to see that W is a subspace of V. Now let $w \in W$ and $g \in G$. For all $u \in U$, we have $ug^{-1} \in U$, so

$$\beta(u, wg) = \beta(ug^{-1}, wgg^{-1}) = \beta(ug^{-1}, w) = 0.$$

Thus $wg \in W$, so W is an $\mathbb{R}G$-submodule of V.   ∎

23.10 Proposition
Suppose that β is a G-invariant symmetric bilinear form on the $\mathbb{R}G$-module V, and that there exist u, $v \in V$ with $\beta(u, u) > 0$ and $\beta(v, v) < 0$. Then V is a reducible $\mathbb{R}G$-module.

Proof Theorem 23.8 supplies us with a G-invariant symmetric bilinear form β_1 on V such that

$$\beta_1(w, w) > 0 \quad \text{for all non-zero } w \in V.$$

By a general result on bilinear forms (see Exercise 23.7), there is a basis $v_1, \ldots, v_n$ of V such that

$$\beta_1(v_i, v_j) = \beta(v_i, v_j) = 0 \quad \text{if } i \neq j,$$

and

$$\beta_1(v_i, v_i) = 1 \quad \text{for all } i,$$

$$\beta(v_1, v_1) > 0,$$

$$\beta(v_2, v_2) < 0.$$

Let $\beta(v_1, v_1) = x$, and define γ by

$$\gamma(u, v) = \beta_1(u, v) - \frac{1}{x}\beta(u, v) \quad (u, v \in V).$$

Since β and β_1 are G-invariant symmetric bilinear forms on V, so is γ. But for all $v = \sum_{i=1}^{n}\lambda_i v_i \in V$ $(\lambda_i \in \mathbb{R})$, we have

$$\gamma(v, v_1) = \lambda_1 \gamma(v_1, v_1) = 0.$$

Therefore, if we define

$$W = \{w \in V: \gamma(v, w) = 0 \text{ for all } v \in V\},$$

then W is non-zero, and is an $\mathbb{R}G$-submodule of V by Proposition 23.9. Moreover,

$$\gamma(v_2, v_2) = 1 - \frac{1}{x}\beta(v_2, v_2) > 0,$$

so $W \neq V$. Therefore V is a reducible $\mathbb{R}G$-module. ∎

We can now relate bilinear forms to the question of whether or not a given character of G can be realized over $\mathbb{R}$.

23.11 Theorem
Let χ be an irreducible character of G. The following two conditions are equivalent:

(1) χ can be realized over $\mathbb{R}$;
(2) there exists a $\mathbb{C}G$-module V with character χ, and a non-zero G-invariant symmetric bilinear form on V.

Proof We first show that (2) implies (1). Let V be a $\mathbb{C}G$-module with character χ, and suppose that β is a non-zero G-invariant symmetric bilinear form on V. There exist $u, v \in V$ with $\beta(u, v) = \beta(v, u) \neq 0$. Since

$$\beta(u + v, \; u + v) = \beta(u, u) + \beta(v, v) + 2\beta(u, v),$$

there exists $w \in V$ with $\beta(w, w) \neq 0$. Let $\beta(w, w) = z$ and $v_1 = z^{-1/2}w$. Then

$$\beta(v_1, v_1) = 1.$$

Extend v_1 to a basis $v_1, \ldots, v_n$ of V. Then $v_1, \ldots, v_n, iv_1, \ldots, iv_n$ is a basis of the $\mathbb{R}G$-module $V_{\mathbb{R}}$.

Define a function ϑ from $V_{\mathbb{R}}$ to V by

$$\vartheta: \sum_{j=1}^{n} \lambda_j v_j + \sum_{j=1}^{n} \mu_j (iv_j) \rightarrow \sum_{j=1}^{n} (\lambda_j + i\mu_j) v_j \quad (\lambda_j, \mu_j \in \mathbb{R}).$$

Then ϑ is a bijection, and for all $w_1, w_2, v \in V_{\mathbb{R}}$, all $\lambda \in \mathbb{R}$ and all $g \in G$, we have

(23.12)
$$(w_1 + w_2)\vartheta = w_1\vartheta + w_2\vartheta,$$

$$(\lambda v)\vartheta = \lambda(v\vartheta),$$

$$(vg)\vartheta = (v\vartheta)g.$$

Now define a function $\tilde{\beta}$ on ordered pairs of elements of $V_{\mathbb{R}}$ by

$$\tilde{\beta}(u, v) = \text{the real part of } \beta(u\vartheta, v\vartheta) \quad (u, v \in V_{\mathbb{R}}).$$

You can readily check, using the properties (23.12), that $\tilde{\beta}$ is a G-invariant symmetric bilinear form on $V_{\mathbb{R}}$. Notice that

$$\tilde{\beta}(v_1, v_1) = 1 \text{ and } \tilde{\beta}(iv_1, iv_1) = -1.$$

Therefore $V_{\mathbb{R}}$ is a reducible $\mathbb{R}G$-module, by Proposition 23.10. It now follows from Proposition 23.6(2) that χ can be realized over $\mathbb{R}$. This establishes that (2) implies (1) in the theorem.

Conversely, suppose that χ can be realized over $\mathbb{R}$, and let U be an $\mathbb{R}G$-module with character χ. By Theorem 23.8, there is a non-zero G-invariant symmetric bilinear form γ on U. Let $u_1, \ldots, u_n$ be a basis of U, and let V be the vector space over $\mathbb{C}$ with basis $u_1, \ldots, u_n$. As explained earlier, V is a $\mathbb{C}G$-module (with $u_i g$ defined as for U). Define $\hat{\gamma}$ on V by

$$\hat{\gamma}\left(\sum_{j=1}^{n} \lambda_j u_j, \; \sum_{k=1}^{n} \mu_k u_k \right) = \sum_{j=1}^{n} \sum_{k=1}^{n} \lambda_j \mu_k \gamma(u_j, u_k)$$

(where λ_j, $\mu_k \in \mathbb{C}$). Then $\hat{\gamma}$ is a non-zero G-invariant symmetric bilinear form on the $\mathbb{C}G$-module V, and V has character χ. Thus (1) implies (2), and the proof of the theorem is complete. ∎

The indicator function

We now associate with each irreducible character χ of G a certain number, called the indicator of χ, which is always 0, 1 or -1. We shall see later that this number tells us whether or not χ can be realized over $\mathbb{R}$.

Observe that

$$\langle \chi^2, 1_G \rangle = \frac{1}{|G|} \sum_{g \in G} \chi(g)\chi(g) = \langle \chi, \bar{\chi} \rangle.$$

Therefore, for irreducible characters χ, we have

$$\langle \chi^2, 1_G \rangle = \begin{cases} 0, & \text{if } \chi \text{ is not real,} \\ 1, & \text{if } \chi \text{ is real.} \end{cases}$$

Let V be a $\mathbb{C}G$-module with character χ. Recall from Chapter 19 that χ^2 is the character of the $\mathbb{C}G$-module $V \otimes V$, and

$$\chi^2 = \chi_S + \chi_A,$$

where χ_S is the character of the symmetric part of $V \otimes V$, and χ_A is the character of the antisymmetric part of $V \otimes V$. Hence if $\langle \chi^2, 1_G \rangle = 1$, then precisely one of χ_S and χ_A has 1_G as a constituent.

23.13 Definition

If χ is an irreducible character of G, then we define the *indicator* $\iota\chi$ of χ by

$$\iota\chi = \begin{cases} 0, & \text{if } \chi \text{ is not a constituent of } \chi_S \text{ or } \chi_A, \\ 1, & \text{if } 1_G \text{ is a constituent of } \chi_S, \\ -1, & \text{if } 1_G \text{ is a constituent of } \chi_A. \end{cases}$$

We call ι the *indicator function* on the set of irreducible characters of G. Note that $\iota\chi \neq 0$ if and only if χ is real.

The next result gives a significant property of the indicator function, relating it to the internal structure of the group G.

23.14 Theorem

For all $x \in G$,

$$\sum_\chi (\iota\chi)\chi(x) = |\{y \in G: y^2 = x\}|,$$

where the sum is over all irreducible characters χ of G.

Proof Define a function $\vartheta: G \to \mathbb{C}$ by

$$\vartheta(x) = |\{y \in G: y^2 = x\}| \quad (x \in G).$$

Note that ϑ is a class function on G, since for $g \in G$ we have

$$y^2 = x \Leftrightarrow (g^{-1}yg)^2 = g^{-1}xg.$$

Therefore by Corollary 15.4, ϑ is a linear combination of the irreducible characters of G.

The definition of $\iota\chi$ gives

$$\iota\chi = \langle \chi_S - \chi_A, 1_G \rangle$$

$$= \frac{1}{|G|} \sum_{g \in G} \chi(g^2) \quad \text{by Proposition 19.14}$$

$$= \frac{1}{|G|} \sum_{x \in G} \sum_{g \in G: g^2 = x} \chi(g^2)$$

$$= \frac{1}{|G|} \sum_{x \in G} \vartheta(x)\chi(x)$$

$$= \langle \vartheta, \chi \rangle.$$

Therefore, $\vartheta = \sum(\iota\chi)\chi$, and the result follows.   ∎

23.15 Example

Let $G = S_3$. The character table of G is

	1	(1 2)	(1 2 3)
χ_1	1	1	1
χ_2	1	-1	1
χ_3	2	0	-1

Using Proposition 19.14 we calculate that $\iota\chi = 1$ for each irreducible character χ of G, so $\sum(\iota\chi)\chi = \chi_1 + \chi_2 + \chi_3$, which takes the following values:

	1	(1 2)	(1 2 3)
$\chi_1 + \chi_2 + \chi_3$	4	0	1

Sure enough, in accordance with Theorem 23.14, four elements of G square to be 1, namely 1, (1 2), (1 3) and (2 3); no elements square to be (1 2); and one element, (1 3 2), squares to be (1 2 3).

Back to reality

We now relate the indicator function to the previous material on bilinear forms. Using this, we show that the indicator of an irreducible character determines whether or not it can be realized over $\mathbb{R}$, and deduce the Frobenius–Schur Count of Involutions.

23.16 Theorem
Let V be an irreducible $\mathbb{C}G$-module with character χ.

(1) *There exists a non-zero G-invariant bilinear form on V if and only if $\iota\chi \neq 0$.*

(2) *There exists a non-zero G-invariant symmetric bilinear form on V if and only if $\iota\chi = 1$.*

(3) *There exists a non-zero G-invariant skew-symmetric bilinear form on V if and only if $\iota\chi = -1$.*

Proof In this proof we regard $\mathbb{C}$ as a 1-dimensional vector space over $\mathbb{C}$, and define a multiplication of $\mathbb{C}$ by elements of G by

$$\lambda g = \lambda \quad (\lambda \in \mathbb{C}, \, g \in G).$$

In this way, $\mathbb{C}$ becomes a trivial $\mathbb{C}G$-module.

(1) Suppose that $\iota\chi \neq 0$. Then 1_G is a constituent of χ^2, and hence the $\mathbb{C}G$-module $V \otimes V$ has a trivial $\mathbb{C}G$-submodule. By Proposition 8.8, there is a non-zero $\mathbb{C}G$-homomorphism from $V \otimes V$ onto this trivial $\mathbb{C}G$-submodule, and hence there is a non-zero $\mathbb{C}G$-homomorphism ϑ from $V \otimes V$ onto the trivial $\mathbb{C}G$-module $\mathbb{C}$.

Now define β by

$$\beta(u, v) = (u \otimes v)\vartheta \quad (u, v \in V).$$

Then β is a non-zero bilinear form on V, and for u, $v \in V$ and $g \in G$, we have

$$\beta(ug, vg) = (ug \otimes vg)\vartheta = ((u \otimes v)g)\vartheta$$

$$= ((u \otimes v)\vartheta)g = (u \otimes v)\vartheta = \beta(u, v).$$

Thus β is G-invariant.

Conversely, suppose that there is a non-zero G-invariant bilinear form β on V. Let $v_1, \ldots, v_n$ be a basis of V, so that $v_i \otimes v_j$ ($1 \leq i \leq n, 1 \leq j \leq n$) form a basis of $V \otimes V$. Define $\vartheta: V \otimes V \to \mathbb{C}$ by putting

$$(v_i \otimes v_j)\vartheta = \beta(v_i, v_j) \quad (1 \leq i \leq n, 1 \leq j \leq n)$$

and extending linearly to the whole of $V \otimes V$. For $g \in G$, we have

$$((v_i \otimes v_j)g)\vartheta = (v_i g \otimes v_j g)\vartheta = \beta(v_i g, v_j g)$$

$$= \beta(v_i, v_j) \quad \text{as } \beta \text{ is } G\text{-invariant}$$

$$= (v_i \otimes v_j)\vartheta.$$

Hence ϑ is a non-zero $\mathbb{C}G$-homomorphism from $V \otimes V$ onto the trivial $\mathbb{C}G$-module $\mathbb{C}$. Thus, by Proposition 10.1, $V \otimes V$ has a trivial $\mathbb{C}G$-submodule. If follows that χ^2 has the trivial character 1_G as a constituent, and therefore $\iota\chi \neq 0$.

(2) Suppose that $\iota\chi = 1$. Then 1_G is a constituent of χ_S, which is the character of the $\mathbb{C}G$-module $S(V \otimes V)$, the symmetric part of $V \otimes V$. As in (1), it follows by Proposition 8.8 that there is a non-zero $\mathbb{C}G$-homomorphism ϑ from $S(V \otimes V)$ onto the trivial $\mathbb{C}G$-module $\mathbb{C}$. Define

$$\beta(u, v) = (u \otimes v + v \otimes u)\vartheta \quad (u, v \in V).$$

Then β is a non-zero G-invariant symmetric bilinear form on V.

Conversely, suppose that there exists a non-zero G-invariant symmetric bilinear form β on V. Let $v_1, \ldots, v_n$ be a basis of V, and define $\vartheta: S(V \otimes V) \to \mathbb{C}$ by putting

$$(v_i \otimes v_j + v_j \otimes v_i)\vartheta = \beta(v_i, v_j) \quad (1 \leq i, j \leq n)$$

and extending linearly. Since β is symmetric, ϑ is well-defined; and ϑ is a non-zero $\mathbb{C}G$-homomorphism from $S(V \otimes V)$ onto the trivial $\mathbb{C}G$-

module $\mathbb{C}$. Hence χ_S has the trivial character 1_G as a constituent, and so $\iota\chi = 1$.

(3) The proof of (3) is very similar to that of (2), and is omitted. ∎

We can now relate real representations of G to *involutions* in G, where by an involution we mean an element of order 2.

23.17 Corollary (The Frobenius–Schur Count of Involutions)
For each irreducible character χ of G, we have

$$\iota\chi = \begin{cases} 0, & \text{if } \chi \text{ is not real,} \\ 1, & \text{if } \chi \text{ can be realized over } \mathbb{R}, \\ -1, & \text{if } \chi \text{ is real, but } \chi \text{ cannot be realized over } \mathbb{R}. \end{cases}$$

Moreover, for all $x \in G$,

$$\sum_\chi (\iota\chi)\chi(x) = |\{y \in G: y^2 = x\}|,$$

where the sum is over all irreducible characters χ of G. In particular,

$$\sum_\chi (\iota\chi)\chi(1) = 1 + t,$$

where t is equal to the number of involutions in G.

Proof When we defined the indicator function, we showed that $\iota\chi \neq 0$ if and only if χ is real. And Theorems 23.11 and 23.16(2) show that χ can be realized over $\mathbb{R}$ if and only if $\iota\chi = 1$. This proves that $\iota\chi$ is determined as in the statement of the corollary.

The expression for $\sum_\chi (\iota\chi)\chi(x)$ was obtained in Theorem 23.14. Putting $x = 1$, we see that $\sum_\chi (\iota\chi)\chi(1)$ is equal to the number of elements y of G satisfying $y^2 = 1$. These elements are just the involutions in G, together with the identity, so the number of them is precisely $1 + t$. ∎

We conclude with some examples illustrating the use of Corollary 23.17.

23.18 Examples
(1) Let χ be a linear character. Then $\iota\chi = 1$ if χ is real, and $\iota\chi = 0$ if χ is non-real.

For an abelian group, the Frobenius–Schur Count of Involutions shows that the number of real irreducible (linear) characters is equal to the number of real conjugacy classes, since in this case g is conjugate to g^{-1} if and only if $g^2 = 1$. This special case of Theorem 23.1 can be proved directly without much difficulty (see Exercise 23.2).

(2) We know that all the irreducible characters of $D_8 = \langle a, b: a^4 = b^2 = 1, b^{-1}ab = a^{-1} \rangle$ can be realized over $\mathbb{R}$ (see Example 23.3(1), and note that all four linear characters are real). Thus $\iota\chi = 1$ for all irreducible characters χ of D_8, and so

$$\sum_\chi (\iota\chi)\chi(1) = 1 + 1 + 1 + 1 + 2 = 6.$$

The five involutions in D_8 which are predicted by the Frobenius–Schur Count of Involutions are a^2, b, ab, a^2b and a^3b.

(3) In the group $Q_8 = \langle a, b: a^4 = 1, a^2 = b^2, b^{-1}ab = a^{-1} \rangle$, there is just one involution, namely a^2. Now $\iota\chi = 1$ for each of the four linear characters, and

$$\sum_\chi (\iota\chi)\chi(1) = 2$$

by the Frobenius–Schur Count of Involutions. Therefore, if ψ is the irreducible character of degree 2, then $\iota\psi = -1$. In particular ψ cannot be realized over $\mathbb{R}$.

(4) The symmetric group S_4 has ten elements whose square is 1, namely the identity, the six elements which are conjugate to $(1\ 2)$, and the three elements which are conjugate to $(1\ 2)(3\ 4)$. Since the degrees 1, 1, 2, 3, 3 of the irreducible characters of S_4 sum to be 10 (see Section 18.1), we see that all the characters of S_4 can be realized over $\mathbb{R}$.

The Brauer–Fowler Theorem

We now apply Corollary 23.17 to give a proof of a famous theorem of Brauer and Fowler.

23.19 Brauer–Fowler Theorem

Let n be a positive integer. Then there exist only finitely many non-isomorphic finite simple groups containing an involution with centralizer of order n.

Despite its fairly elementary proof, this result is of great historical importance in finite group theory. It led Brauer to propose the programme of determining, for each finite group C, all the simple groups G possessing an involution u with $C_G(u) \cong C$. This programme was the start of the modern attempt to classify all finite simple groups, which was finally completed in the early 1980s. For further information about this, see the book by D. Gorenstein listed in the Bibliography.

In Exercise 10 at the end of the chapter you are asked to carry out Brauer's programme in the case where $C \cong C_2$. This should not trouble you too much. In Chapter 30, Theorem 30.8, you will find a much more sophisticated case, in which $C \cong D_8$.

For proof of Theorem 23.19 we require two preliminary lemmas.

23.20 Lemma

If $a_1, \ldots, a_n$ are real numbers, then $\sum a_i^2 \geqslant (\sum a_i)^2 / n$.

Proof This follows from the Cauchy–Schwarz inequality $\|\mathbf{v}\| \|\mathbf{w}\| \geqslant |\mathbf{v}.\mathbf{w}|$, taking $\mathbf{v} = (a_1, \ldots, a_n)$ and $\mathbf{w} = (1, \ldots, 1)$. ∎

23.21 Lemma

Let G be a group of even order m, and let t be the number of involutions in G (so $t > 0$ by Exercise 8 of Chapter 1). Write $a = (m - 1)/t$. Then G contains a non-identity element x such that $|G : C_G(x)| \leqslant a^2$.

Proof By Corollary 23.17, we have

$$t \leqslant \sum_{\chi} \chi(1)$$

where the sum is over all non-trivial irreducible characters χ of G. Writing k for the number of irreducible characters of G, we deduce using Lemma 23.20 and Theorem 11.12 that

$$t^2 \leqslant (\sum_{\chi} \chi(1))^2 \leqslant (k - 1) \sum \chi(1)^2 = (k - 1)(m - 1),$$

and hence $m - 1 \leqslant (k - 1)(m - 1)^2 / t^2 = (k - 1)a^2$. Now $k - 1$ is the number of non-identity conjugacy classes of G. If every non-identity conjugacy class has size more then a^2, then $(k - 1)a^2 >$

$|G| - 1 = m - 1$, a contradiction. Therefore some non-identity class x^G has size at most a^2. Then $|G : C_G(x)| \leqslant a^2$, giving the result.   ∎

Proof of Theorem 23.19 Suppose G is simple and contains an involution u such that $|C_G(u)| = n$. Let $|G| = m$, and let t be the number of involutions in G. By Proposition 12.6, every element of the conjugacy class u^G is an involution, and hence

$$t \geqslant |u^G| = |G : C_G(u)| = m/n.$$

Therefore $(m - 1)/t < n$, and so by Lemma 23.21, there is a non-identity element $x \in G$ such that $|G : C_G(x)| < n^2$.

Let $H = C_G(x)$. If $H = G$ then x lies in $Z(G)$, the centre of G, which is a normal subgroup of G. Since G is simple it follows that $G = Z(G)$, so G is abelian and therefore $G \cong C_2$.

Now suppose that $H \neq G$. Write $r = |G : H|$, so $r < n^2$. By Exercise 9 at the end of the chapter, there is a non-trivial homomorphism θ from G to the symmetric group S_r. As G is simple, the normal subgroup $\text{Ker}\,\theta = \{1\}$. Thus G is isomorphic to a subgroup of S_r, hence of S_{n^2}. In particular, given n, there are only finitely many possibilities for G.   ∎

Summary of Chapter 23

1. The number of real irreducible characters of G is equal to the number of real conjugacy classes of G.

    Let ι be the indicator function, and let χ be an irreducible character of G.

2. $\iota\chi = \begin{cases} 0, & \text{if } \chi \text{ is non-real,} \\ 1, & \text{if there exists an } \mathbb{R}G\text{-module } U \text{ with character } \chi, \\ & \text{with character } \chi \\ -1, & \text{if } \chi \text{ is real, but there does not exist an } \mathbb{R}G\text{-module.} \end{cases}$

3. $\iota\chi = \begin{cases} 0, & \text{if } 1_G \text{ is not a constituent of } \chi_S \text{ or } \chi_A, \\ 1, & \text{if } 1\,G \text{ is a constituent of } \chi_S, \\ -1, & \text{if } 1_G \text{ is a constituent of } \chi_A. \end{cases}$

4. $\sum_\chi (\iota\chi)\chi(1) = |\{g \in G\colon g^2 = 1\}|.$

Exercises for Chapter 23

1. Prove that if G is a group of odd order then no non-identity element of G is real.

2. Let G be a finite abelian group. Use the description of the irreducible characters of G, given in Theorem 9.8, to prove directly that the number of real irreducible characters of G is equal to the number of elements g in G for which $g^2 = 1$.

3. Let $G = D_{2n}$ and consult Section 18.3 for the character table of G. How many elements g of G satisfy $g^2 = 1$? Deduce that $\iota\chi = 1$ for all irreducible characters χ of G.

4. Let ρ be an irreducible representation of degree 2 of a group G, and let χ be the character of ρ. Prove that $\chi_A(g) = \det(g\rho)$ for all $g \in G$. Deduce that $\iota\chi = -1$ if and only if $\det(g\rho) = 1$ for all $g \in G$.

5. Let $G = T_{4n} = \langle a, b: a^{2n} = 1, a^n = b^2, b^{-1}ab = a^{-1}\rangle$, as in Exercise 17.6. Let V be a 2-dimensional vector space over $\mathbb{C}$ with basis v_1, v_2 and let ε be a $(2n)$th root of unity in $\mathbb{C}$ with $\varepsilon \neq \pm 1$. Exercise 17.6 shows that V becomes an irreducible $\mathbb{C}G$-module if we define

$$v_1 a = \varepsilon v_1, \quad v_1 b = v_2,$$
$$v_2 a = \varepsilon^{-1} v_2, \quad v_2 b = \varepsilon^n v_1.$$

Let χ be the character of this $\mathbb{C}G$-module V.
(a) Note that $\varepsilon^n = \pm 1$. Use Exercise 4 to show that $\iota\chi = 1$ if $\varepsilon^n = 1$ and $\iota\chi = -1$ if $\varepsilon^n = -1$.
(b) Let β be the bilinear form on V for which

$$\beta(v_1, v_1) = \beta(v_2, v_2) = 0,$$
$$\beta(v_1, v_2) = 1, \beta(v_2, v_1) = \varepsilon^n.$$

Prove that the bilinear form β is G-invariant, and use Theorem 23.16 to provide a second proof that $\iota\chi = 1$ if $\varepsilon^n = 1$ and $\iota\chi = -1$ if $\varepsilon^n = -1$.
(c) Prove that a^n is the only element of order 2 in T_{4n}.
(d) Use the character table of G, which appears in the solution to Exercise 18.3, to find $\iota\chi$ for each irreducible character χ of G. Check that

$$\sum_\chi (\iota\chi)\chi(1) = 2,$$

in agreement with the Frobenius–Schur Count of Involutions.

6. Prove that if χ is an irreducible character of a group G, and $\iota\chi = -1$, then $\chi(1)$ is even.
   (Hint: the solution uses a well known result about skew-symmetric bilinear forms.)

7. Suppose that V is a vector space over $\mathbb{R}$ and that β_1 and β are symmetric bilinear forms on V. Assume that $\beta_1(w, w) > 0$ for all non-zero w in V. Prove that there is a basis $e_1, \ldots, e_n$ of V such that

   $$\beta_1(e_i, e_i) = 1 \text{ for all } i, \text{ and}$$

   $$\beta_1(e_i, e_j) = \beta(e_i, e_j) = 0 \text{ for all } i \neq j.$$

8. Schur's Lemma is crucial for the development of the theory of $\mathbb{C}G$-modules. This exercise indicates the extent to which results like Schur's Lemma hold for $\mathbb{R}G$-modules.
   Let V and W be irreducible $\mathbb{R}G$-modules.
   (a) Prove that if $\vartheta\colon V \to W$ is an $\mathbb{R}G$-homomorphism then either ϑ is an $\mathbb{R}G$-isomorphism or $v\vartheta = 0$ for all $v \in V$.
   (b) Prove that if $\vartheta\colon V \to V$ is an $\mathbb{R}G$-isomorphism and V remains irreducible as a $\mathbb{C}G$-module, then $\vartheta = \lambda 1_V$ for some real number λ.
   . (c) Give an example of a group G, an irreducible $\mathbb{R}G$-module V and an $\mathbb{R}G$-homomorphism $\vartheta\colon V \to V$ which is not a multiple of 1_V.

9. Let G be a group with a subgroup H of index n. Let Ω be the set of n right cosets Hx of H in G. For $g \in G$, define a function $\rho_g\colon \Omega \to \Omega$ by $(Hx)\rho_g = Hxg$ for all $x \in G$.
   Prove that ρ_g is a permutation of Ω, and that the function $\rho\colon g \to \rho_g$ is a homomorphism from G to the symmetric group on Ω.
   Show that the kernel of ρ is $\bigcap_{x \in G} x^{-1}Hx$.
   Deduce that if a group G has a subgroup H of index n, then there is a homomorphism $G \to S_n$ with kernel contained in H.

10. Suppose that G is a finite group containing and involution t with $C_G(t) \cong C_2$. Prove that $|G:G'| = 2$. Deduce that if G is simple, then $G \cong C_2$.

24
Summary of properties of character tables

In this short chapter we present no new results, but instead we gather together from previous chapters various properties which are helpful when we try to find the character table of a particular group. In the next four chapters we shall calculate several character tables in detail.

Usually we begin by working out the conjugacy classes and centralizer orders of our given finite group G. The size of the character table is determined by the number k of conjugacy classes of G; the character table is then a $k \times k$ matrix, with columns indexed by the conjugacy classes of G (the first column corresponding to the conjugacy class $\{1\}$), and with rows indexed by the irreducible characters of G.

When doing calculations, we commonly come across a new character χ, which may or may not be irreducible. We can then calculate $\langle \chi, \chi \rangle$, which is given by

$$\langle \chi, \chi \rangle = \frac{1}{|G|} \sum_{g \in G} \chi(g) \overline{\chi(g)}.$$

The character χ is irreducible if and only if $\langle \chi, \chi \rangle = 1$ (see Theorem 14.20). If χ is reducible then we calculate $\langle \chi, \chi_i \rangle$ for each of the irreducible characters χ_i which we already know, and then

$$\chi - \sum_i \langle \chi, \chi_i \rangle \chi_i$$

will also be a character. We can thus determine whether χ is a linear combination of the irreducible characters we already know; and if it is not, then we can obtain from χ a linear combination of irreducible characters, all of which are new.

We have developed a number of methods for producing characters χ

283

on which to perform such calculations. For example, every subgroup of S_n has a permutation character (see (13.22)); the product of two characters is again a character (Proposition 19.6); given a character ψ we can form the symmetric and antisymmetric parts of its square, ψ_S and ψ_A (see Proposition 19.14); and if H is a subgroup of G then we can restrict characters of G to H, and induce characters of H to G. These and other properties of characters are summarized in the following list.

Properties of characters

Assume that $\chi_1, \ldots, \chi_k$ are the irreducible characters of G.

(1) (Example 13.8(3)) There is a (trivial) character χ of G which is given by

$$\chi(g) = 1 \quad \text{for all } g \in G.$$

(2) (Theorem 17.11) The group G has precisely $|G/G'|$ linear characters. These are the characters χ given by

$$\chi(g) = \psi(gG') \quad (g \in G)$$

as ψ varies over the irreducible (linear) characters of G/G'.

(3) (Theorem 17.3) As a generalization of (2), if $N \lhd G$ and ψ is an irreducible character of G/N, then we get an irreducible character χ of G which is given by

$$\chi(g) = \psi(gN) \quad (g \in G)$$

(χ is the *lift* of ψ). This method gives precisely those irreducible characters of G which have N contained in their kernel.

(4) (Theorem 19.18) If $G = G_1 \times G_2$ then all the irreducible characters χ of G are given by

$$\chi(g_1, g_2) = \phi_1(g_1)\phi_2(g_2) \quad (g_1 \in G_1, g_2 \in G_2),$$

as ϕ_i varies over the irreducible characters of G_i ($i = 1, 2$).

(5) (Proposition 13.24) If G is a subgroup of S_n, then the function $\nu: G \to \mathbb{C}$ defined by

$$\nu(g) = |\text{fix}(g)| - 1 \quad (g \in G)$$

is a character of G.

(6) (Theorems 11.12 and 22.11) The entries $\chi_i(1)$ $(1 \leqslant i \leqslant k)$ in the first column of the character table of G are positive integers, and satisfy

$$\sum_{i=1}^{k} \chi_i(1)^2 = |G|.$$

Moreover, each integer $\chi_i(1)$ divides $|G|$.

(7) (Row orthogonality relations, Theorem 16.4(1)) For all i, j, we have $\langle \chi_i, \chi_j \rangle = \delta_{ij}$.

(8) (Column orthogonality relations, Theorem 16.4(2)) For all g, $h \in G$, we have

$$\sum_{i=1}^{k} \chi_i(g)\overline{\chi_i(h)} = \begin{cases} |C_G(g)|, & \text{if } g \text{ and } h \text{ are conjugate,} \\ 0, & \text{otherwise.} \end{cases}$$

(9) (Exercise 13.5) If χ is an irreducible character of G and $z \in Z(G)$, then there exists a root of unity ε such that for all $g \in G$,

$$\chi(zg) = \varepsilon\chi(g).$$

(10) (Proposition 13.9(2)) If g is an element of order n in G, and χ is a character of G, then $\chi(g)$ is a sum of nth roots of unity. Moreover, $|\chi(g)| \leqslant \chi(1)$.

(11) (Proposition 13.9(3, 4)) If $g \in G$ and χ is a character of G, then

$$\chi(g^{-1}) = \overline{\chi(g)}.$$

In particular, if g is conjugate to g^{-1} then $\chi(g)$ is real for all characters χ of G.

(12) (Corollary 15.6) If $g \in G$ and g is not conjugate to g^{-1}, then $\chi(g)$ is non-real for some character χ of G.

(13) (Theorem 22.16) Let $g \in G$. If g is conjugate to g^i for all positive integers i which are coprime to the order of g, then $\chi(g)$ is an integer for all characters χ of G.

(14) (Corollary 22.26) Suppose that p is a prime number and that y is the p'-part of the element g of G. If χ is a character of G such that $\chi(g)$ and $\chi(y)$ are both integers, then

$$\chi(g) \equiv \chi(y) \bmod p.$$

In particular, if the order of g is a power of p, then

$$\chi(g) \equiv \chi(1) \bmod p.$$

(15) (Proposition 13.15) If χ is an irreducible character of G, then so is $\bar{\chi}$, where

$$\bar{\chi}(g) = \overline{\chi(g)} \quad (g \in G).$$

(16) (Theorem 23.1) The number of real irreducible characters of G is equal to the number of real conjugacy classes of G.

(17) (Proposition 17.14) If χ is an irreducible character of G and λ is a linear character of G, then $\chi\lambda$ is an irreducible character of G, where

$$\chi\lambda(g) = \chi(g)\lambda(g) \quad (g \in G).$$

(18) (Proposition 19.6) If χ and ψ are characters of G, then so is the product $\chi\psi$, where

$$\chi\psi(g) = \chi(g)\psi(g) \quad (g \in G).$$

(19) (Proposition 19.14) If χ is a character of G, then so are χ_S and χ_A, where for all $g \in G$,

$$\chi_S(g) = \tfrac{1}{2}(\chi^2(g) + \chi(g^2)),$$

$$\chi_A(g) = \tfrac{1}{2}(\chi^2(g) - \chi(g^2)).$$

(20) (Definition 21.13, Proposition 21.23) If H is a subgroup of G and ψ is a character of H, then $\psi \uparrow G$ is a character of G, with values given by Proposition 21.23.

(21) (Chapter 20) If H is a subgroup of G and ψ is a character of G, then $\psi \downarrow H$ is a character of H, where

$$(\psi \downarrow H)(h) = \psi(h) \quad (h \in H).$$

We have seen that the character table of a group G gives group-theoretic information about G. For example, the first column determines $|G|$ and $|G/G'|$ (by (6) and (2)). We can see from the character table whether or not G is simple (Proposition 17.6); indeed, we can find all the normal subgroups of G (Proposition 17.5). Two important normal subgroups are G' and $Z(G)$; these can be determined in the following ways. The derived subgroup G' consists of those elements g in G which satisfy $\chi(g) = 1$ for all linear characters χ of G. The centre $Z(G)$ can be found by noting which elements g of G satisfy

$\sum \chi(g)\overline{\chi(g)} = |G|$, the sum being over all irreducible characters χ of G. In Chapter 30 we shall see some more impressive results about subgroups of G, which can be deduced from the character table.

As a final remark, it is of course true that isomorphic groups have the same character table; however, the converse is false: in Exercise 17.1 we gave examples of non-isomorphic groups, D_8 and Q_8, with the same character table.

25

Characters of groups of order pq

By the end of the next chapter, we shall have determined the character tables of all groups of order less than 32. A number of these groups are so-called Frobenius groups, and in this chapter we shall describe a class of Frobenius groups and find the character tables of the groups in this class. In particular, this will give the character tables of all groups whose order is the product of two prime numbers.

Throughout the chapter, p will denote a prime number.

Primitive roots modulo p

Recall that the set

$$\mathbb{Z}_p = \{0, 1, \ldots, p-1\},$$

with addition and multiplication modulo p, is a field; that is, $\mathbb{Z}_p$ is an abelian group under addition, and $\mathbb{Z}_p^* = \mathbb{Z}_p - \{0\}$ is an abelian group under multiplication. Clearly $\mathbb{Z}_p$ is a cyclic group under addition, generated by 1. It is also true, but not at all obvious, that $\mathbb{Z}_p^*$ is cyclic:

25.1 Theorem
The multiplicative group $\mathbb{Z}_p^*$ is cyclic; that is, there exists an integer n such that

$$n^{p-1} \equiv 1 \bmod p, \text{ and}$$

$$n^r \not\equiv 1 \bmod p \text{ for } 0 < r < p-1.$$

An integer n of order $p-1$ in $\mathbb{Z}_p^*$ is called a *primitive root modulo p*. We shall not provide a proof of Theorem 25.1, but for a good account, we refer you to Theorem 45.3 of the book by J. B. Fraleigh listed in the Bibliography.

25.2 Example
The number 2 is a primitive root modulo 3, 5, 11 and 13, but not modulo 7; and 3 is a primitive root modulo 7.

As an immediate consequence of Theorem 25.1 we have

25.3 Proposition
If $q|p-1$ then there is an integer u such that u has order q modulo p – that is, such that

$$u^q \equiv 1 \bmod p, \text{ and}$$

$$u^r \not\equiv 1 \bmod p \text{ for } 0 < r < q.$$

Frobenius groups of order pq, where $q|p-1$

25.4 Example
Define

$$G = \left\{ \begin{pmatrix} 1 & y \\ 0 & x \end{pmatrix} : x \in \mathbb{Z}_p^*, y \in \mathbb{Z}_p \right\}.$$

Under matrix multiplication, G is a group of order $p(p-1)$ (see Exercise 25.1).

Now let $q|p-1$, and let u be an element of order q in the multiplicative group $\mathbb{Z}_p^*$. Define

$$A = \begin{pmatrix} 1 & 1 \\ 0 & 1 \end{pmatrix}, B = \begin{pmatrix} 1 & 0 \\ 0 & u \end{pmatrix},$$

and let $F = \langle A, B \rangle$, the subgroup of G generated by A and B. Then

$$B^{-1}AB = \begin{pmatrix} 1 & u \\ 0 & 1 \end{pmatrix} = A^u,$$

and so we have the relations

(25.5) $$A^p = B^q = I, \ B^{-1}AB = A^u.$$

Using these relations, we see that every element of F is of the form A^iB^j with $0 \leq i \leq p-1$, $0 \leq j \leq q-1$. These pq elements are distinct, so $|F| = pq$. Moreover the relations (25.5) determine all products in F, so we have the presentation

$$F = \langle A, B: A^p = B^q = I, B^{-1}AB = A^u \rangle.$$

25.6 Definition
If p is a prime and $q|p - 1$, then we write $F_{p,q}$ for the group of order pq with presentation

$$F_{p,q} = \langle a, b: a^p = b^q = 1, b^{-1}ab = a^u \rangle,$$

where u is an element of order q in $\mathbb{Z}_p^*$.

It is not hard to show that, up to isomorphism, $F_{p,q}$ does not depend on which integer u of order q we choose (see Exercise 25.3).

The groups $F_{p,q}$ belong to a wider class of groups known as *Frobenius groups*. We shall not give the general definition of these here, as we shall only be dealing with $F_{p,q}$; the interested reader can find more information in the book by D. S. Passman listed in the Bibliography.

The next result classifies all groups whose order is the product of two distinct prime numbers.

25.7 Proposition
Suppose that G is a group of order pq, where p and q are prime numbers with $p > q$. Then either G is abelian, or q divides $p - 1$ and $G \cong F_{p,q}$.

Proof Assume that G is non-abelian. It follows from Exercise 22.3 that q divides $p - 1$ and G has a normal subgroup H of order p. (Alternatively, these facts follow readily from Sylow's Theorems (see Chapter 18 of the book by J. B. Fraleigh listed in the Bibliography).)

Both H and G/H are cyclic, since they have prime order. Suppose that $H = \langle a \rangle$ and $G/H = \langle Hb \rangle$; then G is generated by a and b. Since $b^q \in H$ but b does not have order pq (as G is non-abelian), it follows that b has order q.

Now $H \lhd G$, so $b^{-1}ab = a^u$ for some integer u. Further,

$$a = b^{-q}ab^q = a^{u^q}$$

and so $u^q \equiv 1 \bmod p$. Thus the order of u in the group $\mathbb{Z}_p^*$ divides q.

If the order of u were 1 then we would have $b^{-1}ab = a$, and G would be abelian. Therefore the order of u is q. We have now established that

$$a^p = b^q = 1, \ b^{-1}ab = a^u, \ \text{order of } u \text{ in } \mathbb{Z}_p^* \text{ is } q.$$

Hence $G \cong F_{p,q}$. ∎

25.8 Example

By Proposition 25.7, every group of order 15 is abelian (indeed, isomorphic to $C_3 \times C_5$); and the groups of order 21 are $C_3 \times C_7$ and $F_{7,3}$.

The character table of $F_{p,q}$

We have, in fact, already found the character tables of certain of the groups $F_{p,q}$: the dihedral group of order $2p$ is the case where $q = 2$, and in Example 21.25 we dealt with $F_{7,3}$. We now construct the character table of $F_{p,q}$ in general. Thus let

$$G = F_{p,q} = \langle a, b: a^p = b^q = 1, \ b^{-1}ab = a^u \rangle$$

where p is prime, $q|p - 1$ (q not necessarily prime), and u has order q modulo p.

Let S be the subgroup of $\mathbb{Z}_p^*$ consisting of the powers of u. Thus $|S| = q$. Write $r = (p - 1)/q$, and choose coset representatives $v_1, \ldots, v_r$ for S in $\mathbb{Z}_p^*$.

25.9 Proposition
The conjugacy classes of $G = F_{p,q}$ are

$$\{1\},$$

$$(a^{v_i})^G = \{a^{v_i s}: s \in S\} \quad (1 \leqslant i \leqslant r),$$

$$(b^n)^G = \{a^m b^n: 0 \leqslant m \leqslant p - 1\} \quad (1 \leqslant n \leqslant q - 1).$$

Proof The equation

$$b^{-j}a^v b^j = a^{vu^j}$$

shows that a^v is conjugate to a^{vs} for all $s \in S$. Therefore the conjugacy class of a^{v_i} has size at least q; also the size of this conjugacy class is equal to $|G: C_G(a^{v_i})|$, and since $\langle a \rangle \leqslant C_G(a^{v_i})$, this size is at most q. Hence $(a^{v_i})^G$ has size q, and has the form stated in the proposition.

Since $C_G(b^n)$ contains $\langle b \rangle$, and $\langle b \rangle$ has index p in G, it follows that for $n \not\equiv 0 \bmod q$, we have $|C_G(b^n)| = q$, and so the conjugacy class of b^n has size p. On the other hand, as $G/\langle a \rangle$ is abelian, every conjugate of b^n has the form $a^m b^n$ for some m. Hence

$$(b^n)^G = \{a^m b^n : 0 \leqslant m \leqslant p - 1\}$$

and the proof is complete. ∎

By Proposition 25.9, G has $q + r$ conjugacy classes, so we seek $q + r$ irreducible characters.

First, observe that the derived subgroup $G' = \langle a \rangle$, so G/G' has order q and therefore by Theorem 17.11, G has precisely q linear characters. These are given by χ_n ($0 \leqslant n \leqslant q - 1$), where

$$\chi_n(a^x b^y) = e^{2\pi i n y / q} \quad (0 \leqslant x \leqslant p - 1, 0 \leqslant y \leqslant q - 1).$$

We shall show that G has r irreducible characters of degree q.

Let $\varepsilon = e^{2\pi i / p}$. For $v \in \mathbb{Z}_p^*$, denote by ψ_v the character of $\langle a \rangle$ which is given by

$$\psi_v(a^x) = \varepsilon^{vx} \quad (0 \leqslant x \leqslant p - 1).$$

We calculate the values of the induced character $\psi_v \uparrow G$, using Proposition 21.23. We obtain

$$(\psi_v \uparrow G)(a^x b^y) = 0 \quad \text{if } 1 \leqslant y \leqslant q - 1, \quad \text{and}$$

$$(\psi_v \uparrow G)(a^x) = \sum_{s \in S} \varepsilon^{vsx} \quad (0 \leqslant x \leqslant p - 1).$$

Note that $\psi_v \uparrow G$ has degree q, and

$$\psi_v \uparrow G = \psi_{vs} \uparrow G \quad \text{if } s \in S.$$

For each coset representative v_j ($1 \leqslant j \leqslant r$) of S in $\mathbb{Z}_p^*$, let

$$\phi_j = \psi_{v_j} \uparrow G.$$

We now prove that each ϕ_j is irreducible. By the Frobenius Reciprocity Theorem 21.16, for all $s \in S$,

$$\langle \phi_j \downarrow \langle a \rangle, \psi_{v_j s} \rangle_{\langle a \rangle} = \langle \phi_j, \psi_{v_j s} \uparrow G \rangle_G = \langle \phi_j, \phi_j \rangle_G.$$

Hence

$$\phi_j \downarrow \langle a \rangle = \langle \phi_j, \phi_j \rangle_G \sum_{s \in S} \psi_{v_j s} + \chi,$$

where χ is either 0 or a character of $\langle a \rangle$. Taking degrees, it follows that

$$\phi_j(1) \geqslant |S| \langle \phi_j, \phi_j \rangle_G.$$

Since $\phi_j(1) = q = |S|$, we deduce that $\langle \phi_j, \phi_j \rangle_G = 1$. This proves that ϕ_j is irreducible, and also that

$$\phi_j \downarrow \langle a \rangle = \langle \phi_j, \phi_j \rangle_G \sum_{s \in S} \psi_{v_j s}.$$

By Theorem 14.23, the characters ψ_v ($v \in \mathbb{Z}_p^*$) are linearly independent, and hence $\phi_1 \downarrow \langle a \rangle, \ldots, \phi_r \downarrow \langle a \rangle$ are distinct. Consequently the irreducible characters $\phi_1, \ldots, \phi_r$ are distinct.

We have now found $q + r$ distinct irreducible characters χ_n, ϕ_j of G ($0 \leqslant n \leqslant q - 1, 1 \leqslant j \leqslant r$), so we have the complete character table of G. We summarize in the following theorem.

25.10 Theorem
Let p be a prime number, $q|p - 1$ and $r = (p - 1)/q$. Then the group

$$F_{p,q} = \langle a, b: a^p = b^q = 1, b^{-1}ab = a^u \rangle$$

$$= \{a^x b^y: 0 \leqslant x \leqslant p - 1, 0 \leqslant y \leqslant q - 1\}$$

has $q + r$ irreducible characters. Of these, q have degree 1 and are given by

$$\chi_n(a^x b^y) = e^{2\pi i ny/q} \quad (0 \leqslant n \leqslant q - 1)$$

and r have degree q and are given by

$$\phi_j(a^x b^y) = 0 \quad \text{if } 1 \leqslant y \leqslant q - 1,$$

$$\phi_j(a^x) = \sum_{s \in S} e^{2\pi i v_j sx/p},$$

for $1 \leqslant j \leqslant r$, where $v_1 S, \ldots, v_r S$ are the cosets in $\mathbb{Z}_p^*$ of the subgroup S generated by u.

We conclude by illustrating Theorem 25.10 in some examples.

25.11 Example
Let

$$G = F_{p,p-1} = \langle a, b: a^p = b^{p-1} = 1, b^{-1}ab = a^u \rangle$$

where u is a primitive root modulo p. Then G has $p - 1$ linear characters, and one irreducible character ϕ of degree $p - 1$, with values given by

$$\phi(a^x b^y) = 0 \qquad \text{if } 1 \leqslant y \leqslant p - 2,$$
$$\phi(a^x) = -1 \qquad \text{if } 1 \leqslant x \leqslant p - 1.$$

25.12 Example

Let $a, b \in S_5$ be the permutations

$$a = (1\,2\,3\,4\,5), \ b = (2\,3\,5\,4).$$

Check that

$$a^5 = b^4 = 1, \ b^{-1}ab = a^2.$$

Hence if $G = \langle a, b \rangle$, then $G \cong F_{5,4}$, and so by the previous example the character table of G is as shown.

Character table of $F_{5,4}$

g_i $|C_G(g_i)|$	1 20	a 5	b 4	b^2 4	b^3 4
χ_0	1	1	1	1	1
χ_1	1	1	i	-1	$-i$
χ_2	1	1	-1	1	-1
χ_3	1	1	$-i$	-1	i
ϕ	4	-1	0	0	0

25.13 Example

We consider the case $p = 13$, $q = 4$. Here

$$F_{13,4} = \langle a, b: a^{13} = b^4 = 1, \ b^{-1}ab = a^5 \rangle.$$

Write $\varepsilon = e^{2\pi i/13}$, and let

$$\alpha = \varepsilon + \varepsilon^5 + \varepsilon^8 + \varepsilon^{12},$$
$$\beta = \varepsilon^2 + \varepsilon^3 + \varepsilon^{10} + \varepsilon^{11},$$
$$\gamma = \varepsilon^4 + \varepsilon^6 + \varepsilon^7 + \varepsilon^9.$$

By Theorem 25.10, the character table of $F_{13,4}$ is as shown opposite.

In Example 21.25 we found the character table of $F_{7,3}$. You may like to check that this agrees with the description of the character table provided by Theorem 25.10.

Character table of $F_{13,4}$

g_i	1	a	a^2	a^4	b	b^2	b^3
$|C_G(g_i)|$	52	13	13	13	4	4	4
χ_0	1	1	1	1	1	1	1
χ_1	1	1	1	1	i	-1	$-i$
χ_2	1	1	1	1	-1	1	-1
χ_3	1	1	1	1	$-i$	-1	i
ϕ_1	4	α	β	γ	0	0	0
ϕ_2	4	β	γ	α	0	0	0
ϕ_3	4	γ	α	β	0	0	0

Summary of Chapter 25

1. Suppose that p is prime and q divides $p - 1$. Let u be an element of order q in $\mathbb{Z}^*_p$. Then

$$F_{p,q} = \langle a, b: a^p = b^q = 1, b^{-1}ab = a^u \rangle.$$

The irreducible characters of $F_{p,q}$ are described in Theorem 25.10.

2. Let p and q be prime numbers with $p > q$. If G has order pq, then either G is abelian or $G \cong F_{p,q}$.

Exercises for Chapter 25

1. Let p be a prime number. Prove that

$$\left\{ \begin{pmatrix} 1 & y \\ 0 & x \end{pmatrix} : x \in \mathbb{Z}^*_p, \, y \in \mathbb{Z}_p \right\},$$

under matrix multiplication, is a group of order $p(p - 1)$.

2. Write down the character table of the non-abelian group $F_{11,5}$ of order 55.

3. Let p and q be positive integers, with p prime and $q|p - 1$. Let u and v be two integers which are of order q modulo p, and define

$$G_1 = \langle a, b: a^p = b^q = 1, b^{-1}ab = a^u \rangle,$$

$$G_2 = \langle a, b: a^p = b^q = 1, b^{-1}ab = a^v \rangle.$$

Prove that $G_1 \cong G_2$.

(This justifies the comment which follows the definition of $F_{p,q}$ in 25.6.)

4. Suppose that p is a prime number, with $p \neq 2$. Let $q = (p - 1)/2$ and let

$$G = F_{p,q} = \langle a, b: a^p = b^q = 1, b^{-1}ab = a^u \rangle,$$

where u is an element of order q modulo p.

(a) Show that there exists an integer m such that $u^m \equiv -1 \bmod p$ if and only if $p \equiv 1 \bmod 4$.

(b) Deduce that a is conjugate to a^{-1} if and only if $p \equiv 1 \bmod 4$.

(c) Using the orthogonality relations, show that the two irreducible characters ϕ_1, ϕ_2 of G of degree q have values

$$\tfrac{1}{2}(-1 \pm \sqrt{(\delta p)})$$

on the element a, where $\delta = 1$ if $p \equiv 1 \bmod 4$, and $\delta = -1$ if $p \equiv -1 \bmod 4$.

(d) Deduce that if $\varepsilon = e^{2\pi i/p}$ then

$$\sum_{s \in Q} \varepsilon^s = (-1 \pm \sqrt{(\delta p)}),$$

where Q is the set of quadratic residues modulo p (that is, $Q = \{1^2, 2^2, \ldots, ((p-1)/2)^2\}$).

5. Let E be the group of order 18 which is given by

$$E = \langle a, b, c: a^3 = b^3 = c^2 = 1, ab = ba, c^{-1}ac = a^{-1},$$

$$c^{-1}bc = b^{-1} \rangle,$$

as in Exercise 5.4. Note that $\langle a, b \rangle$ is a normal subgroup of E which is isomorphic to $C_3 \times C_3$. By inducing linear characters of this subgroup, obtain the character table of E.

6. Show that the group E of Exercise 5 has the properties that $Z(E)$ is cyclic, but E has no faithful irreducible representation. (Thus, E provides a counterexample to the converse of Proposition 9.16.)

7. (a) Find a group whose irreducible character degrees are

$$1, 1, 1, 3, 3, 3, 3.$$

(b) Find a group whose irreducible character degrees are

$$1, 1, 1, 1, 1, 1, 3, 3, 3, 3, 3, 3, 3, 3.$$

(c) Find a group whose irreducible character degrees are

$$1, 1, 1, 1, 1, 1, 2, 2, 2, 3, 3, 3, 3, 3, 3, 3, 3, 3, 6, 6, 6, 6.$$

8. Let G be the group of order 54 which is given by

$$G = \langle a, b: a^9 = b^6 = 1, b^{-1}ab = a^2 \rangle.$$

Find the character table of G.

26

Characters of some p-groups

Throughout this chapter, p will be a prime number. We shall show how to obtain the character tables of all groups of order p^n for $n \leq 4$. The method consists of examining the characters of those p-groups which contain an abelian subgroup of index p, and before explaining the method, we show that all groups of order p^n with $1 \leq n \leq 4$ do, indeed, have an abelian subgroup of index p. We later give explicitly the irreducible characters of all groups of order p^3 and of all groups of order 16. At the end of the chapter we point out, with references, that we have found the character tables of all groups of order less than 32.

Elementary properties of p-groups

A *p-group* is a group whose order is a power of the prime number p. In the first lemma we collect several well known properties of p-groups. Recall that $Z(G)$ denotes the centre of G (see Definition 9.15).

26.1 Lemma
Let G be a group of order p^n with $n \geq 1$.

(1) *If $\{1\} \neq H \triangleleft G$ then $H \cap Z(G) \neq \{1\}$. In particular, $Z(G) \neq \{1\}$.*
(2) *If $K \leq Z(G)$ and G/K is cyclic, then G is abelian.*
(3) *If $n \leq 2$ then G is abelian.*

Proof (1) Since $H \triangleleft G$, H is a union of conjugacy classes of G, all of which have size a power of p; and $H \cap Z(G)$ consists of those conjugacy classes in H which have size 1. Therefore

298

$$|H| = |H \cap Z(G)| + \text{(a multiple of } p\text{)}.$$

As $|H|$ is a multiple of p and $|H \cap Z(G)| \neq 0$, we deduce that $H \cap Z(G) \neq \{1\}$.

(2) Suppose that G/K is cyclic, generated by gK. Let $x_1, x_2 \in G$. Then

$$x_1 = g^i k_1, \; x_2 = g^j k_2$$

for some integers i, j and some $k_1, k_2 \in K$. Since $k_1, k_2 \in Z(G)$, it follows that $x_1 x_2 = x_2 x_1$. Therefore G is abelian.

(3) By (1), $|G/Z(G)| \leq p^{n-1}$. Hence if $n \leq 2$ then $G/Z(G)$ is cyclic and so G is abelian by (2). ∎

26.2 Lemma
Let G be a group of order p^n with $1 \leq n \leq 4$. Then G contains an abelian subgroup of index p.

Proof The result is immediate if $n = 1$, so suppose that $2 \leq n \leq 4$.

Assume that $Z(G)$ contains a subgroup K of order p^{n-2}. Then we can find a subgroup H of G such that $K \leq H$ and $|H| = p^{n-1}$. As $K \leq Z(H)$ and, by Lemma 26.1(2), $H/Z(H)$ is not of order p, we deduce that $Z(H) = H$. Therefore H is an abelian subgroup of index p in G.

Now assume that $Z(G)$ has no subgroup of order p^{n-2}. Since $Z(G) \neq \{1\}$ by Lemma 26.1(1), the only possibility is that $|G| = p^4$ and $|Z(G)| = p$. Then by Exercise 12.7, G has an element x whose conjugacy class x^G is of size p. Let $H = C_G(x)$. Then by Theorem 12.8, $|H| = |G|/|x^G| = p^3$. Moreover, $Z(G)$ and $\langle x \rangle$ are distinct non-identity subgroups of $Z(H)$, and so $Z(H) \geq p^2$. Hence again $Z(H) = H$ by Lemma 26.1(2), and H is an abelian subgroup of index p. ∎

For our final result on the structure of p-groups, recall that the derived subgroup of G is denoted by G' (see Definition 17.7).

26.3 Lemma
Let G be a non-abelian p-group which contains an abelian subgroup H of index p. Then there exists a normal subgroup K of G such that

$$K \leq H \cap G' \cap Z(G) \text{ and } |K| = p.$$

Proof Since G is non-abelian, we have $\{1\} \neq G' \lhd G$, and hence $G' \cap Z(G) \neq \{1\}$ by Lemma 26.1(1). Let K be a subgroup of order p in $G' \cap Z(G)$. Now $K \leqslant Z(G)$ implies that $K \lhd G$ and that KH is an abelian subgroup of G (where $KH = \{kh: k \in K, h \in H\}$). Since G is non-abelian and $|G:H| = p$, we have $KH = H$, and therefore $K \leqslant H$. ∎

Characters of p-groups with an abelian subgroup of index p

In view of Lemma 26.2, the next theorem provides us with all the irreducible characters of non-abelian groups of order p^3 or p^4.

26.4 Theorem
Assume that G is a non-abelian p-group which contains an abelian subgroup H of index p. Let K be a normal subgroup of G as in Lemma 26.3. Then every irreducible character of G is given by either

(1) *the lift of an irreducible character of G/K, or*
(2) $\psi \uparrow G$, *for some linear character ψ of H which satisfies $K \not\leqslant \mathrm{Ker}\,\psi$.*

Proof Let $|G| = p^n$. By Theorem 17.3, the irreducible characters of G/K lift to give precisely those irreducible characters of G which have K in their kernel. The sum of the squares of the degrees of the irreducible characters obtained in this way is $|G/K| = p^{n-1}$, by Theorem 11.12.

We shall construct $p^{n-2} - p^{n-3}$ further irreducible characters of G, each of degree p. Since

$$(*) \qquad p^{n-1} + (p^{n-2} - p^{n-3})p^2 = p^n = |G|,$$

we shall then have obtained all the irreducible characters of G, again by Theorem 11.12.

First note that if χ is a character of G of degree p, then either χ is irreducible or χ is a sum of linear characters (since by Theorem 22.11, the degree of every irreducible character of G is a power of p). In the latter case, we have $G' \leqslant \mathrm{Ker}\,\chi$, as G' is in the kernel of every linear character, and hence $K \leqslant G' \leqslant \mathrm{Ker}\,\chi$. This establishes

(26.5)   if $\chi(1) = p$ and $K \not\leqslant \mathrm{Ker}\,\chi$, then χ is irreducible.

We know by Theorem 9.8 that all the p^{n-1} irreducible characters of the abelian group H are linear. Let Φ denote the set of linear characters of H which do not have K in their kernel. Since the linear characters of H which *do* have K in their kernel are precisely the lifts of linear characters of H/K, we have

$$|\Phi| = p^{n-1} - p^{n-2}.$$

Let $\psi \in \Phi$. By Proposition 21.23, since $K \leqslant Z(G)$,

$$(\psi \uparrow G)(k) = p\psi(k) \quad \text{for all } k \in K.$$

Thus $\psi \uparrow G$ has degree p and does not have K in its kernel. Therefore by (26.5), $\psi \uparrow G$ is an irreducible character of G.

Suppose now that ψ_1 is a linear character of H such that $\psi \uparrow G = \psi_1 \uparrow G$. Then by the Frobenius Reciprocity Theorem 21.16,

$$1 = \langle \psi \uparrow G, \psi_1 \uparrow G \rangle_G = \langle (\psi \uparrow G) \downarrow H, \psi_1 \rangle_H.$$

Since $(\psi \uparrow G) \downarrow H$ has degree p, this implies that there are at most p elements ψ_1 of Φ such that $\psi_1 \uparrow G = \psi \uparrow G$. It follows that

$$\{\psi \uparrow G: \ \psi \in \Phi\}$$

gives at least $|\Phi|/p = (p^{n-1} - p^{n-2})/p$ distinct irreducible characters of G of degree p which do not have K in their kernel. As we saw in (*), G has at most $p^{n-2} - p^{n-3}$ such characters. Therefore $\{\psi \uparrow G: \psi \in \Phi\}$ consists precisely of the $p^{n-2} - p^{n-3}$ irreducible characters we seek, and the proof is complete. ∎

We now use Theorem 26.4 to give an explicit construction of the irreducible characters of the non-abelian groups of order p^3. We shall then illustrate Theorem 26.4 further by constructing the character tables of all the non-abelian groups of order 16.

Groups of order p^3

By Theorem 9.6, the abelian groups of order p^3 are

$$C_{p^3}, \ C_{p^2} \times C_p \text{ and } C_p \times C_p \times C_p.$$

The character tables of these groups are given by Theorem 9.8.

Now let G be a non-abelian group of order p^3. Write $Z = Z(G)$. By Lemma 26.1, $Z \neq \{1\}$ and G/Z is non-cyclic. Hence $G/Z \cong C_p \times C_p$ and $Z = \langle z \rangle \cong C_p$. Choose aZ, bZ such that $G/Z = \langle aZ, bZ \rangle$. Then

$$G/Z = \{a^r b^s Z: \ 0 \leqslant r \leqslant p-1, 0 \leqslant s \leqslant p-1\}$$

and in particular, every element of G is of the form

$$a^r b^s z^t$$

for some r, s, t with $0 \leqslant r, s, t \leqslant p-1$.

26.6 Theorem

Let $G = \{a^r b^s z^t: 0 \leqslant r, s, t \leqslant p-1\}$ be a non-abelian group of order p^3, as above. Write $\varepsilon = e^{2\pi i/p}$. Then the irreducible characters of G are

$$\chi_{u,v} \quad (0 \leqslant u \leqslant p-1, 0 \leqslant v \leqslant p-1), \text{ and}$$

$$\phi_u \quad (1 \leqslant u \leqslant p-1),$$

where for all r, s, t,

$$\chi_{u,v}(a^r b^s z^t) = \varepsilon^{ru+sv},$$

$$\phi_u(a^r b^s z^t) = \begin{cases} p\varepsilon^{ut}, & \text{if } r = s = 0, \\ 0, & \text{otherwise.} \end{cases}$$

Proof By Theorem 9.8, the irreducible characters of G/Z are $\psi_{u,v}$ $(0 \leqslant u, v \leqslant p-1)$, where

$$\psi_{u,v}(a^r b^s Z) = \varepsilon^{ru+sv}.$$

The lift to G of $\psi_{u,v}$ is the linear character $\chi_{u,v}$ which appears in the statement of the theorem.

Let $H = \langle a, z \rangle$, so that H is an abelian subgroup of order p^2. For $1 \leqslant u \leqslant p-1$, choose a character ψ_u of H which satisfies

$$\psi_u(z^t) = \varepsilon^{ut} \quad (0 \leqslant t \leqslant p-1).$$

We shall calculate $\psi_u \uparrow G$.

Let r be an integer with $1 \leqslant r \leqslant p-1$. If a^r is conjugate to an element g of G, then $a^r Z$ is conjugate to gZ in the abelian group G/Z, so $a^r Z = gZ$, and therefore $g = a^r z^t$ for some t. Since $a^r \notin Z$, the conjugacy class $(a^r)^G$ does not have size 1, and hence

$$(a^r)^G = \{a^r z^t: \ 0 \leqslant t \leqslant p-1\}.$$

Then by Proposition 21.23,

$$(\psi_u \uparrow G)(a^r z^t) = \psi_u(a^r) + \psi_u(a^r z) + \ldots + \psi_u(a^r z^{p-1})$$

$$= \psi_u(a^r) \sum_{s=0}^{p-1} \psi_u(z^s)$$

$$= \psi_u(a^r) \sum_{s=0}^{p-1} \varepsilon^{us}$$

$$= 0.$$

Also,

$$(\psi_u \uparrow G)(z^t) = p\psi_u(z^t) = p\varepsilon^{ut}, \text{ and}$$

$$(\psi_u \uparrow G)(g) = 0 \text{ if } g \notin H.$$

We have now established that if $\phi_u = \psi_u \uparrow G$, then ϕ_u takes the values stated in the theorem.

We find that

$$\langle \phi_u, \phi_u \rangle_G = \frac{1}{p^3} \sum_{g \in G} \phi_u(g) \overline{\phi_u(g)}$$

$$= \frac{1}{p^3} \sum_{g \in Z} \phi_u(g) \overline{\phi_u(g)}$$

$$= \frac{1}{p^3} \sum_{g \in Z} p^2$$

$$= 1.$$

Therefore ϕ_u is irreducible.

Clearly the irreducible characters $\chi_{u,v}$ $(0 \leqslant u, v \leqslant p - 1)$ and ϕ_u $(1 \leqslant u \leqslant p - 1)$ are all distinct, and the sum of the squares of their degrees is

$$p^2 \cdot 1^2 + (p - 1) \cdot p^2 = |G|.$$

Hence we have found all the irreducible characters of G.   ∎

Notice that the calculation in the proof of Theorem 26.6 is a special case of the proof of Theorem 26.4 (with $K = Z(G)$).

In fact, up to isomorphism, there are precisely two non-abelian groups of order p^3. If $p = 2$, they are D_8 and Q_8. And if p is odd, they are

(26.7) $H_1 = \langle a, b \colon a^{p^2} = b^p = 1, b^{-1}ab = a^{p+1} \rangle$, and

$H_2 = \langle a, b, z \colon a^p = b^p = z^p = 1, az = za, bz = zb,$

$b^{-1}ab = az \rangle.$

We have $Z(H_1) = \langle a^p \rangle$, $Z(H_2) = \langle z \rangle$. The elements a, b in H_1 and H_2 will serve for the elements a, b chosen in the statement of Theorem 26.6.

26.8 The groups of order 16

It is known that, up to isomorphism, there are precisely fourteen groups of order 16 (see p. 134 of the book by Coxeter and Moser listed in the Bibliography). We shall describe all these groups and their character tables.

By Theorem 9.6, the abelian groups of order 16 are

$$C_{16}, C_8 \times C_2, C_4 \times C_4, C_4 \times C_2 \times C_2 \text{ and } C_2 \times C_2 \times C_2 \times C_2,$$

and their character tables are given by Theorem 9.8.

For each of the nine non-abelian groups G of order 16 it is the case that $|G' \cap Z(G)| = 2$ (see Exercise 26.7), so the subgroup K described in Lemma 26.3 is given by

$$K = G' \cap Z(G).$$

Now G/K is a group of order 8. It is not C_8 by Lemma 26.1(2), and it is not Q_8 by Exercise 26.8. Hence

$$G/K \cong D_8, C_4 \times C_2 \text{ or } C_2 \times C_2 \times C_2.$$

We shall divide our descriptions into three parts, according to these three possibilities for G/K. Our descriptions will be in terms of presentations; it is possible to see, using Exercise 26.5, that all the nine groups $G_1, \ldots, G_9$ given below do indeed have order 16.

(A) There are three non-abelian groups G of order 16 with $G/K \cong D_8$. These are

$$G_1 = \langle a, b \colon a^8 = b^2 = 1, b^{-1}ab = a^{-1} \rangle = D_{16},$$

$$G_2 = \langle a, b \colon a^8 = 1, b^2 = a^4, b^{-1}ab = a^{-1} \rangle,$$

$$G_3 = \langle a, b \colon a^8 = b^2 = 1, b^{-1}ab = a^3 \rangle.$$

In each case $K = \langle a^4 \rangle$. Each of the groups has seven conjugacy classes $C_1, \ldots, C_7$, and these are given in the following table.

	C_1	C_2	C_3	C_4	C_5	C_6	C_7
G_1, G_2	1	a^4	a^2, a^6	a, a^7	a^3, a^5	$a^i b\,(i\text{ even})$	$a^i b\,(i\text{ odd})$
G_3	1	a^4	a^2, a^6	a, a^3	a^5, a^7	$a^i b\,(i\text{ even})$	$a^i b\,(i\text{ odd})$

Using Theorem 26.4, we obtain the character tables of G_1, G_2 and G_3:

Class Centralizer order	C_1	C_2	C_3	C_4	C_5	C_6	C_7
	16	16	8	8	8	4	4
	1	1	1	1	1	1	1
	1	1	1	1	1	-1	-1
	1	1	1	-1	-1	1	-1
	1	1	1	-1	-1	-1	1
	2	2	-2	0	0	0	0
	2	-2	0	α	β	0	0
	2	-2	0	β	α	0	0

where

$$\alpha = \sqrt{2} = -\beta \quad \text{for } G_1, G_2,$$

$$\alpha = i\sqrt{2} = -\beta \quad \text{for } G_3.$$

The first five characters are the lifts of the irreducible characters of $G/K \cong D_8$. The last two characters can be obtained as in Theorem 26.4(2) as induced characters $\psi \uparrow G$, where ψ is a linear character of the abelian subgroup $\langle a \rangle$ of index 2; alternatively, they can be found by using the column orthogonality relations. Note that a is conjugate to a^{-1} in G_1 and G_2, but not in G_3; hence the values in the columns C_4 and C_5 are all real for G_1 and G_2, but not for G_3 (see Corollary 15.6).

(B) There are three non-abelian groups G of order 16 with $G/K \cong C_4 \times C_2$ (where, as before, $K = G' \cap Z(G)$, of order 2). These are

$$G_4 = \langle a, b, z: \ a^4 = z, \ b^2 = z^2 = 1, \ b^{-1}ab = az \rangle,$$

$$G_5 = \langle a, b, z: \ a^4 = 1, \ b^2 = z, \ z^2 = 1, \ b^{-1}ab = az \rangle,$$

$$G_6 = \langle a, b, z: \ a^4 = 1, \ b^2 = z^2 = 1, \ b^{-1}ab = az, \ az = za, \ bz = zb \rangle.$$

The given presentations are somewhat cumbersome (for example, since $a^4 = z$ in G_4, z is redundant), but they are in a form which allows us to describe the conjugacy classes $C_1, \ldots, C_{10}$ of all three groups G_4, G_5, G_6 simultaneously:

C_1	C_2	C_3	C_4	C_5	C_6	C_7	C_8	C_9	C_{10}
1	z	a^2	a^2z	a, az	a^3, a^3z	b, bz	a^2b, a^2bz	ab, abz	a^3b, a^3bz

In each case, $K = \langle z \rangle$. The character tables of G_4, G_5 and G_6 can again be found using Theorem 26.4:

Class Centralizer order	C_1 16	C_2 16	C_3 16	C_4 16	C_5 8	C_6 8	C_7 8	C_8 8	C_9 8	C_{10} 8
	1	1	1	1	1	1	1	1	1	1
	1	1	-1	-1	i	$-i$	1	-1	i	$-i$
	1	1	1	1	-1	-1	1	1	-1	-1
	1	1	-1	-1	$-i$	i	1	-1	$-i$	i
	1	1	1	1	1	1	-1	-1	-1	-1
	1	1	-1	-1	i	$-i$	-1	1	$-i$	i
	1	1	1	1	-1	-1	-1	-1	1	1
	1	1	-1	-1	$-i$	i	-1	1	i	$-i$
	2	-2	α	β	0	0	0	0	0	0
	2	-2	β	α	0	0	0	0	0	0

where

$$\alpha = 2i = -\beta \quad \text{for } G_4,$$

$$\alpha = 2 = -\beta \quad \text{for } G_5, G_6.$$

(C) Finally, there are three non-abelian groups G of order 16 with $G/K \cong C_2 \times C_2 \times C_2$ (where $K = G' \cap Z(G)$, of order 2). These are

$$G_7 = \langle a, b, z: \ a^4 = b^2 = z^2 = 1, \ b^{-1}ab = a^{-1}, \ az = za, \ bz = zb \rangle$$
$$\cong D_8 \times C_2,$$

$$G_8 = \langle a, b, z: \ a^4 = z^2 = 1, \ a^2 = b^2, \ b^{-1}ab = a^{-1}, \ az = za, \ bz = zb \rangle$$
$$\cong Q_8 \times C_2,$$

$$G_9 = \langle a, b, z: \ a^2 = b^2 = z^4 = 1, \ b^{-1}ab = az^2, \ az = za, \ bz = zb \rangle.$$

Each of these groups has ten conjugacy classes, which are given by

	C_1	C_2	C_3	C_4	C_5	C_6	C_7	C_8	C_9	C_{10}
G_7, G_8	1	a^2	z	a^2z	a, a^3	az, a^3z	b, a^2b	bz, a^2bz	ab, a^3b	abz, a^3bz
G_9	1	z^2	z	z^3	a, az^2	az, az^3	b, bz^2	bz, bz^3	ab, abz^2	abz, abz^3

We have

$$K = \begin{cases} \langle a^2 \rangle & \text{for } G_7, G_8, \\ \langle z^2 \rangle & \text{for } G_9, \end{cases}$$

and the character tables of G_7, G_8 and G_9, given by Theorem 26.4, are as follows.

Class	C_1	C_2	C_3	C_4	C_5	C_6	C_7	C_8	C_9	C_{10}
Centralizer order	16	16	16	16	8	8	8	8	8	8
	1	1	1	1	1	1	1	1	1	1
	1	1	1	1	1	1	−1	−1	−1	−1
	1	1	1	1	−1	−1	1	1	−1	−1
	1	1	−1	−1	1	−1	1	−1	1	−1
	1	1	1	1	−1	−1	−1	−1	1	1
	1	1	−1	−1	1	−1	−1	1	−1	1
	1	1	−1	−1	−1	1	1	−1	−1	1
	1	1	−1	−1	−1	1	−1	1	1	−1
	2	−2	α	β	0	0	0	0	0	0
	2	−2	β	α	0	0	0	0	0	0

where

$$\alpha = 2 = -\beta \quad \text{for } G_7, G_8,$$

$$\alpha = 2\mathrm{i} = -\beta \quad \text{for } G_9.$$

26.9 The groups of order less than 32

At this point, we have in fact found the character tables of all groups of order 31 or less. Apart from abelian groups and dihedral groups, whose character tables are given by Theorem 9.8 and Section 18.3, the groups, with references for their character tables, are as follows:

$|G|$	G	Reference for character table
8	Q_8	Exercise 17.1
12	A_4	Section 18.2
	T_{12}	Exercise 18.3
16	$G_1, \ldots, G_9$	Section 26.8
18	$D_6 \times C_3$	Theorem 19.18
	E	Exercise 25.5
20	T_{20}	Exercise 18.3
	$F_{5,4}$	Theorem 25.10
21	$F_{7,3}$	Theorem 25.10
24	$D_{12} \times C_2,\ A_4 \times C_2,\ T_{12} \times C_2$	Theorem 19.18
	$D_8 \times C_3,\ Q_8 \times C_3,\ D_6 \times C_4$	Theorem 19.18
	S_4	Section 18.1
	SL$(2, 3)$	Exercise 27.2
	T_{24}	Exercise 18.3
	U_{24}	Exercise 18.4
	V_{24}	Exercise 18.5
27	H_1, H_2	Theorem 26.6
28	T_{28}	Exercise 18.3
30	$D_6 \times C_5,\ D_{10} \times C_3$	Theorem 19.18

Summary of Chapter 26

In this chapter, we gave the irreducible characters of various non-abelian p-groups G, as follows.

1. Theorem 26.4: p-groups which contain an abelian subgroup of index p.

2. Theorem 26.6: groups of order p^3.

3. Section 26.8: groups of order 16.

Exercises for Chapter 26

1. Suppose that G is a group of order p^n (p prime, $n \geqslant 2$), with an abelian subgroup H of index p. Show that for some integer $m \geqslant 2$, G has p^m linear characters and $p^{n-2} - p^{m-2}$ irreducible characters of degree p.

2. Let H be the group of order 27 which is given by

$$H = \langle a, b, z: a^3 = b^3 = z^3 = 1, az = za, bz = zb, b^{-1}ab = az \rangle$$

(see (26.7)).

Find the conjugacy classes of H, and use Theorem 26.6 to write down the character table of H.

3. Let G be the group of order 32 which is given by

$$G = \langle a, b: a^{16} = 1, b^2 = a^8, b^{-1}ab = a^{-1} \rangle.$$

Using Theorem 26.4, or otherwise, find the character table of G.

4. Let A, B, C, D be the following 4×4 matrices:

$$A = \begin{pmatrix} -1 & 0 & 0 & 0 \\ 0 & 1 & 0 & 0 \\ 0 & 0 & 1 & 0 \\ 0 & 0 & 0 & -1 \end{pmatrix}, \quad B = \begin{pmatrix} 0 & 0 & i & 0 \\ 0 & 0 & 0 & -i \\ i & 0 & 0 & 0 \\ 0 & -i & 0 & 0 \end{pmatrix},$$

$$C = \begin{pmatrix} 0 & i & 0 & 0 \\ i & 0 & 0 & 0 \\ 0 & 0 & 0 & -i \\ 0 & 0 & -i & 0 \end{pmatrix}, \quad D = \begin{pmatrix} 0 & 0 & 0 & 1 \\ 0 & 0 & 1 & 0 \\ 0 & 1 & 0 & 0 \\ 1 & 0 & 0 & 0 \end{pmatrix},$$

and let $G = \langle A, B, C, D \rangle$. Write $Z = -I$.

(a) Prove that all pairs of generators commute modulo $\langle Z \rangle$, and deduce that $G' = \langle Z \rangle$.

(b) Show that for all g in G, $g^2 \in \langle Z \rangle$, and deduce that G is a 2-group of order at most 32.

(c) Prove that the given representation of G of degree 4 is irreducible. (Hint: use Corollary 9.3.)

(d) Show that $|G| = 32$, and find all the irreducible representations of G.

5. Let $G_1, \ldots, G_9$ be the non-abelian groups of order 16 with presentations as given in the text.

(a) Find faithful irreducible representations of degree 2 for G_1, G_2, G_3, G_4 and G_9.

(b) Why do the remaining groups G_5, G_6, G_7 and G_8 have no faithful irreducible representations?

(c) Check that the following give faithful representations of G_5 and G_6:

$$G_5: a \to \begin{pmatrix} 0 & 1 & 0 \\ 1 & 0 & 0 \\ 0 & 0 & i \end{pmatrix}, b \to \begin{pmatrix} i & 0 & 0 \\ 0 & -i & 0 \\ 0 & 0 & 1 \end{pmatrix},$$

$$z \to \begin{pmatrix} -1 & 0 & 0 \\ 0 & -1 & 0 \\ 0 & 0 & 1 \end{pmatrix};$$

$$G_6: a \to \begin{pmatrix} i & 0 & 0 \\ 0 & -i & 0 \\ 0 & 0 & i \end{pmatrix}, b \to \begin{pmatrix} 0 & 1 & 0 \\ 1 & 0 & 0 \\ 0 & 0 & 1 \end{pmatrix},$$

$$z \to \begin{pmatrix} -1 & 0 & 0 \\ 0 & -1 & 0 \\ 0 & 0 & 1 \end{pmatrix}.$$

(d) Find faithful representations of degree 3 for $G_7 \cong D_8 \times C_2$ and $G_8 \cong Q_8 \times C_2$.

(Note: This exercise can be used to confirm that the presentations of $G_1, \ldots, G_9$ given in the text do indeed give groups of order 16.)

6. Prove that no two of the groups $G_1, \ldots, G_9$ are isomorphic.

7. Let G be a non-abelian group of order p^4.
   (a) Prove that $|Z(G)| = p$ or p^2, and that if $|Z(G)| = p^2$ then G has $p^3 + p^2 - p$ conjugacy classes.
   (b) Prove that $|G'| = p$ or p^2, and that if $|G'| = p^2$ then G has $2p^2 - 1$ conjugacy classes.
   (c) Deduce that $|G' \cap Z(G)| = p$.

8. (a) Prove that if G is any group, then $G/Z(G) \not\cong Q_8$.
       (Hint: assume that $G/Z = \langle aZ, bZ: a^4 \in Z, a^2 \equiv b^2 \bmod Z, b^{-1}ab \equiv a^{-1} \bmod Z \rangle$. Prove that a^2 commutes with b, and hence that $a^2 \in Z$.)
   (b) Deduce from the result of Exercise 7 that if G is a group of order 16, then $G/(G' \cap Z(G)) \not\cong Q_8$.

27
Character table of the simple group
of order 168

Recall that a *simple* group is a non-trivial group G such that the only normal subgroups of G are $\{1\}$ and G itself. We discussed briefly in Chapter 1 the significance of simple groups in the theory of finite groups. Examples of simple groups which we have met so far are cyclic groups of prime order, A_5 and A_6. In fact the group A_5, of order 60, is the smallest non-abelian simple group. The next smallest is a certain group of order 168, and in this chapter we shall describe this group and find its character table. The group belongs to a whole family of simple groups, and we begin with a description of this family.

Special linear groups

Let p be a prime number, and recall that $\mathbb{Z}_p$ is the field which consists of the numbers $0, \ldots, p - 1$, with addition and multiplication modulo p. Denote by $SL(2, p)$ the set of all 2×2 matrices M with entries in $\mathbb{Z}_p$ such that $\det M = 1$. Then $SL(2, p)$ is a group under matrix multiplication, and is called the 2-dimensional *special linear group* over $\mathbb{Z}_p$.

To calculate the order of the group $SL(2, p)$, we count the matrices

$$\begin{pmatrix} a & b \\ c & d \end{pmatrix} \quad (a, b, c, d \in \mathbb{Z}_p, ad - bc = 1).$$

If $c = 0$, then there are $p(p - 1)$ choices for a, b, d which make $ad - bc = 1$ (since a, b are arbitrary, except that $a \neq 0$; and d is determined by a). And there are $p^2(p - 1)$ choices for a, b, c, d

with $c \neq 0$, such that $ad - bc = 1$ (since a, d may be chosen arbitrarily, c is any non-zero element of $\mathbb{Z}_p$; and then b is determined). Therefore

$$|\mathrm{SL}\,(2,\,p)| = p(p-1) + p^2(p-1)$$
$$= p(p^2 - 1).$$

If $p = 2$ then $\mathrm{SL}\,(2, p)$ has order 6, and it is easy to see that this group is isomorphic to S_3; so assume that p is an odd prime. By Exercise 27.1, the centre of $\mathrm{SL}\,(2, p)$ is

$$Z = \{I,\, -I\}$$

(where I is the 2×2 identity matrix). The factor group $\mathrm{SL}\,(2, p)/Z$ is called the 2-dimensional *projective special linear group*, and is written as $\mathrm{PSL}\,(2, p)$. Thus

$$\mathrm{PSL}\,(2, p) = \mathrm{SL}\,(2, p)/\{\pm I\}.$$

Since $|\mathrm{SL}\,(2, p)| = p(p^2 - 1)$, we have

$$|\mathrm{PSL}(2, p)| = p(p^2 - 1)/2.$$

It is known that $\mathrm{PSL}\,(2, 3) \cong A_4$, $\mathrm{PSL}\,(2, 5) \cong A_5$, and that for $p \geqslant 5$, the group $\mathrm{PSL}\,(2, p)$ is simple (see Theorem 8.19 of the book by J. J. Rotman listed in the Bibliography).

The simple group $G = \mathrm{PSL}\,(2, 7)$ has order 168, and we shall construct the character table of this group. After finding the conjugacy classes of G, we shall find the character table using only numerical calculations, notably the orthogonality relations (Theorem 16.4) and congruence properties (Corollary 22.26). The power of these techniques is therefore well illustrated. In the exercises, we indicate other ways of obtaining characters of G, using information about subgroups.

The conjugacy classes of PSL (2, 7)

27.1 Lemma

The group $\mathrm{PSL}\,(2, 7)$ *has exactly six conjugacy classes. The following table records representatives* g_i ($1 \leqslant i \leqslant 6$) *for the conjugacy classes, together with the order of* g_i, *the order of* $C_G(g_i)$, *and the size of the conjugacy class containing* g_i.

	Order of g_i	$|C_G(g_i)|$	$|g_i^G|$
$g_1 = \begin{pmatrix} 1 & 0 \\ 0 & 1 \end{pmatrix} Z$	1	168	1
$g_2 = \begin{pmatrix} 0 & 1 \\ -1 & 0 \end{pmatrix} Z$	2	8	21
$g_3 = \begin{pmatrix} 2 & -2 \\ 2 & 2 \end{pmatrix} Z$	4	4	42
$g_4 = \begin{pmatrix} 2 & 0 \\ 0 & 4 \end{pmatrix} Z$	3	3	56
$g_5 = \begin{pmatrix} 1 & 1 \\ 0 & 1 \end{pmatrix} Z$	7	7	24
$g_6 = \begin{pmatrix} 1 & -1 \\ 0 & 1 \end{pmatrix} Z$	7	7	24

Proof For each i, we verify that g_i has the stated order, and then use direct calculation to find all the elements of G which commute with g_i. Consider, for example, g_4. Suppose that

$$\begin{pmatrix} a & b \\ c & d \end{pmatrix} Z$$

commutes with g_4. Then

$$\begin{pmatrix} a & b \\ c & d \end{pmatrix} \begin{pmatrix} 2 & 0 \\ 0 & 4 \end{pmatrix} = \pm \begin{pmatrix} 2 & 0 \\ 0 & 4 \end{pmatrix} \begin{pmatrix} a & b \\ c & d \end{pmatrix},$$

and hence $b = c = 0$. Consequently

$$C_G(g_4) = \left\{ \begin{pmatrix} 1 & 0 \\ 0 & 1 \end{pmatrix} Z, \begin{pmatrix} 2 & 0 \\ 0 & 4 \end{pmatrix} Z, \begin{pmatrix} 4 & 0 \\ 0 & 2 \end{pmatrix} Z \right\}.$$

Similarly

$$C_G(g_2) = \left\{ MZ: M = \begin{pmatrix} 1 & 0 \\ 0 & 1 \end{pmatrix}, \begin{pmatrix} 0 & -1 \\ 1 & 0 \end{pmatrix}, \begin{pmatrix} 3 & 2 \\ 2 & 4 \end{pmatrix}, \begin{pmatrix} 3 & -2 \\ -2 & 4 \end{pmatrix}, \right.$$
$$\left. \begin{pmatrix} 2 & 3 \\ 3 & -2 \end{pmatrix}, \begin{pmatrix} 2 & 4 \\ 4 & -2 \end{pmatrix}, \begin{pmatrix} 2 & 2 \\ -2 & 2 \end{pmatrix}, \begin{pmatrix} 2 & -2 \\ 2 & 2 \end{pmatrix} \right\}.$$

Also, $C_G(g_i) = \langle g_i \rangle$ for $i = 3, 5, 6$.

Among $g_1, \ldots, g_6$, the only elements with the same order are g_5

and g_6; so no two of these six elements are conjugate, except possibly g_5 and g_6. Suppose that $g^{-1}g_6g = g_5$ with

$$g = \begin{pmatrix} a & b \\ c & d \end{pmatrix} Z \in G.$$

Then $gg_5 = g_6g$, and so

$$\begin{pmatrix} a & a+b \\ c & c+d \end{pmatrix} = \pm \begin{pmatrix} a-c & b-d \\ c & d \end{pmatrix} \text{ with } ad - bc = 1.$$

It follows that $c = 0$, $a \neq 0$, $d = a^{-1}$ and

$$\begin{pmatrix} a & a+b \\ 0 & a^{-1} \end{pmatrix} = \begin{pmatrix} a & b-a^{-1} \\ 0 & a^{-1} \end{pmatrix}.$$

Therefore $a^2 = -1$, which is impossible for $a \in \mathbb{Z}_7$. Thus g_5 is not conjugate to g_6, and we have established that no two of the elements $g_1, \ldots, g_6$ are conjugate.

The size of the conjugacy class g_i^G is obtained by dividing 168 by $|C_G(g_i)|$ (Theorem 12.8). Since the sum of the sizes of the six conjugacy classes g_i^G $(1 \leq i \leq 6)$ is 168, these exhaust the conjugacy classes of G.   ∎

Notice that using Lemma 27.1, it is easy to check that G is indeed simple, since any normal subgroup is a union of conjugacy classes (see Proposition 12.19).

27.2 Corollary
(1) If $1 \leq i \leq 4$ and χ is a character of G, then $\chi(g_i)$ is an integer.
(2) For some character χ of G, $\chi(g_5)$ is non-real.

Proof (1) By Lemma 27.1, for $1 \leq i \leq 4$, g_i is conjugate to $(g_i)^k$ whenever g_i and $(g_i)^k$ have the same order. Hence the conclusion follows from Theorem 22.16.

(2) Notice that $g_6 = g_5^{-1}$, so g_5 is not conjugate to its inverse. Therefore (2) follows from Corollary 15.6.   ∎

The character table of $G = \text{PSL}(2, 7)$

Since G has six conjugacy classes, it also has six irreducible characters. Let $\chi_1, \ldots, \chi_6$ be the irreducible characters of G, where χ_1 is the trivial character (so that $\chi_1(g) = 1$ for all $g \in G$). Recall that the character table is the 6×6 matrix with ij-entry $\chi_i(g_j)$.

We shall repeatedly exploit the column orthogonality relations, Theorem 16.4(2), and the congruence properties given by Corollaries 22.26 and 22.27 for the elements g_2, g_3, g_4, for which the character values are known to be integers, by Corollary 27.2.

The entries in the column of g_4 are integers, and the sum of the squares of these integers is equal to $|C_G(g_4)| = 3$. The entries must therefore be 1, ± 1, ± 1, 0, 0, 0 in some order. (We know that the entry in the first row is $\chi_1(g_4) = 1$.)

Similarly the entries in the column of g_3 are 1, ± 1, ± 1, ± 1, 0, 0 in some order, and the entries in column g_2 are 1, ± 1, ± 1, ± 1, ± 2, 0 in some order. Now for all characters χ of G, we have by Corollary 22.27,

$$\chi(g_2) \equiv \chi(1) \bmod 2, \text{ and}$$

$$\chi(g_3) \equiv \chi(1) \bmod 2,$$

and so

$$\chi(g_2) \equiv \chi(g_3) \bmod 2.$$

Since we also know that

$$\sum_{i=1}^{6} \chi_i(g_3)\chi_i(g_4) = 0,$$

we see that, with a suitable ordering of $\chi_2, \ldots, \chi_6$, part of the character table of G is as follows:

Class representative Centralizer order	g_2 8	g_3 4	g_4 3
χ_1	1	1	1
χ_2	± 1	± 1	± 1
χ_3	0	0	± 1
χ_4	± 1	± 1	0
χ_5	± 1	± 1	0
χ_6	± 2	0	0

We shall determine the signs later. For the moment we concentrate on the entries in the first column of the character table (i.e. the degrees $\chi_i(1)$). Let $d_i = \chi_i(1)$, so d_i is the entry on row i of column 1. By

Corollary 22.27, Theorem 22.11 and the fact that $\sum_{i=1}^{6} d_i^2 = 168$, we have

$$d_4 \equiv 0 \bmod 3,$$

$$d_4 \equiv 1 \bmod 2,$$

$$d_4 \text{ divides } |G| = 168, \quad \text{and}$$

$$d_4^2 \leqslant 168.$$

The only positive integer d_4 which satisfies these conditions is $d_4 = 3$. In the same way, $d_5 = 3$.

Next,

$$d_6 \equiv 0 \bmod 3,$$

$$d_6 \equiv 0 \bmod 2,$$

$$d_6 \text{ divides } 168, \quad \text{and}$$

$$d_6^2 \leqslant 168.$$

Therefore d_6 is 6 or 12. But

$$0 = \sum_{i=1}^{6} \chi_i(g_2) d_i = 1 \pm d_2 \pm 3 \pm 3 \pm 2 d_6,$$

so as $d_2^2 \leqslant 168$, we have $d_6 \neq 12$, and hence $d_6 = 6$.

Now

$$1 + d_2^2 + d_3^2 + 3^2 + 3^2 + 6^2 = 168,$$

so $d_2^2 + d_3^2 = 113$. The only solutions to this equation with d_2, d_3 positive integers have d_2, d_3 equal to 7, 8 in some order. Since $d_2 \equiv 1 \bmod 2$, we have $d_2 = 7$ and $d_3 = 8$.

We have now found the first column of the character table, and have the following portion:

Class representative Centralizer order	g_1 168	g_2 8	g_3 4	g_4 3
χ_1	1	1	1	1
χ_2	7	± 1	± 1	± 1
χ_3	8	0	0	± 1
χ_4	3	± 1	± 1	0
χ_5	3	± 1	± 1	0
χ_6	6	± 2	0	0

The equations

$$\sum_{i=1}^{6} \chi_i(g_1)\chi_i(g_j) = 0 \quad \text{for } j = 2, 3, 4$$

now enable us to determine the signs in the columns for g_2, g_3, g_4. We obtain

Class representative Centralizer order	g_1 168	g_2 8	g_3 4	g_4 3
χ_1	1	1	1	1
χ_2	7	-1	-1	1
χ_3	8	0	0	-1
χ_4	3	-1	1	0
χ_5	3	-1	1	0
χ_6	6	2	0	0

Next, the equation

$$1 = \langle \chi_2, \chi_2 \rangle = \sum_{i=1}^{6} \frac{\chi_2(g_i)\overline{\chi_2(g_i)}}{|C_G(g_i)|}$$

$$= \frac{7 \cdot 7}{168} + \frac{1}{8} + \frac{1}{4} + \frac{1}{3} + \frac{\chi_2(g_5)\overline{\chi_2(g_5)}}{7} + \frac{\chi_2(g_6)\overline{\chi_2(g_6)}}{7}$$

gives $\chi_2(g_5) = \chi_2(g_6) = 0$. (Note that $\chi_2(g_5) \equiv \chi_2(1) \bmod 7$, but we could not use this fact as we were not sure that $\chi_2(g_5)$ was an integer.) Also, for $j = 5, 6$,

$$0 = \sum_{i=1}^{6} \chi_i(g_4)\chi_i(g_j) = 1 - \chi_3(g_j)$$

and so $\chi_3(g_5) = \chi_3(g_6) = 1$.

By Corollary 27.2, there is an irreducible character χ of G such that $\chi(g_5)$ is non-real. For this character χ, the complex conjugate $\overline{\chi}$ will be a different character of the same degree. Hence χ_4 and χ_5 (being the only two irreducible characters with the same degree) must be complex conjugates of each other.

Let $\chi_4(g_5) = \overline{\chi_5(g_5)} = z$, and let $\chi_6(g_5) = t$. Thus the column for g_5 is

Class representative Centralizer order	g_5 7
χ_1	1
χ_2	0
χ_3	1
χ_4	z
χ_5	$\bar{z}$
χ_6	t

Now

$$0 = \sum_{i=1}^{6} \chi_i(g_2)\chi_i(g_5) = 1 - z - \bar{z} + 2t,$$

$$0 = \sum_{i=1}^{6} \chi_i(g_3)\chi_i(g_5) = 1 + z + \bar{z},$$

$$7 = \sum_{i=1}^{6} \chi_i(g_5)\overline{\chi_i(g_5)} = 2 + 2z\bar{z} + t\bar{t}.$$

Solving these equations, we obtain

$$t = -1, \; z = (-1 \pm i\sqrt{7})/2.$$

Since $g_6 = g_5^{-1}$, we have $\chi(g_6) = \overline{\chi(g_5)}$ for all characters χ of G.

We have now completely determined the character table of $G = \text{PSL}(2, 7)$, as shown.

Character table of PSL $(2, 7)$

Class representative Centralizer order	g_1 168	g_2 8	g_3 4	g_4 3	g_5 7	g_6 7
χ_1	1	1	1	1	1	1
χ_2	7	-1	-1	1	0	0
χ_3	8	0	0	-1	1	1
χ_4	3	-1	1	0	α	$\bar{\alpha}$
χ_5	3	-1	1	0	$\bar{\alpha}$	α
χ_6	6	2	0	0	-1	-1

where $\alpha = (-1 + i\sqrt{7})/2$.

It is known that there are precisely five non-abelian simple groups of order less than 1000. We give you the character tables of all of these, as follows:

G	Order of G	Reference for character table
A_5	60	Example 20.13
PSL $(2, 7)$	168	This chapter
A_6	360	Exercise 20.2
PSL $(2, 8)$	504	Exercise 28.3
PSL $(2, 11)$	660	Exercise 27.6

Summary of Chapter 27

1. $SL(2, p) = \left\{ \begin{pmatrix} a & b \\ c & d \end{pmatrix} : a, b, c, d \in \mathbb{Z}_p, ad - bc = 1 \right\}.$

   $|SL(2, p)| = p(p^2 - 1).$

2. $PSL(2, p) = SL(2, p)/\{\pm I\}.$

   $|PSL(2, p)| = p(p^2 - 1)/2$ (p odd).

3. We constructed the character table of PSL$(2, 7)$, the simple group of order 168.

Exercises for Chapter 27

1. Prove that $Z(SL(2, p)) = \{\pm I\}$.

2. Find the character table of SL$(2, 3)$.

3. Deduce directly from the character table of PSL$(2, 7)$ that this group is simple.

4. In this exercise we present an alternative construction of the character table of $G = PSL(2, 7)$, given the conjugacy classes of G, as in Lemma 27.1.

   (a) Define the subgroup T of G, of order 21, as follows:

$$T = \left\{ \begin{pmatrix} a & b \\ 0 & a^{-1} \end{pmatrix} Z : a \in \mathbb{Z}_7^*, b \in \mathbb{Z}_7 \right\}$$

   (where $Z = \{\pm I\}$). Calculate the values of the induced character $(1_T) \uparrow G$, and show that

$$(1_T) \uparrow G = 1_G + \chi,$$

   where χ is an irreducible character of G.

(b) Let λ be a non-trivial linear character of T. Calculate the values of $\lambda \uparrow G$ and prove that this is an irreducible character of G.

(c) By considering χ_S (see Proposition 19.14), obtain an irreducible character of G of degree 6.

(d) From (a), (b), (c), we now have irreducible characters of G of degrees 1, 7, 8 and 6. Use orthogonality relations to complete the character table of G.

5. *The character table of* SL $(2, 7)$.

Let $G = \mathrm{SL}\,(2, 7)$, the group of all 2×2 matrices of determinant 1, with entries in the field $\mathbb{Z}_7$.

(a) Show that G has 11 conjugacy classes with representatives g_i as follows:

g_i	Order of g_i	$|C_G(g_i)|$	$|g_i^G|$
$g_1 = \begin{pmatrix} 1 & 0 \\ 0 & 1 \end{pmatrix}$	1	336	1
$g_2 = \begin{pmatrix} -1 & 0 \\ 0 & -1 \end{pmatrix}$	2	336	1
$g_3 = \begin{pmatrix} 0 & 1 \\ -1 & 0 \end{pmatrix}$	4	8	42
$g_4 = \begin{pmatrix} 2 & -2 \\ 2 & 2 \end{pmatrix}$	8	8	42
$g_5 = \begin{pmatrix} -2 & 2 \\ -2 & -2 \end{pmatrix}$	8	8	42
$g_6 = \begin{pmatrix} 2 & 0 \\ 0 & 4 \end{pmatrix}$	3	6	56
$g_7 = \begin{pmatrix} -2 & 0 \\ 0 & -4 \end{pmatrix}$	6	6	56
$g_8 = \begin{pmatrix} 1 & 1 \\ 0 & 1 \end{pmatrix}$	7	14	24
$g_9 = \begin{pmatrix} -1 & -1 \\ 0 & -1 \end{pmatrix}$	14	14	24
$g_{10} = \begin{pmatrix} 1 & -1 \\ 0 & 1 \end{pmatrix}$	7	14	24
$g_{11} = \begin{pmatrix} -1 & 1 \\ 0 & -1 \end{pmatrix}$	14	14	24

(b) Use the character table of PSL(2,7) to write down the six irreducible characters of G with kernel containing $Z = \{\pm I\}$.

(c) Let χ_7, χ_8, χ_9, χ_{10}, χ_{11} be the remaining irreducible characters of G. Show that for any j with $7 \le j \le 11$ and any $g \in G$, we have $\chi_j(g) = -\chi_j(-g)$.

(d) Prove that $\chi_j(g_3) = 0$ for $7 \le j \le 11$, and deduce that $\chi_j(1)$ is even.

(e) By considering the column of g_6 in the character table, and congruences modulo 3, show that the degrees of $\chi_7, \ldots, \chi_{11}$ are 4, 4, 6, 6, 8, and find $\chi_j(g_6)$ for $7 \le j \le 11$.

(f) Let ψ be one of the irreducible characters of degree 4. By considering the values of ψ_A on g_1, g_2, g_3 and g_6 (see Proposition 19.14), prove that ψ_A is equal to the irreducible character of G of degree 6 whose kernel contains Z. Deduce the values of the irreducible characters of degree 4 on all g_i.

(g) Complete the character table of G.

6. *The character table of* PSL(2, 11).

   Let $G = \text{PSL}(2, 11)$. This group has eight conjugacy classes with representatives $g_1, \ldots, g_8$ having orders and centralizer orders as follows:

g_i	g_1	g_2	g_3	g_4	g_5	g_6	g_7	g_8
Order of g_i	1	2	3	6	5	5	11	11
$|C_G(g_i)|$	660	12	6	6	5	5	11	11

Also, g_5^{-1}, g_6^{-1}, g_7^{-1}, g_8^{-1} are conjugate to g_5, g_6, g_8, g_7, respectively.

   Find the character table of G.

28
Character table of GL(2, q)

We are now going to calculate the character tables of an important infinite series of groups, and one of the exercises will show you how to use the results of this chapter to determine the character tables of infinitely many simple groups. In the last chapter and its exercises, we found the character tables of certain groups of 2×2 matrices with entries in $\mathbb{Z}_7$ and $\mathbb{Z}_{11}$. We shall determine the character tables of some matrix groups with entries from an arbitrary finite field. At first sight, this is a daunting task, since the number of irreducible characters increases with the size of the field. However, we shall see that the conjugacy classes of our groups fall into four families, as do the irreducible characters. Consequently, we can display the character values in a 4×4 matrix.

The fields $\mathbb{F}_q$ and $\mathbb{F}_{q^2}$

We consider finite fields, and we shall tell you which properties of these fields we will use; if you are unfamiliar with finite fields then you might like to consult the book by J. B. Fraleigh listed in the Bibliography.

Recall that a field $(F, +, \times)$ is a set F with two binary operations $+$ and $\times$ such that the following properties hold. First, $(F, +)$ is an abelian group, with identity element 0. Secondly, if we write $F^* = F \backslash \{0\}$, then $(F^*, \times)$ is an abelian group, with identity element 1. Finally, the distributive law holds; that is $(a + b)c = ac + bc$ for all a, b, $c \in F$. For example, $\mathbb{R}$, $\mathbb{C}$ and $\mathbb{Z}_p$ (p prime) are fields, with the usual definitions of $+$ and $\times$.

The basic properties of finite fields which we will use without proof are these:

(28.1) *Let p be a prime and n be a positive integer, and write*
$q = p^n$. *Then there exists a field* $\mathbb{F}_q$ *of order q and*
every field of order q is isomorphic to $\mathbb{F}_q$.
For every $s \in \mathbb{F}_q$ *the sum of s with itself p times is*
zero; in short, ps = 0.
The group $(\mathbb{F}_q^*, \times)$ *is cyclic.*

Notice that the binomial coefficients $\binom{p}{i}$ with $1 \leqslant i \leqslant p - 1$ are all
divisible by p; it follows that $(s + t)^p = s^p + t^p$ for all $s, t \in \mathbb{F}_q$, and
hence $(s + t)^{p^k} = s^{p^k} + t^{p^k}$ for all positive integers k. We use this
remark in the proof of the next proposition.

28.2 Proposition
Let $F = \mathbb{F}_{q^2}$ and $S = \{s \in F : s^q = s\}$.

(1) *The set S is a subfield of F of order q, and hence* $S \cong \mathbb{F}_q$.
(2) *If* $r \in F$ *then* $r + r^q, r^{1+q} \in S$.

Proof (1) Suppose that $s, t \in S$. Then $(s + t)^q = s^q + t^q = s + t$, so
$s + t \in S$. It is now easy to check that $(S, +)$ and $(S \setminus \{0\}, \times)$ are
abelian groups, so S is a field.
(2) Since $\mathbb{F}_{q^2}^*$ is a group of order $q^2 - 1$, we see that $r^{q^2} = r$ for all
$r \in F$. This implies that $(r + r^q)^q = r^q + r^{q^2} = r + r^q$ and $(r^{1+q})^q = r^{1+q}$, so $r + r^q, r^{1+q} \in S$. $\blacksquare$

Hereafter, we shall identify the subfield S of $\mathbb{F}_{q^2}$ in Proposition 28.2
with the field $\mathbb{F}_q$.
We introduce the following useful notation.

(28.3) *Let* ε *be a generator of the cyclic group* $\mathbb{F}_{q^2}^*$ *and let*
$\omega = e^{(2\pi i/(q^2-1))}$. *Suppose that* $r \in \mathbb{F}_{q^2}^*$. *We may write*
$r = \varepsilon^m$ *for some m and we let* $\bar{r} = \omega^m$.
Then $r \rightarrow \bar{r}$ *is an irreducible character of* $\mathbb{F}_{q^2}^*$. *More-*
over, every irreducible character of $\mathbb{F}_{q^2}^*$ *has the form*
$r \rightarrow \bar{r}^j$ *for some integer j.*

You are now in a position to appreciate the statement of the main
result of this chapter, namely Theorem 28.5.

The conjugacy classes of GL(2, *q*)

The general linear group $GL(2, q)$ is defined to be the group of invertible 2×2 matrices with entries in $\mathbb{F}_q$. The subgroup consisting of all matrices of determinant 1 is the special linear group $SL(2, q)$, and we talked about some special linear groups in the last chapter. Here, we are going to calculate the character table of $GL(2, q)$.

Let $G = GL(2, q)$, and remember that the matrix

$$\begin{pmatrix} a & b \\ c & d \end{pmatrix}$$

belongs to G if and only if its rows are linearly independent. The number of such matrices is found by noting that (a, b) can be any non-zero row vector, giving us $q^2 - 1$ choices; and once (a, b) has been chosen, (c, d) can be any row vector which is not a multiple of (a, b), giving us $q^2 - q$ choices. Therefore,

$$|G| = (q^2 - 1)(q^2 - q) = q(q - 1)^2(q + 1).$$

There are four families of conjugacy classes of G, of which three are easy to describe.

First,

$$\begin{pmatrix} a & b \\ 0 & c \end{pmatrix}$$

can be conjugate to

$$\begin{pmatrix} a' & b' \\ 0 & c' \end{pmatrix}$$

only if $\{a, c\} = \{a', c'\}$, since conjugate matrices have the same eigenvalues. Keep this in mind during the following discussion.

The matrices

$$sI = \begin{pmatrix} s & 0 \\ 0 & s \end{pmatrix} \ (s \in \mathbb{F}_q^*)$$

belong to the centre of G. They give us $q - 1$ conjugacy classes of size 1.

Next, consider the matrices

$$u_s = \begin{pmatrix} s & 1 \\ 0 & s \end{pmatrix} \ (s \in \mathbb{F}_q^*).$$

Let

$$g = \begin{pmatrix} a & b \\ c & d \end{pmatrix} \in G.$$

Then

$$gu_s = \begin{pmatrix} as & a+bs \\ cs & c+ds \end{pmatrix} \text{ and } u_s g = \begin{pmatrix} as & d+bs \\ cs & ds \end{pmatrix}$$

so g belongs to the centralizer of u_s if and only if $c = 0$ and $a = d$. Thus, the matrices u_s $(s \in \mathbb{F}_q^*)$ give us $q - 1$ conjugacy classes; the centralizer order is $(q - 1)q$, so, by Theorem 12.8, each conjugacy class contains $q^2 - 1$ elements.

Now, let

$$d_{s,t} = \begin{pmatrix} s & 0 \\ 0 & t \end{pmatrix} \in G \; (s, t \in \mathbb{F}_q^*)$$

and note that

$$\begin{pmatrix} 0 & 1 \\ 1 & 0 \end{pmatrix}^{-1} d_{s,t} \begin{pmatrix} 0 & 1 \\ 1 & 0 \end{pmatrix} = d_{t,s}.$$

On the other hand, if $s \neq t$, then we have that $gd_{s,t} = d_{s,t}g$ if and only if $b = c = 0$. Thus, the matrices $d_{s,t}$ $(s, t \in \mathbb{F}_q^*, s \neq t)$ give us $(q - 1)(q - 2)/2$ conjugacy classes; the centralizer order is $(q - 1)^2$, so each conjugacy class contains $q(q + 1)$ elements.

Finally, consider

$$v_r = \begin{pmatrix} 0 & 1 \\ -r^{1+q} & r+r^q \end{pmatrix} \; (r \in \mathbb{F}_{q^2} \setminus \mathbb{F}_q).$$

By Proposition 28.2, $v_r \in G$. The characteristic polynomial of v_r is

$$\det(xI - v_r) = x(x - (r + r^q)) + r^{1+q} = (x - r)(x - r^q),$$

so v_r has eigenvalues r and r^q. Since $r \notin \mathbb{F}_q$ we see that v_r lies in none of the conjugacy classes we have constructed so far.

Now,

$$gv_r = \begin{pmatrix} -br^{1+q} & a+b(r+r^q) \\ -dr^{1+q} & c+d(r+r^q) \end{pmatrix} \text{ and }$$

$$v_r g = \begin{pmatrix} c & d \\ -ar^{1+q}+c(r+r^q) & -br^{1+q}+d(r+r^q) \end{pmatrix}.$$

Hence $gv_r = v_r g$ only if $c = -br^{1+q}$ and $d = a + b(r + r^q)$. If these conditions hold, then

$$ad - bc = a^2 + ab(r + r^q) + b^2 r^{1+q} = (a + br)(a + br^q).$$

Since $(a, b) \neq (0, 0)$ and $r, r^q \notin \mathbb{F}_q$, we see that $a + br$ and $a + br^q$ are non-zero. Therefore, $g \in C_G(v_r)$ if and only if

$$g = \begin{pmatrix} a & b \\ -br^{1+q} & a + b(r + r^q) \end{pmatrix}.$$

Thus, $|C_G(v_r)| = q^2 - 1$, and the conjugacy class containing v_r has size $q^2 - q$.

The matrix v_t has eigenvalues t and t^q, so it is not conjugate to v_r unless $t = r$ or $t = r^q$. We therefore partition $\mathbb{F}_{q^2} \backslash \mathbb{F}_q$ into subsets $\{r, r^q\}$; each subset gives us a conjugacy class representative v_r and different subsets give us representatives of different conjugacy classes.

We have now found all the conjugacy classes of G.

28.4 Proposition
There are $q^2 - 1$ conjugacy classes in $\mathrm{GL}(2, q)$, described as follows.

Class rep. g	sI	u_s	$d_{s,t}$	v_r
$|C_G(g)|$	$(q^2 - 1)(q^2 - q)$	$(q - 1)q$	$(q - 1)^2$	$q^2 - 1$
No. of classes	$q - 1$	$q - 1$	$(q - 1)(q - 2)/2$	$(q^2 - q)/2$

The families of conjugacy class representatives sI and u_s are indexed by elements s of $\mathbb{F}_q^*$.

The family of conjugacy class representatives $d_{s,t}$ is indexed by unordered pairs $\{s, t\}$ of distinct elements of $\mathbb{F}_q^*$.

The family of conjugacy class representatives v_r is indexed by unordered pairs $\{r, r^q\}$ of elements of $\mathbb{F}_{q^2}^* \backslash \mathbb{F}_q^*$.

Proof The conjugacy classes we have found account for

$$(q - 1) + (q - 1)(q^2 - 1) + (q - 1)(q - 2)q(q + 1)/2 + (q^2 - q)(q^2 - q)/2$$

elements altogether. But this sum is equal to the order of $\mathrm{GL}(2, q)$, so we have found all the conjugacy classes. ∎

The characters of GL(2,*q*)

We are now in a position to describe the character table of *G*.

28.5 Theorem
Label the conjugacy classes of GL(2, *q*) as in Proposition 28.4, and let $r \to \bar{r}$ be the function from $\mathbb{F}_{q^2}^*$ to $\mathbb{C}$ described in (28.3). Then the irreducible characters of GL(2, *q*) are given by $\lambda_i, \psi_i, \psi_{i,j}, \chi_i$ as follows.

	sI	u_s	$d_{s,t}$	v_r
λ_i	$\bar{s}^{2i}$	$\bar{s}^{2i}$	$(\overline{st})^i$	$\bar{r}^{i(1+q)}$
ψ_i	$q\bar{s}^{2i}$	0	$(\overline{st})^i$	$-\bar{r}^{i(1+q)}$
$\psi_{i,j}$	$(q+1)\bar{s}^{i+j}$	$\bar{s}^{i+j}$	$\bar{s}^i\bar{t}^j + \bar{s}^j\bar{t}^i$	0
χ_i	$(q-1)\bar{s}^i$	$-\bar{s}^i$	0	$-(\bar{r}^i + \bar{r}^{iq})$

Here, we have the following restrictions on the subscripts.
(a) For λ_i we have $0 \leqslant i \leqslant q-2$. Thus, there are $q-1$ characters λ_i, each of degree 1.
(b) For ψ_i we have $0 \leqslant i \leqslant q-2$. Thus, there are $q-1$ characters ψ_i, each of degree q.
(c) For $\psi_{i,j}$ we have $0 \leqslant i < j \leqslant q-2$. Thus, there are $(q-1)(q-2)/2$ characters $\psi_{i,j}$, each of degree $q+1$.
(d) For χ_i, we first consider the set of integers j with $0 \leqslant j \leqslant q^2-1$ and $(q+1) \nmid j$; if j_1 and j_2 belong to this set and $j_1 \equiv j_2 q \bmod(q^2-1)$ then we choose precisely one of j_1 and j_2 to belong to the indexing set for the characters χ_i. Hence, there are $(q^2-q)/2$ characters χ_i, each of degree $q-1$.

Before we embark upon the task of calculating the irreducible characters of *G*, we present a proposition which will be useful later. Recall that ε is our chosen generator for $\mathbb{F}_{q^2}^*$ and

$$v_\varepsilon = \begin{pmatrix} 0, & 1 \\ -\varepsilon^{1+q} & \varepsilon + \varepsilon^q \end{pmatrix}.$$

28.6 Proposition
Let $K = \langle v_\varepsilon \rangle$. Then $|K| = q^2 - 1$. The group *K* contains the $q-1$ scalar matrices *sI* in *G*, and of the remaining $q^2 - q$ elements of *K*,

precisely two belong to each of the conjugacy classes represented by v_r *with* $r \in \mathbb{F}_{q^2}^* \backslash \mathbb{F}_q^*$.

Proof The eigenvalues of v_ε are ε and ε^q, so v_ε has order $q^2 - 1$.

The eigenvalues of v_ε^i and of v_ε^{iq} are ε^i and ε^{iq}. If $\varepsilon^i \neq \varepsilon^{iq}$ then $\varepsilon^i \notin \mathbb{F}_q$ and v_ε^i and v_ε^{iq} must be conjugate to v_{ε^i}. Hence two elements of K belong to the conjugacy class of v_{ε^i}.

If $\varepsilon^i = \varepsilon^{iq}$ then $v_\varepsilon^i = \varepsilon^i I$, since v_ε^i has eigenvalues in $\mathbb{F}_q$ and v_ε^i is diagonalizable, and this case accounts for the $q - 1$ scalar matrices. ∎

We shall construct, in turn, the irreducible characters λ_i, ψ_i, $\psi_{i,j}$ and χ_i which appear in Theorem 28.5.

28.7 Proposition
There are $q - 1$ *linear characters* λ_i *of* G, *and they are given in Theorem 28.5.*

Proof The map $\det: g \to \det g$ is a homomorphism from G onto $\mathbb{F}_q^*$. As i varies between 0 and $q - 2$ inclusive, the functions

$$\lambda_i: g \to (\overline{\det g})^i \quad (g \in G)$$

give $q - 1$ distinct linear characters of G, whose values appear in Theorem 28.5. ∎

We will see later that the linear characters λ_i $(0 \leqslant i \leqslant q - 2)$ which appear in Proposition 28.7 are all the linear characters of G.

28.8 Proposition
For all integers i, j *there is a character* $\psi_{i,j}$ *of* G *whose values on the conjugacy class representatives, as described in Proposition 28.4, are as follows.*

	sI	u_s	$d_{s,t}$	v_r
$\psi_{i,j}$	$(q+1)\bar{s}^{i+j}$	$\bar{s}^{i+j}$	$\bar{s}^i \bar{t}^j + \bar{s}^j \bar{t}^i$	0

Proof Let

$$B = \left\{ \begin{pmatrix} a & b \\ 0 & c \end{pmatrix} \in G \right\}.$$

Then B is a subgroup of G with $|B| = (q-1)^2 q$. Define $\lambda_{i,j} : B \to \mathbb{C}$ by

$$\lambda_{i,j} : \begin{pmatrix} s & r \\ 0 & t \end{pmatrix} \to \bar{s}^i \bar{t}^j.$$

Then $\lambda_{i,j}$ is a character of B. We let $\psi_{i,j} = \lambda_{i,j} \uparrow G$.

We use Proposition 21.23 to calculate $\psi_{i,j}(g)$ for each conjugacy class representative g, as follows.

$$g = sI : \quad \psi_{i,j}(g) = \frac{|C_G(g)|}{|C_B(g)|} \lambda_{i,j}(g)$$

$$g = u_s : \quad \psi_{i,j}(g) = \frac{|C_G(g)|}{|C_B(g)|} \lambda_{i,j}(g)$$

$$g = d_{s,t} : \quad \psi_{i,j}(g) = |C_G(g)| \left(\frac{\lambda_{i,j}(g)}{|C_B(g)|} + \frac{\lambda_{i,j}(g')}{|C_B(g')|} \right), \text{ where } g' = d_{t,s}$$

$$g = v_r : \quad \psi_{i,j}(g) = 0.$$

Hence, the values of $\psi_{i,j}$ are as stated in the proposition. ∎

28.9 Proposition
For each integer i, there is an irreducible character ψ_i of G whose values are given in Theorem 28.5. The characters ψ_i for $0 \leqslant i \leqslant q - 2$ are all different.

Proof We shall demonstrate that the character $\psi_{i,i}$ which appears in Proposition 28.8 gives us $\psi_{i,i} = \lambda_i + \psi_i$. To this end, we calculate $\langle \psi_{i,i}, \psi_{i,i} \rangle$ and $\langle \psi_{i,i}, \lambda_i \rangle$. Remember that the complex conjugate of $\bar{s}^i$ is $\bar{s}^{-i}$. We have

$$\langle \psi_{i,i}, \psi_{i,i} \rangle = \frac{(q+1)^2}{(q^2-1)(q^2-q)}(q-1) + \frac{1}{(q-1)q}(q-1)$$

$$+ \frac{4}{(q-1)^2} \frac{(q-1)(q-2)}{2}$$

$$= 2.$$

Here, the first term corresponds to the conjugacy classes of elements

sI, and the calculation of this first term involves the following three observations.

(1) $\psi_{i,i}(sI)\overline{\psi_{i,i}(sI)} = (q+1)^2$.

(2) $|C_G(sI)| = (q^2-1)(q^2-q)$.

(3) There are $q-1$ conjugacy classes with representatives of the forms sI.

The remaining terms in $\langle \psi_{i,i}, \psi_{i,i} \rangle$ are calculated in a similar fashion.

Next,

$$\langle \psi_{i,i}, \lambda_i \rangle = \frac{(q+1)}{(q^2-1)(q^2-q)}(q-1) + \frac{1}{(q-1)q}(q-1)$$
$$+ \frac{2}{(q-1)^2}\frac{(q-1)(q-2)}{2}$$
$$= 1.$$

The facts that $\langle \psi_{i,i}, \lambda_i \rangle = 1$ and $\langle \psi_{i,i}, \psi_{i,i} \rangle = 2$ imply that $\psi_{i,i} = \lambda_i + \psi_i$ for some irreducible character ψ_i. Subtract λ_i from $\psi_{i,i}$ to get the values of ψ_i as given in Theorem 28.5.

Let s be an element of $\mathbb{F}_q^*$ of order $q-1$. Then $\psi_i: d_{s,1} \to \bar{s}^i$. Hence the characters ψ_i for $0 \leq i \leq q-2$ are all different. ∎

28.10 Proposition

Suppose that $0 \leq i < j \leq q-2$. Then the character $\psi_{i,j}$ which appears in Proposition 28.8 is irreducible.

Proof We shall show that $\langle \psi_{i,j}, \psi_{i,j} \rangle = 1$. Using the values of $\psi_{i,j}$ which are given in Proposition 28.8, we obtain $\langle \psi_{i,j}, \psi_{i,j} \rangle = A + B + C$, where

$$A = \frac{(q+1)^2}{(q^2-1)(q^2-q)}(q-1), \qquad B = \frac{1}{(q-1)q}(q-1) \qquad \text{and}$$

$$C = \frac{1}{(q-1)^2}\frac{1}{2}\sum_{s \neq t}(\bar{s}^i\bar{t}^j + \bar{s}^j\bar{t}^i)(\bar{s}^{-i}\bar{t}^{-j} + \bar{s}^{-j}\bar{t}^{-i}).$$

The coefficent $\frac{1}{2}$ appears in C because we have just one conjugacy class for each unordered pair $\{s, t\}$ of distinct elements of $\mathbb{F}_q^*$.

To evaluate C, note that $\{d_{s,t}: s, t \in \mathbb{F}_q^*\}$ is an abelian group of order $(q-1)^2$, and if $\sigma: d_{s,t} \to \bar{s}^i\bar{t}^j + \bar{s}^j\bar{t}^i$ then σ is a sum of two inequivalent irreducible characters of this group. Thus, $\langle \sigma, \sigma \rangle = 2$. That is,

$$\frac{1}{(q-1)^2}\left(4(q-1) + \sum_{s \neq t}(\bar{s}^i \bar{t}^j + \bar{s}^j \bar{t}^i)(\bar{s}^{-i}\bar{t}^{-j} + \bar{s}^{-j}\bar{t}^{-i})\right) = 2.$$

Hence,

$$C = \frac{q-3}{q-1}.$$

And now we find that $A + B + C = 1$. Therefore, $\langle \psi_{i,j}, \psi_{i,j} \rangle = 1$, and $\psi_{i,j}$ is irreducible. ∎

28.11 Corollary
The characters $\psi_{i,j}$ for $0 \leqslant i < j \leqslant q - 2$ are distinct irreducible characters of G.

Proof Suppose that $0 \leqslant i < j \leqslant q - 2$ and $0 \leqslant i' < j' \leqslant q - 2$, and $(i, j) \neq (i', j')$. We must prove that $\psi_{i,j} \neq \psi_{i',j'}$.

Consider the group B and its linear characters $\lambda_{i,j}$ which were used in the proof of Proposition 28.8. We have

$$\lambda_{i,j} + \lambda_{j,i} : \begin{pmatrix} s & b \\ 0 & t \end{pmatrix} \to \bar{s}^i \bar{t}^j + \bar{s}^j \bar{t}^i.$$

Since $\lambda_{i,j} + \lambda_{j,i} \neq \lambda_{i',j'} + \lambda_{j',i'}$, there exists complex $(q-1)th$ roots of unity $\bar{s}$ and $\bar{t}$ such that

either $\bar{s} \neq \bar{t}$ and $\bar{s}^i \bar{t}^j + \bar{s}^j \bar{t}^i \neq \bar{s}^{i'} \bar{t}^{j'} + \bar{s}^{j'} \bar{t}^{i'}$
or $\bar{s} = \bar{t}$ and $\bar{s}^{i+j} \neq \bar{s}^{i'+j'}$.

In either case, we see that $\psi_{i,j}$ differs from $\psi_{i',j'}$ on a conjugacy class of G. Therefore, $\psi_{i,j} \neq \psi_{i',j'}$. ∎

28.12 Proposition
For each integer i, there exists a character ϕ_i of G which takes the following values.

	sI	u_s	$d_{s,t}$	v_r
ϕ_i	$q(q-1)\bar{s}^i$	0	0	$\bar{r}^i + \bar{r}^{iq}$

Proof Let $K = \langle v_\varepsilon \rangle$, as in Proposition 28.6, and consider the linear character α_i of K which sends the generator v_ε of K to $\bar{\varepsilon}^i$. Suppose that $g \in K$ and g is conjugate in G to v_r. Then g has eigenvalues r and r^q. Hence $\alpha_i(g) = \bar{r}^i$ or $\bar{r}^{iq}$ and

$$\alpha_i(g) + \alpha_i(g^q) = \bar{r}^i + \bar{r}^{iq}.$$

Let $\phi_i = \alpha_i \uparrow G$.

In order to calculate ϕ_i, first recall that $\alpha_i \uparrow G$ is zero on all elements which are not conjugate to an element of K. Thus, by Proposition 28.6, ϕ_i is zero on the elements of the form u_s and $d_{s,t}$ ($s \neq t$).

If $g = sI$ with $s \in \mathbb{F}_q^*$ then $g \in K$ and

$$\phi_i(g) = \frac{|C_G(g)|}{|C_K(g)|} \alpha_i(g) = q(q-1)\bar{s}^i.$$

Suppose that $r \in \mathbb{F}_{q^2} \backslash \mathbb{F}_q$. Then, by Proposition 28.6, v_r is conjugate to an element of g of K. Also,

$$\phi_i(g) = |C_G(g)| \left(\frac{\alpha_i(g)}{|C_K(g)|} + \frac{\alpha_i(g^q)}{|C_K(g)|} \right)$$

$$= \alpha_i(g) + \alpha_i(g^q)$$

$$= \bar{r}^i + \bar{r}^{iq}.$$

Thus, ϕ_i has the values stated in the proposition. ∎

To be able to work out certain inner products involving our characters ϕ_i, we shall the use the following lemma.

28.13 Lemma

Assume that i is an integer and $(q + 1) \nmid i$. Then

$$\sum_{r \in \mathbb{F}_{q^2} \backslash \mathbb{F}_q} (\bar{r}^i + \bar{r}^{iq})(\bar{r}^{-i} + \bar{r}^{-iq}) = 2(q-1)^2.$$

Proof Note that

$$G_1 = \left\{ \begin{pmatrix} r & 0 \\ 0 & r^q \end{pmatrix} : r \in \mathbb{F}_{q^2}^* \right\} \text{ and } G_2 = \left\{ \begin{pmatrix} r & 0 \\ 0 & r_q \end{pmatrix} : r \in \mathbb{F}_q^* \right\}$$

are abelian groups of orders $q^2 - 1$ and $q - 1$, respectively. Now,

$$\begin{pmatrix} r & 0 \\ 0 & r^q \end{pmatrix} \rightarrow \bar{r}^i + \bar{r}^{iq}$$

gives a character χ of degree 2 for each group. For G_1, the character χ is a sum of two inequivalent irreducible characters, since $(q+1) \nmid i$ implies that $\bar{\varepsilon}^i \neq \bar{\varepsilon}^{iq}$; and for G_2, the character χ is twice an irreducible character, since $r^q = r$ for $r \in \mathbb{F}_q^*$. Taking the inner product of the character χ of G_1 with itself, we get

$$\frac{1}{q^2 - 1} \sum_{r \in \mathbb{F}_{q^2}^*} (\bar{r}^i + \bar{r}^{iq})(\bar{r}^{-i} + \bar{r}^{-iq}) = 2$$

and doing the same for the character χ of G_2, we get

$$\frac{1}{q - 1} \sum_{r \in \mathbb{F}_q^*} (\bar{r}^i + \bar{r}^{iq})(\bar{r}^{-i} + \bar{r}^{-iq}) = 4.$$

Hence

$$\sum_{r \in \mathbb{F}_{q^2} \backslash \mathbb{F}_q} (\bar{r}^i + \bar{r}^{iq})(\bar{r}^{-i} + \bar{r}^{-iq}) = 2(q^2 - 1) - 4(q - 1)$$

$$= 2(q - 1)^2. \qquad \blacksquare$$

28.14 Proposition
For each integer i, let χ_i be the class function on G with the following values.

	sI	u_s	$d_{s,t}$	v_r
χ_i	$(q-1)\bar{s}^i$	$-\bar{s}^i$	0	$-(\bar{r}^i + \bar{r}^{iq})$

If $(q+1) \nmid i$ then χ_i is an irreducible character of G.

Proof We can justify the manoeuvre which we now perform only by saying that it gives the correct answer.

Recall the characters $\psi_{i,j}$, ψ_i and ϕ_i given in Propositions 28.8, 28.9 and 28.12. Now, χ_i is the class function on G which is given by

$$\chi_i = \psi_{0,-i}\psi_i - \psi_{0,i} - \phi_i.$$

The table below allows us to verify this.

	sI	u_s	$d_{s,t}$	v_r
$\psi_{0,-i}$	$(q+1)\bar{s}^{-i}$	$\bar{s}^{-i}$	$\bar{s}^{-i}+\bar{t}^{-i}$	0
ψ_i	$q\bar{s}^{2i}$	0	$(\bar{s}\bar{t})^i$	$-\bar{r}^{i(1+q)}$
$\psi_{0,-i}\psi_i$	$q(q+1)\bar{s}^i$	0	$\bar{s}^i+\bar{t}^i$	0
$\psi_{0,i}$	$(q+1)\bar{s}^i$	$\bar{s}^i$	$\bar{s}^i+\bar{t}^i$	0
ϕ_i	$q(q-1)\bar{s}^i$	0	0	$\bar{r}^i+\bar{r}^{iq}$
χ_i	$(q-1)\bar{s}^i$	$-\bar{s}^{-i}$	0	$-(\bar{r}^i+\bar{r}^{iq})$

Next, assume that $(q+1) \nmid i$. We work out $\langle \chi_i, \chi_i \rangle$ using Lemma 28.13.

$$\langle \chi_i, \chi_i \rangle = \frac{(q-1)^2}{(q^2-1)(q^2-q)}(q-1) + \frac{1}{q^2-q}(q-1) + \frac{(q-1)^2}{q^2-1} = 1.$$

Since χ_i is a linear combination of irreducible characters of G, with integer coefficients, and $\langle \chi_i, \chi_i \rangle = 1$ and $\chi_i(1) > 0$, it follows that χ_i is an irreducible character of G. ∎

28.15 Proposition
Suppose that i and j are integers with $(q+1) \nmid i$ and $(q+1) \nmid j$ and $j \not\equiv i, iq \bmod(q^2-1)$. Then the characters χ_i and χ_j of G are different.

Proof Let $K = \langle v_\varepsilon \rangle$, as in Proposition 28.6, and consider the linear character α_i of K which sends the generator v_ε of K to $\bar{\varepsilon}^i$. Suppose that $g \in K$.

If $g = sI$ where $s \in \mathbb{F}_q^*$ then $(\alpha_i + \alpha_{iq})(g) = 2\bar{s}^i$.

If g is conjugate to v_r where $r \in \mathbb{F}_{q^2} \backslash \mathbb{F}_q$ then $(\alpha_i + \alpha_{iq})(g) = \bar{r}^i + \bar{r}^{iq}$.

Since $j \not\equiv i, iq \bmod(q^2-1)$, the characters $\alpha_i + \alpha_{iq}$ and $\alpha_j + \alpha_{jq}$ of K are different, so either $\bar{s}^i \neq \bar{s}^j$ for some $s \in \mathbb{F}_q^*$ or $\bar{r}^i + \bar{r}^{iq} \neq \bar{r}^j + \bar{r}^{jq}$ for some $r \in \mathbb{F}_{q^2} \backslash \mathbb{F}_q$. Therefore, $\chi_i \neq \chi_j$, as we wished to show. ∎

We have now completed the proof of Theorem 28.5, since we have shown that the class functions given in the theorem are inequivalent irreducible characters; and the number of them is $q^2 - 1$, which is the same as the number of conjugacy classes of G.

It is possible to use the character table of $GL(2, q)$ to find the character table of SL(2, q), since most of the irreducible characters remain irreducible when restricted. We do not go fully into this, since the answers are quite complicated, and they depend upon whether q is a power of 2 or $q \equiv 1 \bmod 4$ or $q \equiv 3 \bmod 4$. In Exercise 28.2, though, you are asked to consider the easiest case, namely that where q is a power of 2. Since SL(2, q) $\cong$ PSL(2, q) when q is a power of 2, this gives the character tables of an infinite series of simple groups PSL(2, q).

Among the characters of SL(2, q), those with kernel containing the centre of SL(2, q) provide the characters of the groups PSL(2, q) – compare Chapter 27 – and so the character table of PSL(2, q) is rather easier to find than that of SL(2, q). A challenging exercise is to determine the character table of PSL(2, q) when $q \equiv 1 \bmod 4$ or $q \equiv 3 \bmod 4$ from the character table of GL(2, q).

Although the character table of GL(2, q) was first given in 1907, it was not until the 1950's that the character table of GL(3, q) was found. Then, in 1955, J. A. Green determined the character table of GL(n, q) for all positive integers n.

Summary of Chapter 28

The character table of GL(2, q) has the following properties.

(a) There are $q - 1$ conjugacy classes with representatives of the form $sI = \begin{pmatrix} s & 0 \\ 0 & s \end{pmatrix}$, and there are $q - 1$ irreducible characters of degree 1.

(b) There are $q - 1$ conjugacy classes with representatives of the form $u_s = \begin{pmatrix} s & 1 \\ 0 & s \end{pmatrix}$, and there are $q - 1$ irreducible characters of degree q.

(c) There are $(q - 1)(q - 2)/2$ conjugacy classes with representatives of the form $d_{s,t} = \begin{pmatrix} s & 0 \\ 0 & t \end{pmatrix}$ $(s \neq t)$, and there are $(q - 1)(q - 2)/2$ irreducible characters of degree $q + 1$.

(d) There are $(q^2 - q)/2$ conjugacy classes with representatives of the form $v_r = \begin{pmatrix} 0 & 1 \\ -r^{1+q} & r + r^q \end{pmatrix}$, and there are $(q^2 - q)/2$ irreducible characters of degree $q - 1$.

Exercises for Chapter 28

1. Use Theorem 28.5 to write down explicitly the character table of GL(2, 3).

2. Suppose that q is a power of 2. Let $Z = \{sI : s \in \mathbb{F}_q^*\}$. Prove that
$$GL(2, q) \cong Z \times SL(2, q).$$

Deduce the character table of SL(2, q) from that of GL(2, q). Prove that if $q \neq 2$ then SL(2, q) is simple.

3. Use your solution to Exercise 28.2 to write down explicitly the character table of PSL(2, 8).

29

Permutations and characters

We have already seen in Chapter 13 that if G is a *permutation group*, i.e. a subgroup of S_n for some n, then G has a permutation character π defined by $\pi(g) = |\text{fix}(g)|$ for $g \in G$, a fact which proved useful in many of our subsequent character table calculations. In this chapter we take the theory of permutation groups and characters somewhat further, and develop some useful results, particularly about irreducible characters of symmetric groups (see Theorem 29.12 below).

Group actions

We begin with a more general notion than that of a permutation group. If Ω is a set, denote by $\text{Sym}(\Omega)$ the group of all permutations of Ω. In particular, if $\Omega = \{1, 2, \ldots, n\}$ then $\text{Sym}(\Omega) = S_n$.

Definition
Let G be a group and Ω a set. An *action* of G on Ω is a homomorphism $\phi: G \to \text{Sym}(\Omega)$. We also say that G *acts on* Ω (via ϕ).

29.1 Examples
(1) If $G \leq S_n$ then the identity map is an action of G on $\{1, \ldots, n\}$.
(2) Let $G = S_n$ and let Ω be the set consisting of all pairs $\{i, j\}$ of elements of $\{1, \ldots, n\}$. Define $\phi: G \to \text{Sym}(\Omega)$ by setting

$$\{i, j\}(g\phi) = \{ig, jg\}$$

for all $g \in S_n$ and $1 \leq i < j \leq n$. (So for example, $(1\ 2)\phi$ sends $\{1, 3\} \to \{2, 3\}$.) Check that ϕ is an action of S_n; it is called the action of S_n on pairs.

337

(3) Let $G = GL(2, q)$, the group of invertible 2×2 matrices over the finite field $\mathbb{F}_q$, as defined in Chapter 28. Let V be the 2-dimensional vector space over $\mathbb{F}_q$ consisting of all row vectors (a, b) with $a, b \in \mathbb{F}_q$, and let Ω be the set of all 1-dimensional subspaces $\langle v \rangle$ of V. Define $\phi: G \to \mathrm{Sym}(\Omega)$ by setting

$$\langle v \rangle (g\phi) = \langle vg \rangle$$

for all $\langle v \rangle \in \Omega$ and $g \in G$. For example, if

$$g = \begin{pmatrix} 1 & 1 \\ 0 & 1 \end{pmatrix}$$

then $g\phi$ sends

$$\langle (a, b) \rangle \to \langle (a, a + b) \rangle.$$

Then ϕ is an action of G on Ω.

(4) Let G be a group with a subgroup H of index n, and let Ω be the set of all right cosets Hx of H in G (so $|\Omega| = n$). Define ϕ: $G \to \mathrm{Sym}(\Omega)$ by

$$(Hx)(g\phi) = Hxg$$

for all $x, g \in G$. Then by Exercise 9 of Chapter 23, ϕ is an action of G, and $\mathrm{Ker}\,\phi = \bigcap_{x \in G} x^{-1} Hx \leqslant H$.

To simplify notation, if $\phi: G \to \mathrm{Sym}(\Omega)$ is an action, for $\omega \in \Omega$ and $g \in G$ we usually just write ωg instead of $\omega(g\phi)$. With this notation, the fact that ϕ is a homomorphism simply says that $\omega(gh) = (\omega g)h$ for all $\omega \in \Omega$ and $g, h \in G$.

Adopting this notation, define a relation $\sim$ on Ω as follows: for $\alpha, \beta \in \Omega$, we have $\alpha \sim \beta$ if and only if there exists $g \in G$ such that $\alpha g = \beta$. It is easy to see that $\sim$ is an equivalence relation on Ω. The equivalence classes are called the *orbits* of G on Ω. Thus Ω is the disjoint union of the orbits of G. Write

$$\mathrm{orb}(G, \Omega)$$

for the number of orbits of G on Ω. The group G is said to be *transitive* on Ω if $\mathrm{orb}(G, \Omega) = 1$; in other words, G is transitive if, given any $\alpha, \beta \in \Omega$, there exists $g \in G$ such that $\alpha g = \beta$.

29.2 Examples

(1) Let $G = C_4$, generated by x, say, and let $\phi: G \to S_8$ be the action

defined by $x\phi = (1\ 2\ 3\ 4)(5\ 6)(7\ 8)$ (and of course $x^k\phi = ((1\ 2\ 3\ 4)(5\ 6)(7\ 8))^k$ for any k). Then G has three orbits on $\Omega = \{1, \ldots, 8\}$, namely $\{1, 2, 3, 4\}$, $\{5, 6\}$ and $\{7, 8\}$.

(2) The group G is transitive on the set Ω in each of Examples 29.1(2, 3, 4). This is clear in Example (2); to verify it for Example (3) you need to convince yourself that for any two non-zero row vectors $v, w \in V$ there is an invertible 2×2 matrix $A \in \mathrm{GL}(2, q)$ such that $vA = w$; and in Example (4), simply observe that, given two right cosets Hx, $Hy \in \Omega$, the element $g = x^{-1}y \in G$ has the property that $(Hx)g = Hy$.

Let G be a group acting on a set Ω. For $\omega \in \Omega$, write ω^G for the orbit of G which contains ω, so $\omega^G = \{\omega g : g \in G\}$; and define

$$G_\omega = \{g \in G : \omega g = \omega\}.$$

We call G_ω the *stabilizer* of ω in G.

29.3 Proposition
The stabilizer G_ω is a subgroup of G. Moreover, the size of the orbit ω^G is equal to the index of G_ω in G; that is,

$$|\omega^G| = |G : G_\omega|.$$

Proof If $g, h \in G_\omega$ then $\omega(gh) = (\omega g)h = \omega h = \omega$, hence $gh \in G_\omega$. Also $g^{-1} \in G_\omega$, and G_ω contains the identity, so G_ω is a subgroup.

Now let Δ be the set of right cosets $G_\omega x$ of G_ω in G. Observe that for $x, y \in G$,

$$G_\omega x = G_\omega y \Leftrightarrow xy^{-1} \in G_\omega \Leftrightarrow \omega xy^{-1} = \omega \Leftrightarrow \omega x = \omega y.$$

Hence we can define an injective function $\gamma : \Delta \rightarrow \omega^G$ by

$$\gamma(G_\omega x) = \omega x$$

for all $x \in G$. Clearly γ is also surjective, and hence $|\Delta| = |\omega^G|$, as required. ∎

Permutation characters

Let G be a group acting on a finite set Ω. Denote by $\mathbb{C}\Omega$ the vector space over $\mathbb{C}$ for which Ω is a basis. In other words, $\mathbb{C}\Omega$ consists of all expressions of the form

$$\sum_{\omega \in \Omega} \lambda_\omega \omega \qquad (\lambda_\omega \in \mathbb{C})$$

with the obvious addition and scalar multiplication. As in Chapter 13, we can make $\mathbb{C}\Omega$ into a $\mathbb{C}G$-module, called the *permutation module*, by defining

$$\left(\sum \lambda_\omega \omega \right) g = \sum \lambda_\omega (\omega g)$$

for all $g \in G$. We see just as in Chapter 13 (p. 129) that if π is the character of this permutation module, then for $g \in G$,

$$\pi(g) = |\mathrm{fix}_\Omega(g)|,$$

where $\mathrm{fix}_\Omega(g) = \{\omega \in \Omega : \omega g = \omega\}$. We call π the permutation character of G on Ω.

The next result, though elementary, is rather famous, and provides a basic link between the permutation character and the action of G. It is often referred to as "Burnside's Lemma", but is in fact due to Cauchy and Frobenius.

29.4 Proposition
Let G be a group acting on a finite set Ω, and let π be the permutation character. Then

$$\langle \pi, 1_G \rangle = \frac{1}{|G|} \sum_{g \in G} |\mathrm{fix}_\Omega(g)| = \mathrm{orb}(G, \Omega).$$

Proof First note that

$$\langle \pi, 1_G \rangle = \frac{1}{|G|} \sum_{g \in G} \pi(g) = \frac{1}{|G|} \sum_{g \in G} |\mathrm{fix}_\Omega(g)|.$$

Let $\Delta_1, \ldots, \Delta_t$ be the orbits of G on Ω, and for each i, pick $\omega_i \in \Delta_i$. By Proposition 29.3, for $1 \leq i \leq t$ we have

$$|\Delta_i| = |\omega_i^G| = |G : G_{\omega_i}|.$$

Hence $|\Delta_i| |G_{\omega_i}| = |G|$. Now define $\Phi = \{(\omega, g) : \omega \in \Omega, \ g \in G, \ \omega g = \omega\}$. We calculate $|\Phi|$ in two different ways. First, for each g, the number of $\omega \in \Omega$ such that $\omega g = \omega$ is equal to $|\mathrm{fix}_\Omega(g)|$, hence

$$|\Phi| = \sum_{g \in G} |\text{fix}_\Omega(g)|.$$

Secondly, for each ω, the number of $g \in G$ such that $\omega g = \omega$ is equal to $|G_\omega|$, hence

$$|\Phi| = \sum_{\omega \in \Omega} |G_\omega| = \sum_{i=1}^{t} |\Delta_i| |G_{\omega_i}| = \sum_{1}^{t} |G| = t|G|.$$

Therefore $\sum_{g \in G} |\text{fix}_\Omega(g)| = t|G|$, and the conclusion follows. ∎

29.5 Corollary
G *is transitive on* Ω *if and only if* $\langle \pi, 1_G \rangle = 1$.

Now let G be a group, and suppose that G acts on two sets Ω_1 and Ω_2, with corresponding permutation characters π_1 and π_2 respectively. Then we can define an action of G on the Cartesian product $\Omega_1 \times \Omega_2$ by setting

$$(\omega_1, \omega_2)g = (\omega_1 g, \omega_2 g)$$

for all $\omega_i \in \Omega_i$, $g \in G$. It is clear that $\text{fix}_{\Omega_1 \times \Omega_2}(g) = \text{fix}_{\Omega_1}(g) \times \text{fix}_{\Omega_2}(g)$ for any $g \in G$. Hence if π is the permutation character of G on $\Omega_1 \times \Omega_2$, then

$$\pi(g) = \pi_1(g)\pi_2(g)$$

for all $g \in G$.

29.6 Proposition
Let G act on Ω_1 and Ω_2, with permutation characters π_1 and π_2 respectively. Then

$$\langle \pi_1, \pi_2 \rangle = \text{orb}(G, \Omega_1 \times \Omega_2).$$

Proof We have

$$\langle \pi_1, \pi_2 \rangle = \frac{1}{|G|} \sum_{g \in G} |\text{fix}_{\Omega_1}(g)||\text{fix}_{\Omega_2}(g)| = \frac{1}{|G|} \sum_{g \in G} |\text{fix}_{\Omega_1 \times \Omega_2}(g)|,$$

which is equal to $\text{orb}(G, \Omega_1 \times \Omega_2)$ by Proposition 29.4. ∎

In the rest of the chapter we apply Proposition 29.6 in a number of situations, the first being the case where $\Omega_1 = \Omega_2$.

Suppose G acts on Ω. Then G also acts on $\Omega \times \Omega$ in the way defined above, namely $(\omega_1, \omega_2)g = (\omega_1 g, \omega_2 g)$ for all $\omega_1, \omega_2 \in \Omega$, $g \in G$.

29.7 Definition
The number of orbits of G on $\Omega \times \Omega$ is called the *rank* of G on Ω, written $r(G, \Omega)$. Thus

$$r(G, \Omega) = \mathrm{orb}(G, \Omega \times \Omega).$$

The next result is immediate from Proposition 29.6.

29.8 Proposition
Let G act on Ω, with permutation character π. Then

$$r(G, \Omega) = \langle \pi, \pi \rangle.$$

Now suppose G is transitive on Ω and $|\Omega| > 1$. Then

$$\Delta = \{(\omega, \omega) : \omega \in \Omega\}$$

is an orbit of G on $\Omega \times \Omega$, and hence certainly $r(G, \Omega) \geqslant 2$. The case where equality holds is of particular interest.

29.9 Definition
Let G be transitive on Ω. Then G is said to be *2-transitive* on Ω if $r(G, \Omega) = 2$.

In other words, G is 2-transitive if, for any ordered pairs (α_1, α_2) and (β_1, β_2) in $\Omega \times \Omega$, with $\alpha_1 \neq \alpha_2$, $\beta_1 \neq \beta_2$, there exists $g \in G$ such that $\alpha_1 g = \beta_1$ and $\alpha_2 g = \beta_2$.

29.10 Corollary
If G is 2-transitive on Ω, with permutation character π, then

$$\pi = 1_G + \chi,$$

where χ is an irreducible character of G.

Proof We have $\langle \pi, 1_G \rangle = 1$ by Corollary 29.5, and $\langle \pi, \pi \rangle = 2$ by Proposition 29.8. The result follows, using Theorem 14.17. ∎

29.11 Examples

(1) The symmetric group S_n is 2-transitive on $\{1, \ldots, n\}$. Also A_n is 2-transitive, provided $n \geq 4$. Hence these groups have an irreducible character χ given by

$$\chi(g) = |\mathrm{fix}(g)| - 1.$$

We have seen this irreducible character in a number of previous examples (see 18.1, 19.16, 19.17).

(2) Consider the action of $G = \mathrm{GL}(2, q)$ given in Example 29.1(3). Here Ω is the set of all 1-dimensional subspaces of the 2-dimensional vector space V. We claim that G is 2-transitive on Ω. To see this, let $(\langle v_1 \rangle, \langle v_2 \rangle)$ and $(\langle w_1 \rangle, \langle w_2 \rangle)$ be two pairs of distinct 1-spaces in Ω. Then v_1, v_2 and w_1, w_2 are both bases of V. The linear transformation from V to V which sends $v_1 \rightarrow w_1$, $v_2 \rightarrow w_2$ is therefore invertible, giving an element of $\mathrm{GL}(2, q)$ which sends $\langle v_1 \rangle \rightarrow \langle w_1 \rangle$, $\langle v_2 \rangle \rightarrow \langle w_2 \rangle$. Hence G is 2-transitive on Ω, as claimed. Since $|\Omega| = q + 1$ (see Exercise 1 at the end of the chapter), the irreducible character χ of G given by Corollary 29.10 is the character ψ_0 of degree q in Theorem 28.5.

(3) Consider the action of S_n on pairs defined in Example 29.1(2), with $n \geq 4$. This action is not 2-transitive, since, for example, there is no element of S_n which sends $(\{1, 2\}, \{3, 4\})$ to $(\{1, 2\}, \{2, 3\})$. In fact it is easy to see that the orbits of $G = S_n$ on $\Omega \times \Omega$ are Δ, Δ_1 and Δ_2, where Δ is as above, and

$$\Delta_1 = \{(\{i, j\}, \{k, l\}) : |\{i, j\} \cap \{k, l\}| = 1\},$$
$$\Delta_2 = \{(\{i, j\}, \{k, l\}) : |\{i, j\} \cap \{k, l\}| = 0\}.$$

Thus $\langle \pi, \pi \rangle = r(G, \Omega) = 3$, and so $\pi = 1_G + \chi + \psi$, where χ and ψ are irreducible characters of S_n.

Some irreducible characters of S_n

By Theorem 12.15 we know that the conjugacy classes of S_n are in bijective correspondence with the set of all possible cycle-shapes of permutations. Each cycle-shape (including 1-cycles) is a sequence $\lambda = (\lambda_1, \ldots, \lambda_s)$ of positive integers λ_i such that $\lambda_1 \geq \lambda_2 \geq \ldots \geq \lambda_s$ and $\lambda_1 + \ldots + \lambda_s = n$, and we call such a sequence a *partition* of n.

By Theorem 15.3, the irreducible characters of S_n are also in bijective correspondence with the partitions λ of n. A key aim is therefore to construct, for each partition λ, an irreducible character χ^λ

of S_n, in a natural way. We shall apply the material of this chapter to carry out this aim in the case where $\lambda = (n - k, k)$, a *2-part partition* (see Theorem 29.13 below). The ideas can be developed to carry out the aim in general, but we do not do this; if you want to see this, and much more, on the character theory of S_n, we refer you to the book by G. James listed in the Bibliography.

Let $G = S_n$ and $I = \{1, 2, \ldots, n\}$. For an integer $k \leqslant n/2$, define I_k to be the set consisting of all subsets of I of size k. Just as in Example 29.1(2) we can define an action of G on I_k as follows: for any subset $A = \{i_1, \ldots, i_k\} \in I_k$ and any $g \in G$, let

$$Ag = \{i_1 g, \ldots, i_k g\}.$$

Let π_k be the permutation character of G in its action on I_k. Observe that

$$\pi_k(1) = |I_k| = \binom{n}{k}.$$

29.12 Proposition
If $l \leqslant k \leqslant n/2$, then $\langle \pi_k, \pi_l \rangle = l + 1$.

Proof By Proposition 29.6, $\langle \pi_k, \pi_l \rangle = \mathrm{orb}(G, I_k \times I_l)$. The orbits of $G = S_n$ on $I_k \times I_l$ are easily seen to be $J_0, J_1, \ldots J_l$, where for $0 \leqslant s \leqslant l$,

$$J_s = \{(A, B) \in I_k \times I_l : |A \cap B| = s\}.$$

Hence $\mathrm{orb}(G, I_k \times I_l) = l + 1$, giving the conclusion. ∎

29.13 Theorem
Let $m = n/2$ if n is even, and $m = (n - 1)/2$ if n is odd. Then S_n has distinct irreducible characters $\chi^{(n)} = 1_G$, $\chi^{(n-1,1)}$, $\chi^{(n-2,2)}$, $\ldots, \chi^{(n-m,m)}$ such that for all $k \leqslant m$,

$$\pi_k = \chi^{(n)} + \chi^{(n-1,1)} + \ldots + \chi^{(n-k,k)}.$$

In particular, $\chi^{(n-k,k)} = \pi_k - \pi_{k-1}$.

Proof We prove the existence of irreducible characters $\chi^{(n)}$, $\chi^{(n-1,1)}, \ldots, \chi^{(n-k,k)}$ such that $\pi_k = \chi^{(n)} + \chi^{(n-1,1)} + \ldots + \chi^{(n-k,k)}$, by induction on k. This holds for $k = 1$ by Corollary 29.10.

Now assume the statement holds for all values less than k. Then there exist irreducible characters $\chi^{(n)}$, $\chi^{(n-1,1)}$, $\ldots$, $\chi^{(n-k+1,k-1)}$ such that

$$\pi_i = \chi^{(n)} + \chi^{(n-1,1)} + \ldots + \chi^{(n-i,i)}$$

for all $i < k$. Now by Proposition 29.12,

$$\langle \pi_k, 1_G \rangle = 1, \ \langle \pi_k, \pi_1 \rangle = 2, \ \ldots, \ \langle \pi_k, \pi_{k-1} \rangle = k, \ \langle \pi_k, \pi_k \rangle = k+1.$$

It follows that $\pi_k = \pi_{k-1} + \chi$ for some irreducible character χ. Writing $\chi = \chi^{(n-k,k)}$, we have $\pi_k = \chi^{(n)} + \chi^{(n-1,1)} + \ldots + \chi^{(n-k,k)}$, as required. ∎

29.14 Examples

(1) The formula $\chi^{(n-k,k)} = \pi_k - \pi_{k-1}$ makes it easy to calculate the values of the characters $\chi^{(n-k,k)}$. For example, the degree is

$$\chi^{(n-k,k)}(1) = \pi_k(1) - \pi_{k-1}(1) = \binom{n}{k} - \binom{n}{k-1}.$$

As another example, suppose $n = 7$ and let us calculate the value of the irreducible character $\chi^{(5,2)}$ on a 3-cycle:

$$\chi^{(5,2)}(123) = \pi_2(123) - \pi_1(123) = |\mathrm{fix}_{I_2}(123)| - |\mathrm{fix}_{I_1}(123)| = 6 - 4 = 2.$$

(2) In the character table of S_6 given in Example 19.17, the irreducible characters $\chi_1, \chi_3, \chi_7, \chi_9$ are equal to $\chi^{(6)}, \chi^{(5,1)}, \chi^{(4,2)}, \chi^{(3,3)}$, respectively.

Summary of Chapter 29

1. An action of G on Ω is a homomorphism $G \to \mathrm{Sym}(\Omega)$. The orbits are the equivalence classes in Ω of the relation defined by $\alpha \sim \beta \Leftrightarrow \alpha g = \beta$ for some $g \in G$. The size of the orbit ω^G containing ω is $|\omega^G| = |G : G_\omega|$.

2. If G acts on Ω then $\mathbb{C}\Omega$ is the permutation module, and the corresponding character of G is π, where $\pi(g) = |\mathrm{fix}_\Omega(g)|$. The number of orbits is equal to $\langle \pi, 1_G \rangle$.

3. The rank $r(G, \Omega)$ is the number of orbits of G on $\Omega \times \Omega$, and $r(G, \Omega) = \langle \pi, \pi \rangle$. If G is 2-transitive then $r(G, \Omega) = 2$ and $\pi = 1_G + \chi$ with χ irreducible.

4. The irreducible characters of S_n are in bijective correspondence with partitions of n. The irreducible characters $\chi^{(n-k,k)}$ corresponding to 2-part partitions have values given by Theorem 29.13.

Exercises for Chapter 29

1. Let G be a finite group, and define a function $\phi: G \times G \to \mathrm{Sym}(G)$ by $x((g, h)\phi) = g^{-1}xh$ for all $x, g, h \in G$.
   (a) Show that ϕ is an action of $G \times G$ on G, which is transitive.
   (b) Find the stabilizer $(G \times G)_1$ of the identity $1 \in G$, and find the kernel of ϕ.
   (c) Show that the rank of this action $r(G \times G, G)$ is equal to the number of conjugacy classes of G, and the permutation character π is

   $$\pi = \sum_\chi \chi \times \bar{\chi},$$

   where the sum is over all irreducible characters χ of G, and $\chi \times \bar{\chi}$ is the irreducible character of $G \times G$ given by Theorem 19.18.

2. Show that if Ω is the set of all 1-dimensional subspaces of a 2-dimensional vector space over $\mathbb{F}_q$ (as in Example 29.1(3)), then $|\Omega| = q + 1$.

3. Let $G = \mathrm{GL}(2, q)$ and let $V = \mathbb{F}_q^2$ as in Example 29.1(2). Let $V^* = V - \{0\}$, and define an action $\phi : G \to \mathrm{Sym}(V^*)$ by $v(g\phi) = vg$ for $v \in V^*$, $g \in G$. Let π be the permutation character of G in this action.

   Decompose π as a sum of irreducible characters of $\mathrm{GL}(2, q)$ (the latter are given by Theorem 28.5).

   (*Hint:* one way to do this is to write down the values of π on the conjugacy classes of G, and take inner products with the irreducible characters of G given in 28.5.)

4. Let G be a finite group, and let H_1, H_2 be subgroups of G. For $i = 1, 2$ define Ω_i to be the set of right cosets of H_i in G, so that G acts on Ω_i as in Example 29.1(4); let π_i be the permutation character of G in the action on Ω_i. Suppose that $\pi_1 = \pi_2$.

   Prove that if G is abelian, then $H_1 = H_2$.

   Give an example to show that this need not be the case in general.

5. Let G be a finite group acting transitively on a set Ω of size greater than 1. Prove that G contains an element g such that $|\mathrm{fix}_\Omega(g)| = 0$. (Such an element is called a *fixed-point-free* element of G.)

6. Let n be a positive integer, and let Ω be the set of all ordered pairs (i, j) with $i, j \in \{1, \dots, n\}$ and $i \neq j$. Let S_n act on Ω in the obvious way (namely, $(i, j)g = (ig, jg)$ for $g \in S_n$), and let the permutation character of S_n in this action be $\pi^{(n-2,1,1)}$.

By considering inner products as in the proof of Theorem 29.13, prove that

$$\pi^{(n-2,1,1)} = 1 + 2\chi^{(n-1,1)} + \chi^{(n-2,2)} + \chi,$$

where χ is an irreducible character.

Writing $\chi = \chi^{(n-2,1,1)}$, calculate the degree of $\chi^{(n-2,1,1)}$, and calculate its value on the elements (12) and (123) of S_n.

In the character table of S_6 given in Example 19.17, which irreducible character is equal to $\chi^{(4,1,1)}$?

30
Applications to group theory

There are several ways of using the character theory of a group to determine information about the structure of the group. The examples which we have come across so far – finding the centre of the group, seeing whether or not the group is simple, and so on – require little calculation. In this chapter we present some rather deeper applications. The first involves doing arithmetic with character values to determine certain numbers, known as the class algebra constants. These constants carry information about the multiplication in G, and they can be used to investigate the subgroup structure of G, as we shall demonstrate. The second application takes this much further: the Brauer–Fowler Theorem 23.19 motivates the study of simple groups containing an involution with centralizer isomorphic to a given group C. Using a little group theory and a lot of character theory we shall carry out such a study in the case where $C = D_8$, the dihedral group of order 8.

Class algebra constants

Let G be a finite group and let $C_1, \ldots, C_l$ be the distinct conjugacy classes of G. Recall from Proposition 12.22 that the class sums $\bar{C}_1, \ldots, \bar{C}_l$ form a basis for the centre of the group algebra $\mathbb{C}G$ (where $\bar{C}_i = \sum_{g \in C_i} g$).

30.1 Proposition
There exist non-negative integers a_{ijk} such that for $1 \leq i \leq l$ and $1 \leq j \leq l$,

348

$$\overline{C}_i\overline{C}_j = \sum_{k=1}^{l} a_{ijk}\overline{C}_k.$$

Proof For $g \in C_k$ the coefficient of g in the product $\overline{C}_i\overline{C}_j$ is equal to the number of pairs (a, b) with $a \in C_i$, $b \in C_j$ and $ab = g$. This number is a non-negative integer, and is independent of the chosen element g of C_k. The result follows. ∎

Another way of looking at Proposition 30.1 is to note that $\overline{C}_i\overline{C}_j$ belongs to $Z(\mathbb{C}G)$, so it must be a linear combination of $\overline{C}_1, \ldots, \overline{C}_l$.

30.2 Definition
The integers a_{ijk} in the formula

$$\overline{C}_i\overline{C}_j = \sum_{k=1}^{l} a_{ijk}\overline{C}_k$$

are the *class algebra constants* of G.

From their very definition, the numbers a_{ijk} carry information about the multiplication in G:

(30.3) For all $g \in C_k$ and all i, j we have

$a_{ijk} = $ the number of pairs (a, b) with $a \in C_i$, $b \in C_j$ and $ab = g$.

Also, the constants a_{ijk} determine the product of any two elements in the centre $Z(\mathbb{C}G)$ of the group algebra, since $\overline{C}_1, \ldots, \overline{C}_l$ is a basis of $Z(\mathbb{C}G)$. As the centre of the group algebra plays an important role in representation theory, you might suspect that the class algebra constants are determined by the character table of G. Our next theorem shows that this is indeed the case.

30.4 Theorem
Let $g_i \in C_i$ for $1 \le i \le l$. Then for all i, j, k, we have

$$a_{ijk} = \frac{|G|}{|C_G(g_i)|\,|C_G(g_j)|} \sum_{\chi} \frac{\chi(g_i)\chi(g_j)\overline{\chi(g_k)}}{\chi(1)}$$

where the sum is over all the irreducible characters χ of G.

Proof Let χ be an irreducible character of G, and let U be a $\mathbb{C}G$-module with character χ. Then by Lemma 22.7, for all $u \in U$ we have

$$u\bar{C}_i = \frac{|G|\chi(g_i)}{|C_G(g_i)|\chi(1)}u.$$

Therefore

$$u\bar{C}_i\bar{C}_j = \frac{|G|^2}{|C_G(g_i)|\,|C_G(g_j)|}\frac{\chi(g_i)\chi(g_j)}{(\chi(1))^2}u$$

and

$$\sum_{m=1}^{l} a_{ijm}u\bar{C}_m = \sum_{m=1}^{l} a_{ijm}\frac{|G|\chi(g_m)}{|C_G(g_m)|\chi(1)}u.$$

Since $\bar{C}_i\bar{C}_j = \sum_m a_{ijm}\bar{C}_m$, we deduce that

(30.5) $$\sum_{m=1}^{l} a_{ijm}\frac{\chi(g_m)}{|C_G(g_m)|} = \frac{|G|}{|C_G(g_i)|\,|C_G(g_j)|}\frac{\chi(g_i)\chi(g_j)}{\chi(1)}.$$

Pick k with $1 \leqslant k \leqslant l$. Multiply both sides of equation (30.5) by $\overline{\chi(g_k)}$ and sum over all irreducible characters χ of G, to obtain

$$\sum_{m=1}^{l} a_{ijm}\sum_{\chi}\frac{\chi(g_m)\overline{\chi(g_k)}}{|C_G(g_m)|} = \frac{|G|}{|C_G(g_i)|\,|C_G(g_j)|}\sum_{\chi}\frac{\chi(g_i)\chi(g_j)\overline{\chi(g_k)}}{\chi(1)}.$$

By the column orthogonality relations, Theorem 16.4(2), this yields

$$a_{ijk} = \frac{|G|}{|C_G(g_i)|\,|C_G(g_j)|}\sum_{\chi}\frac{\chi(g_i)\chi(g_j)\overline{\chi(g_k)}}{\chi(1)}. \qquad \blacksquare$$

Examples

30.6 Example

In this example we shall use the class algebra constants to prove some facts about the elements and subgroups of the symmetric group S_4; these results can readily be proved directly, but they serve as a useful illustration of the method.

Let $G = S_4$. By Section 18.1, the character table of G is as shown:

Character table of S_4

Class C_i	C_1	C_2	C_3	C_4	C_5
g_i	1	$(1\,2)$	$(1\,2\,3)$	$(1\,2)(3\,4)$	$(1\,2\,3\,4)$
$|C_G(g_i)|$	24	4	3	8	4
χ_1	1	1	1	1	1
χ_2	1	-1	1	1	-1
χ_3	2	0	-1	2	0
χ_4	3	1	0	-1	-1
χ_5	3	-1	0	-1	1

(1) We use Theorem 30.4 to calculate the class algebra constant a_{555}:

$$a_{555} = \frac{24}{4\cdot 4}\left(\frac{1}{1}+\frac{-1}{1}+\frac{0}{2}+\frac{-1}{3}+\frac{1}{3}\right) = 0.$$

Hence, by (30.3), S_4 does not possess elements a, b of order 4 such that the product ab also has order 4. We deduce from this that S_4 does not have a subgroup which is isomorphic to the quaternion group Q_8: for Q_8 does have two elements of order 4 with product of order 4.

(2) By Theorem 30.4,

$$a_{245} = \frac{24}{4\cdot 8}\left(1+1+\frac{1}{3}+\frac{1}{3}\right) = 2.$$

Hence S_4 has elements a, b of order 2 such that ab has order 4. Writing $x = ab$, we have

$$x^4 = 1, \quad a^{-1}xa = ba = (ab)^{-1} = x^{-1},$$

so $\langle a, b\rangle \cong D_8$. We deduce the fact (which we already know from Exercise 18.1) that S_4 has a subgroup which is isomorphic to D_8.

(3) Finally,

$$a_{235} = \frac{24}{4\cdot 3}(1+1) = 4,$$

so S_4 has elements a of order 2 and b of order 3 with ab of order 4. In fact, it can be shown that S_4 has a *presentation* as follows:

$$S_4 = \langle a, b:\ a^2 = b^3 = (ab)^4 = 1\rangle.$$

In other words, S_4 is generated by a and b, and *all* products of elements of S_4 are determined by the given relations. We supply a

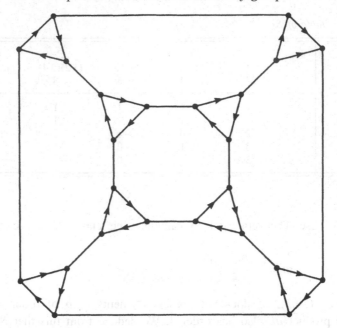

proof in the solution to Exercise 30.6 – in the meantime, you may wish to puzzle out the relevance of the figure above.

30.7 Example
We use Theorem 30.4 to find a subgroup H of the simple group PSL $(2, 7)$ with H isomorphic to S_4. That such a subgroup exists is not obvious, and it is quite tricky to construct directly.

We found in Chapter 27 that the character table of $G = $ PSL $(2, 7)$ is as follows.

Character table of PSL $(2, 7)$

Class rep. g_i	g_1	g_2	g_3	g_4	g_5	g_6
Order of g_i	1	2	4	3	7	7
$|C_G(g_i)|$	168	8	4	3	7	7
χ_1	1	1	1	1	1	1
χ_2	7	-1	-1	1	0	0
χ_3	8	0	0	-1	1	1
χ_4	3	-1	1	0	α	$\bar{\alpha}$
χ_5	3	-1	1	0	$\bar{\alpha}$	α
χ_6	6	2	0	0	-1	-1

where $\alpha = (-1 + i\sqrt{7})/2$.

We calculate the class algebra constant a_{243}. By Theorem 30.4,

$$a_{243} = \frac{168}{8 \cdot 3} \left(1 + \frac{1}{7} + 0 + 0 + 0 + 0 \right) = 8.$$

Hence, by (30.3), G contains elements x and y such that x has order 2, y has order 3 and xy has order 4. Let H be the subgroup $\langle x, y \rangle$ of G. From Example 30.6, we know that

$$S_4 = \langle a, b: \; a^2 = b^3 = (ab)^4 = 1 \rangle.$$

Hence there is a homomorphism ϕ from S_4 onto H (ϕ sends a to x and b to y). By Theorem 1.10, $S_4/\mathrm{Ker}\,\phi \cong H$. Now $\mathrm{Ker}\,\phi$, being a normal subgroup of S_4, is $\{1\}$, V_4, A_4 or S_4 (see Example 12.20), so H is isomorphic to S_4, S_3, C_2 or $\{1\}$. Since H has an element of order 4, namely xy, we conclude that

$$H \cong S_4.$$

Thus we have shown that $\mathrm{PSL}(2, 7)$ has a subgroup which is isomorphic to S_4.

The Brauer programme

The Brauer–Fowler Theorem 23.19 states that there are only finitely many non-isomorphic finite simple groups containing an involution with a given centralizer. This fact led Brauer to initiate a programme to find, given a finite group C, all finite simple groups G possessing an involution t such that $C_G(t) \cong C$. This programme formed an important part of the effort of many mathematicians to classify all the finite simple groups, an effort which was finally completed in the early 1980s (see the book by D. Gorenstein listed in the Bibliography).

The next result carries out part of Brauer's programme in the case where $C = D_8$, a dihedral group of order 8. It determines the possible orders of simple groups G having an involution t such that $C_G(t) \cong D_8$. We have chosen to present this result because it provides a wonderful illustration of the use of character theory in the service of group theory.

30.8 Theorem
Let G be a finite non-abelian simple group which has an involution t such that $C_G(t) \cong D_8$. Then G has order 168 or 360.

Observe that PSL(2, 7) is a simple group of order 168 having an involution with centralizer D_8 (see Lemma 27.1); and A_6 is a simple group of order 360 with this property (see Exercise 7 at the end of the chapter). Using some rather more sophisticated group theory than that covered in this book, one can show that PSL(2, 7) and A_6 are the only simple groups of order 168 or 360.

Before embarking upon the proof of Theorem 30.8, we require a couple of preliminary results. The first is *Sylow's Theorem*, a basic result in finite group theory. We shall not prove this, but refer you to Theorems 18.3 and 18.4 of the book by J. Fraleigh listed in the Bibliography.

30.9 Sylow's Theorem

Let p be a prime number, and let G be a finite group of order $p^a b$, where a, b are positive integers and $p \nmid b$. Then

(1) *G contains a subgroup of order p^a; such a subgroup is call a Sylow p-subgroup of G;*

(2) *all Sylow p-subgroups are conjugate in G (i.e. if P, Q are Sylow p-subgroups, then there exists $g \in G$ such that $Q = g^{-1} Pg$);*

(3) *if R is a subgroup of G with $|R| = p^c$ for some c, then there is a Sylow p-subgroup of G containing R.*

30.10 Lemma

Let G be a finite non-abelian simple group, and let P be a Sylow 2-subgroup of G. Suppose Q is a subgroup of P with $|P : Q| = 2$. If u is an involution in G, then u is conjugate to an element of Q.

Proof Suppose u is not conjugate to an element of Q. Let Ω be the set of right cosets Qx of Q in G, and define an action of G on Ω by $(Qx)g = Qxg$ for $x, g \in G$ (see Example 29.2(4)). Observe that $|\Omega| = 2|G : P| = 2m$, where m is odd since P is a Sylow 2-subgroup.

Now consider $\text{fix}_\Omega(u) = \{\omega \in \Omega : \omega u = \omega\}$. If $Qx \in \text{fix}_\Omega(u)$, then $Qxu = Qx$ and hence $xux^{-1} \in Q$, contrary to assumption. Hence $\text{fix}_\Omega(u) = \varnothing$. This means that in its action on Ω, the involution u is a product of m disjoint 2-cycles, hence is an odd permutation. Hence the subgroup

$$\{g \in G : g \text{ acts as an even permutation on } \Omega\}$$

is a normal subgroup of index 2 in G. This is impossible since G is non-abelian and simple. This contradiction completes the proof. ∎

We also need to introduce the idea of a *generalized character* of a group H. This is simply a class function of the form

$$\psi = \sum_x n_\chi \chi$$

where the sum is over all the irreducible characters of H, and each $n_\chi \in \mathbb{Z}$. If $n_\chi \geqslant 0$ for all χ then of course ψ is a character, but this need not be the case for a generalized character. In particular, the degree $\psi(1)$ can be 0 or negative for a generalized character ψ. Notice also that the orthogonality relations give the usual inner products

$$\langle \psi, \chi \rangle = n_\chi, \quad \langle \psi, \psi \rangle = \sum n_\chi^2$$

for a generalized character ψ as above.

The generalized character ψ can be expressed as a difference $\alpha - \beta$, where α and β are characters of H: take

$$\alpha = \sum_{n_\chi \geqslant 0} n_\chi \chi, \quad \beta = -\sum_{n_\chi < 0} n_\chi \chi.$$

Finally, if H is a subgroup of a group G, we define the induced generalized character $\psi \uparrow G$ by

$$\psi \uparrow G = (\alpha \uparrow G) - (\beta \uparrow G)$$

where $\psi = \alpha - \beta$ as above. It is clear from this definition that the formulae for the values of $\psi \uparrow G$ given in Proposition 21.19 and Corollary 21.20 hold for generalized characters ψ.

Proof of Theorem 30.8

Let G be a finite non-abelian simple group with an involution t such that $C_G(t) = D \cong D_8$. Certainly t commutes with itself, so $t \in D$; and as t commutes with all elements of D, we have $t \in Z(D)$, the centre of D. The centre of D_8 is a cyclic group of order 2 (see (12.12)), and hence $Z(D) = \langle t \rangle$.

By Theorem 30.9(3), there is a Sylow 2-subgroup P of G such that $D \leqslant P$. Then $Z(P) \leqslant C_G(t) = D$, so $Z(P) \leqslant Z(D) = \langle t \rangle$. By Lemma 26.1(1) we have $Z(P) \neq 1$, and hence $Z(P) = \langle t \rangle$. Therefore $P \leqslant$

$C_G(t) = D$, and so $P = D$. In other words, D is a Sylow 2-subgroup of G.

Write $D = \langle a, b \rangle$ where $a^4 = b^2 = 1$ and $b^{-1}ab = a^{-1}$. Then $t = a^2$. Let $C = \langle a \rangle$ be the cyclic subgroup of index 2 in D. By Lemma 30.10, every involution of G is conjugate to an involution in C. As $t = a^2$ is the only such involution, we conclude that t^G is the unique conjugacy class of involutions in G.

Next, let $g \in G$ and suppose that $g^{-1}cg \in C$ for some non-identity element $c \in C$. Since $t = c$ or c^2, we must have $g^{-1}tg = t$, hence $g \in C_G(t) = D$ and so $g^{-1}Cg = C$.

We summarise what we have proved so far:

(30.11)   *D is a Sylow 2-subgroup of G; t^G is the unique conjugacy class of involutions in G; for any $g \in G$ we have $C \cap g^{-1}Cg = \{1\}$ or C; and if $C \cap g^{-1}Cg = C$ then $g \in D$.*

This is all the group theory we will need for the proof. The rest is character theory.

Let λ be the linear character of C such that $\lambda(a) = i$, and define

$$\theta = (1_C \uparrow D) - (\lambda \uparrow D),$$

a generalized character of D. Then θ takes the value 2 on a, a^{-1}, the value 4 on t, and 0 elsewhere. Referring to the character table of D_8 in Example 16.3(3), we have $\theta = \chi_1 + \chi_2 - \chi_5$. (In particular, $\theta(1) = 0$.) Hence $\langle \theta, \theta \rangle = 3$.

We next establish

(30.12) $\langle \theta \uparrow G, \theta \uparrow G \rangle = 3.$

To see this, observe first that by Frobenius reciprocity, $\langle \theta \uparrow G, \theta \uparrow G \rangle = \langle (\theta \uparrow G) \downarrow D, \theta \rangle$ Now for $1 \neq c \in C$, Proposition 21.19 gives

$$(\theta \uparrow G)(c) = \frac{1}{8} \sum_{y \in G} \dot\theta(y^{-1}cy).$$

By (30.11), if $y^{-1}cy \in C$ then $y \in D$, whence $y^{-1}cy = c^{\pm 1}$ and $\theta(y^{-1}cy) = \theta(c)$. And if $y^{-1}cy \in D - C$ then $\theta(y^{-1}cy) = 0$. It follows that $(\theta \uparrow G)(c) = \theta(c)$. Since θ vanishes on $D - C$, we therefore have $\langle (\theta \uparrow G) \downarrow D, \theta \rangle = \langle \theta, \theta \rangle = 3$, giving (30.12).

Now $\langle \theta \uparrow G, 1_G \rangle = \langle 1_C - \lambda, 1_C \rangle = 1$. Also $(\theta \uparrow G)(1) = 0$ (see Corollary 21.20), and so it follows from (30.12) that

$$\theta \uparrow G = 1_G + \alpha - \beta,$$

where α, β are irreducible characters of G. Since we have shown that $(\theta \uparrow G)(t) = \theta(t) = 4$, we have now proved the following.

(30.13) *We have* $\theta \uparrow G = 1_G + \alpha - \beta$, *where* α, β *are irreducible*, $1 + \alpha(1) - \beta(1) = 0$ *and* $1 + \alpha(t) - \beta(t) = 4$.

Note that by Corollary 13.10, $\alpha(t)$ and $\beta(t)$ are integers.

We now introduce another class function of G into the picture. For $g \in G$, define $\gamma(g)$ to be number of ordered pairs $(x, y) \in t^G \times t^G$ such that $g = xy$. If we write $t^G = C_i$ and g lies in the conjugacy class C_k of G, then $\gamma(g) = a_{iik}$ in the notation of (30.3). Hence Theorem 30.4 yields the following.

(30.14) *We have*

$$\gamma = \frac{|G|}{|D|^2} \sum_\chi \frac{\chi(t)^2}{\chi(1)} \chi,$$

where the sum is over all irreducible characters χ *of* G.

We shall calculate the inner product of γ and $\theta \uparrow G$ in two ways. First, from (30.13) and (30.14) we have

(30.15) $$\langle \theta \uparrow G, \gamma \rangle = \frac{|G|}{64} \left(1 + \frac{\alpha(t)^2}{\alpha(1)} - \frac{\beta(t)^2}{\beta(1)} \right).$$

On the other hand, by Frobenius Reciprocity, $\langle \theta \uparrow G, \gamma \rangle = \langle 1_C - \lambda, \gamma \downarrow C \rangle$. Consider $\gamma(c)$ for $1 \neq c \in C$. If $c = xy$ with $x, y \in t^G$, then $x^{-1}cx = yx = c^{-1}$, and hence $x \in D$ by (30.11); similarly $y \in D$. Now calculation in D_8 shows that $\gamma(c) = 4$. Therefore

$$\langle 1_C - \lambda, \gamma \downarrow C \rangle = \frac{1}{|C|} \cdot 4 \cdot ((1 - i) + 2 + (1 + i)) = 4.$$

Hence from (30.15) we deduce

(30.16) $$|G| \left(1 + \frac{\alpha(t)^2}{\alpha(1)} - \frac{\beta(t)^2}{\beta(1)} \right) = 2^8.$$

This equation gives us enough number-theoretic information about $|G|$ to finish the proof fairly quickly. Write $d = \alpha(1)$ and $e = \alpha(t) \in \mathbb{Z}$. By (30.13) we have

$$\beta(1) = d + 1, \quad \beta(t) = e - 3.$$

From the column orthogonality relations 16.4(2), we have

$$8 = |C_G(t)| \geqslant 1 + \alpha(t)^2 + \beta(t)^2 = 1 + e^2 + (e - 3)^2,$$

from which it follows that $e = 1$ or 2.

Suppose now that $e = 1$. Then (30.16) gives

$$|G|\left(1 + \frac{1}{d} - \frac{4}{d + 1}\right) = 2^8,$$

whence

$$|G| = 2^8 \frac{d(d + 1)}{(d - 1)^2}.$$

Now the highest common factor $\mathrm{hcf}(d - 1, d + 1)$ is 1 or 2, and $\mathrm{hcf}(d - 1, d) = 1$. Hence $(d - 1)^2$ must divide 2^{10}, and so $d - 1 = 2^r$ with $r \leqslant 5$. Moreover, a Sylow 2-subgroup of G has order 8. It follows that $r = 3$ and $d = 9$, giving $|G| = 360$, one of the possibilities in the conclusion of Theorem 30.8.

Finally, suppose that $e = 2$. Then (30.16) yields

$$|G| = 2^8 \frac{d(d + 1)}{(d + 2)^2}.$$

Reasoning as above, we deduce that $d + 2 = 2^3$, giving $d = 6$ and $|G| = 168$. This completes the proof of Theorem 30.8. ∎

Summary of Chapter 30

1. The class algebra constants a_{ijk} are given by

$$\bar{C}_i \bar{C}_j = \sum_k a_{ijk} \bar{C}_k.$$

They can be calculated from the character table, by using the formula

$$a_{ijk} = \frac{|G|}{|C_G(g_i)| \, |C_G(g_j)|} \sum_\chi \frac{\chi(g_i)\chi(g_j)\overline{\chi(g_k)}}{\chi(1)}.$$

2. Given groups G and H, the class algebra constants of G can sometimes be used to determine whether or not G has a subgroup which isomorphic to H.

3. Using Sylow's Theorem, together with lots of ingenious character theory, it can be shown that any simple group possessing an involution with centralizer isomorphic to D_8 must have order 168 or 360.

Exercises for Chapter 30

1. Use the character table of PSL$(2, 7)$, given at the end of Chapter 27, to prove that PSL$(2, 7)$ contains elements a and b such that a has order 2, b has order 3 and ab has order 7.

2. Does PSL$(2, 7)$ contain a subgroup isomorphic to D_{14}?
   (Hint: $D_{14} = \langle a, b: a^2 = b^2 = 1, (ab)^7 = 1 \rangle$.)

   For the next three exercises, you may assume that A_5 is a simple group, and that A_5 has the following presentation:

   $$A_5 = \langle a, b: a^2 = b^3 = (ab)^5 = 1 \rangle.$$

3. The character table of PSL$(2, 11)$ is given in the solution to Exercise 27.6. Does PSL$(2, 11)$ contain a subgroup which is isomorphic to A_5?

4. Prove that A_5 is characterized by its character table – that is, if G is a group with the same character table as A_5 (see Example 20.13), then $G \cong A_5$.

	g_1	g_2	g_3	g_4	g_5	g_6	g_7
χ_1	1	1	1	1	1	1	1
χ_2	5	1	-1	2	-1	0	0
χ_3	5	1	-1	-1	2	0	0
χ_4	8	0	0	-1	-1	α	β
χ_5	8	0	0	-1	-1	β	α
χ_6	9	1	1	0	0	-1	-1
χ_7	10	-2	0	1	1	0	0

where $\alpha = (1 + \sqrt{5})/2, \beta = (1 - \sqrt{5})/2$.

5. Suppose that G is a group, and that G has the character table shown.
   (a) Show that G is a simple group of order 360.
   (b) Use the Frobenius–Schur Count of Involutions to obtain an

upper bound for the number of involutions in G, and deduce that g_2 has order 2 and g_3 has order 4.

(c) Prove that G has a subgroup H which is isomorphic to A_5.

(d) Using Exercise 23.9, show that $G \cong A_6$.

6. Use the figure which appears in Example 30.6(3) to show that every group G which is generated by two elements a and b which satisfy

$$a^2 = b^3 = (ab)^4 = 1$$

has order at most 24.

7. Prove that PSL(2, 7) and A_6 are simple groups of order 168, 360 respectively, both of which contain an involution with centralizer isomorphic to D_8.

8. Find a simple group G having an involution t such that $C_G(t) \cong D_{16}$.

(Hint: look for a suitable simple group PSL(2, p).)

31
Burnside's Theorem

One of the most famous applications of representation theory is Burnside's Theorem, which states that if p and q are prime numbers and a and b are positive integers, then no group of order $p^a q^b$ is simple. In the first edition of his book *Theory of groups of finite order* (1897), Burnside presented group-theoretic arguments which proved the theorem for many special choices of the integers a, b, but it was only after studying Frobenius's new theory of group representations that he was able to prove the theorem in general. Indeed, many later attempts to find a proof which does not use representation theory were unsuccessful, until H. Bender found one in 1972.

A preliminary lemma

We prepare for the proof of Burnside's Theorem with a lemma (31.2) which is concerned with character values. In order to establish this lemma we require some basic facts about algebraic integers and algebraic numbers, which we now describe. We omit proofs of these – for a good account, see for instance the book by Pollard and Diamond listed in the Bibliography.

An *algebraic number* is a complex number which is a root of some non-zero polynomial over $\mathbb{Q}$. We call a polynomial in x *monic* if the coefficient of the highest power of x in it is 1.

Let α be an algebraic number; and let $p(x)$ be a monic polynomial over $\mathbb{Q}$ of smallest possible degree having α as a root. Then $p(x)$ is unique and irreducible; it is called the *minimal polynomial* of α. The roots of $p(x)$ are called the *conjugates* of α.

For example, if ω is an nth root of unity then the minimal polynomial of ω divides $x^n - 1$, and so every conjugate of ω is also an nth root of unity.

If α is an algebraic integer, then α is a root of a monic polynomial with integer coefficients (see Chapter 22), and it turns out that the minimal polynomial of α also has integer coefficients.

We shall require the following fact about conjugates:

(31.1) Let α and β be algebraic numbers. Then every conjugate of $\alpha + \beta$ is of the form $\alpha' + \beta'$, where α' is a conjugate of α and β' is a conjugate of β. Moreover, if $r \in \mathbb{Q}$ then every conjugate of $r\alpha$ is of the form $r\alpha'$, where α' is a conjugate of α.

For an elementary proof of this, see Pollard and Diamond, Chapter V, Section 3. Alternatively, (31.1) can be proved easily using some Galois theory.

31.2 Lemma
Let χ be a character of a finite group G, and let $g \in G$. Then $|\chi(g)/\chi(1)| \le 1$, and if

$$0 < |\chi(g)/\chi(1)| < 1$$

then $\chi(g)/\chi(1)$ is not an algebraic integer.

Proof Let $\chi(1) = d$. By Proposition 13.9 we have $\chi(g) = \omega_1 + \ldots + \omega_d$, where each ω_i is a root of unity, so

$$\chi(g)/\chi(1) = (\omega_1 + \ldots + \omega_d)/d.$$

Since $|\chi(g)| = |\omega_1 + \ldots + \omega_d| \le |\omega_1| + \ldots + |\omega_d| = d$, it follows that $|\chi(g)/\chi(1)| \le 1$.

Now suppose that $\chi(g)/\chi(1)$ is an algebraic integer and $|\chi(g)/\chi(1)| < 1$. We prove that $\chi(g) = 0$.

Write $\gamma = \chi(g)/\chi(1)$, and let $p(x)$ be the minimal polynomial of γ, so that

$$p(x) = x^n + a_{n-1}x^{n-1} + \ldots + a_1 x + a_0$$

where $a_i \in \mathbb{Z}$ for all i. By (31.1), each conjugate of γ is of the form

$$(\omega'_1 + \ldots + \omega'_d)/d$$

where $\omega'_1, \ldots, \omega'_d$ are roots of unity. Hence each conjugate of γ has modulus at most 1. It follows that if λ is the product of all the conjugates of γ (including γ), then

$$|\lambda| < 1.$$

But the conjugates of γ are, by definition, the roots of the polynomial $p(x)$, and the product of all these roots is equal to $\pm a_0$. Thus

$$\lambda = \pm a_0.$$

Since $a_0 \in \mathbb{Z}$ and $|\lambda| < 1$, it follows that $a_0 = 0$. As $p(x)$ is irreducible, this implies that

$$p(x) = x,$$

which in turn forces $\gamma = 0$. Thus $\chi(g) = 0$, and the proof is complete. ∎

Burnside's $p^a q^b$ Theorem

We deduce the main result, Theorem 31.4, from another interesting theorem of Burnside.

31.3 Theorem
Let p be a prime number and let r be an integer with $r \geqslant 1$. Suppose that G is a finite group with a conjugacy class of size p^r. Then G is not simple.

Proof Let $g \in G$ with $|g^G| = p^r$. Since $p^r > 1$, G is not abelian and $g \neq 1$. As usual, denote the irreducible characters of G by $\chi_1, \ldots, \chi_k$, and take χ_1 to be the trivial character.

The column orthogonality relations, Theorem 16.4(2), applied to the columns corresponding to 1 and g in the character table of G, give

$$1 + \sum_{i=2}^{k} \chi_i(g)\chi_i(1) = 0.$$

Therefore

$$\sum_{i=2}^{k} \chi_i(g) \cdot \frac{\chi_i(1)}{p} = -\frac{1}{p}.$$

Now $-1/p$ is not an algebraic integer, by Proposition 22.5. Therefore, for some $i \geqslant 2$, $\chi_i(g)\chi_i(1)/p$ is not an algebraic integer (see Theorem 22.3). Since $\chi_i(g)$ is an algebraic integer (Corollary 22.4), it follows that $\chi_i(1)/p$ is not an algebraic integer; in other words, p does not divide $\chi_i(1)$. Thus

$$\chi_i(g) \neq 0 \quad \text{and} \quad p \nmid \chi_i(1).$$

As $|g^G| = p^r$, this means that $\chi_i(1)$ and $|g^G|$ are coprime integers, and so there are integers a and b such that

$$a|G{:}C_G(g)| + b\chi_i(1) = 1.$$

Hence

$$a\frac{|G|\chi_i(g)}{|C_G(g)|\chi_i(1)} + b\chi_i(g) = \frac{\chi_i(g)}{\chi_i(1)}.$$

By Corollaries 22.10 and 22.4, the left-hand side of this equation is an algebraic integer; and since $\chi_i(g) \neq 0$, it is non-zero. Now Lemma 31.2 implies that

$$|\chi_i(g)/\chi_i(1)| = 1.$$

Let ρ be a representation of G with character χ_i. By Theorem 13.11(1), there exists $\lambda \in \mathbb{C}$ such that

$$g\rho = \lambda I.$$

Let $K = \operatorname{Ker} \rho$, so that K is a normal subgroup of G. Since χ_i is not the trivial character, $K \neq G$. If $K \neq \{1\}$ then G is not simple, as required; so assume that $K = \{1\}$, that is, ρ is a faithful representation.

Since $g\rho$ is a scalar multiple of the identity, $g\rho$ commutes with $h\rho$ for all $h \in G$. As ρ is faithful, it follows that g commutes with all $h \in G$; in other words,

$$g \in Z(G).$$

Therefore $Z(G) \neq \{1\}$. As $Z(G)$ is a normal subgroup of G and $Z(G) \neq G$, we conclude that G is not simple. ∎

We now come to the main result of the chapter, Burnside's Theorem.

31.4 Burnside's $p^a q^b$ Theorem
Let p and q be prime numbers, and let a and b be non-negative integers with $a + b \geq 2$. If G is a group of order $p^a q^b$, then G is not simple.

Proof First suppose that either $a = 0$ or $b = 0$. Then the order of G is a power of a prime, so by Lemma 26.1(1) we have $Z(G) \neq \{1\}$.

Choose $g \in Z(G)$ of prime order. Then $\langle g \rangle \lhd G$ and $\langle g \rangle$ is not equal to $\{1\}$ or G. Hence G is not simple.

Now assume that $a > 0$ and $b > 0$. By Sylow's Theorem 30.9, G has a subgroup Q of order q^b. We have $Z(Q) \neq \{1\}$ by Lemma 26.1(1). Let $g \in Z(Q)$ with $g \neq 1$. Then $Q \leqslant C_G(g)$, so

$$|g^G| = |G{:}C_G(g)| = p^r$$

for some r. If $p^r = 1$ then $g \in Z(G)$, so $Z(G) \neq \{1\}$ and G is not simple as before. And if $p^r > 1$ then G is not simple, by Theorem 31.3. ∎

In fact Burnside's $p^a q^b$ Theorem leads to a somewhat more informative result about groups of order $p^a q^b$:

(31.5) *Every group of order $p^a q^b$ is soluble.*

Here, by a *soluble* group we mean a group G which has subgroups $G_0, G_1, \ldots, G_r$ with

$$1 = G_0 < G_1 < \ldots < G_r = G$$

such that for $1 \leqslant i \leqslant r$, $G_{i-1} \lhd G_i$ and the factor group G_i/G_{i-1} is cyclic of prime order.

We sketch a proof of (31.5), using induction on $a + b$. The result is clear if $a + b \leqslant 1$, so assume that $a + b \geqslant 2$ and let G be a group of order $p^a q^b$. By Burnside's Theorem 31.4, G has a normal subgroup H such that H is not $\{1\}$ or G. Both H and the factor group G/H have order equal to a product of powers of p and q, and these orders are less than $p^a q^b$. Hence by induction, H and G/H are both soluble. Therefore there are subgroups

$$1 = G_0 \lhd G_1 \lhd \ldots \lhd G_s = H,$$

$$1 = G_s/H \lhd G_{s+1}/H \lhd \ldots \lhd G_r/H = G/H$$

with all factor groups G_i/G_{i-1} of prime order. Then the series

$$1 = G_0 \lhd G_1 \lhd \ldots \lhd G_r = G$$

shows that G is soluble.

Summary of Chapter 31

1. If G has a conjugacy class of size p^r (p prime, $r \geqslant 1$), then G is not simple.

2. If $|G| = p^a q^b$ (p, q primes, $a + b \geqslant 2$), then G is not simple.

Exercises for Chapter 31

1. Show that a non-abelian simple group cannot have an abelian subgroup of prime power index.

2. Prove that if G is a non-abelian simple group of order less than 80, then $|G| = 60$.
   (Hint: use Exercise 13.8.)

32

An application of representation theory to molecular vibration

Representation theory is used extensively in many of the physical sciences. Such applications come about because every physical system has a symmetry group G, and certain vector spaces associated with the system turn out to be $\mathbb{R}G$-modules. For example, the vibration of a molecule is governed by various differential equations, and the symmetry group of the molecule acts on the space of solutions of these equations. It is on this application – the theory of molecular vibrations – that we concentrate in this final chapter. In order to keep our treatment elementary, we stay within the framework of classical mechanics throughout. (Quantum mechanical effects can be incorporated subsequently, but we shall not go into this – for an account, consult the book by D. S. Schonland listed in the Bibliography.)

Symmetry groups

Let V be $\mathbb{R}^2$ or $\mathbb{R}^3$, and for v, $w \in V$, let $d(v, w)$ denote the distance between v and w – in other words, if $v = (x_1, x_2, \ldots)$ and $w = (y_1, y_2, \ldots)$, then

$$d(v, w) = \sqrt{\left(\sum (x_i - y_i)^2 \right)}.$$

An *isometry* of V is an invertible endomorphism ϑ of V such that

$$d(v\vartheta, w\vartheta) = d(v, w) \quad \text{for all } v, w \in V.$$

The set of all isometries of V forms a group under composition, called the *orthogonal group of V,* and denoted by $O(V)$.

Any rotation of $\mathbb{R}^3$ about an axis through the origin is an example

367

of an isometry; so is any reflection in a plane through the origin. The endomorphism $-1_{\mathbb{R}^3}$ which sends every vector v to $-v$ is another example of an element in the orthogonal group $O(\mathbb{R}^3)$. It turns out that the composition of two rotations is again a rotation, and that for every isometry g in $O(\mathbb{R}^3)$, either g or $-g$ is a rotation (see Exercise 32.1). The orthogonal group $O(\mathbb{R}^3)$ therefore contains a subgroup of index 2 which consists of the rotations. The same is true of the group $O(\mathbb{R}^2)$.

If Δ is a subset of V, where $V = \mathbb{R}^2$ or $\mathbb{R}^3$, then we define $G(\Delta)$ to be the set of isometries which leave Δ invariant – that is,

$$G(\Delta) = \{g \in O(V): \Delta g = \Delta\}$$

(where $\Delta g = \{vg: v \in \Delta\}$). Then $G(\Delta)$ is a subgroup of $O(V)$, and is called the *symmetry group* of Δ. The subgroup of $G(\Delta)$ consisting of the rotations in $G(\Delta)$ is called the *rotation group* of Δ. The index of the rotation group of Δ in the symmetry group $G(\Delta)$ is 1 or 2.

32.1 Example

Let $V = \mathbb{R}^2$, and let Δ be a regular n-sided polygon, with $n \geqslant 3$, centred at the origin. The symmetry group of Δ is easily seen to be the dihedral group D_{2n}, which was defined as a group of n rotations and n reflections preserving Δ (see Example 1.1(3)).

Now let $V = \mathbb{R}^3$, and again let Δ be a regular n-sided polygon ($n \geqslant 3$) centred at the origin. This time, $G(\Delta) = D_{2n} \times C_2$; the extra elements arise because there is an isometry which fixes all points of Δ, namely the reflection in the plane of Δ.

32.2 Example

Let Δ be a regular tetrahedron in $\mathbb{R}^3$ centred at the origin:

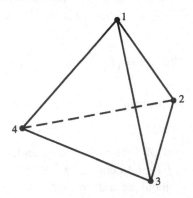

Label the corners of the tetrahedron 1, 2, 3, 4. We claim that each permutation of the numbers 1, 2, 3, 4 corresponds to an isometry of Δ. For example, the 2-cycle (1 2) corresponds to a reflection in the plane which contains the origin and the edge 34; similarly each 2-cycle corresponds to a reflection. Since S_4 is generated by the 2-cycles, each of the 24 permutations of 1, 2, 3, 4 corresponds to an isometry, as claimed.

No non-identity endomorphism of $\mathbb{R}^3$ fixes all the corners of Δ, since Δ contains three linearly independent vectors. Therefore we have found all the isometries, and $G(\Delta) \cong S_4$.

Notice that the rotation group of Δ is isomorphic to A_4; for example, (1 2)(3 4) corresponds to a rotation through π about the axis through the mid-points of the edges 12 and 34, and (1 2 3) corresponds to a rotation through $2\pi/3$ about the axis through the origin and the corner 4.

Finally, observe that the group $G(\Delta)$ is unchanged if we take Δ to consist of just the four corners of the tetrahedron.

32.3 Example

In this example we describe the symmetry groups of the molecules H_2O (water), CH_3Cl (methyl chloride) and CH_4 (methane). The *symmetry group of a molecule* is defined to be the group of isometries which not only preserve the position of the molecule in space, but also send each atom to an atom of the same kind.

The shapes of the three molecules are as follows.

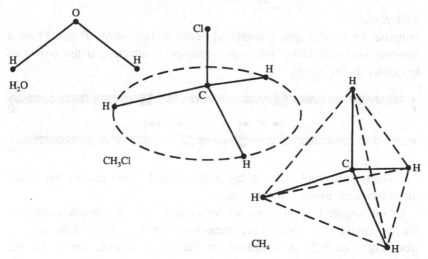

We always assume that the centroid of our molecule lies at the origin in $\mathbb{R}^3$.

The CH_4 molecule has four hydrogen atoms at the corners of a regular tetrahedron, and a carbon atom at the centre of the tetrahedron. So the symmetry group of the molecule CH_4 is equal to the symmetry group of the tetrahedron, as given in Example 32.2. This group is isomorphic to S_4, permuting the four hydrogen atoms among themselves and fixing the carbon atom.

As for the CH_3Cl molecule, this possesses a rotation symmetry a of order 3 about the vertical axis, and three reflection symmetries in the planes containing the C, Cl and one of the H atoms. If b is one of these reflections, then the symmetry group is

$$\{1, a, a^2, b, ab, a^2b\}$$

and is isomorphic to S_3, permuting the three H atoms and fixing the C and Cl atoms.

Finally, the H_2O molecule possesses two reflection symmetries, one in the plane of the molecule, and one in a plane perpendicular to this one passing through the O atom; and it has a rotation symmetry of order 2. Hence the symmetry group is isomorphic to $C_2 \times C_2$.

Vibration of a physical system

We prepare for a description of the general problem with an example.

32.4 Example
Suppose we have a spring stretched between two points P and Q on a smooth horizontal table, with equal masses m attached at the points of trisection of the spring:

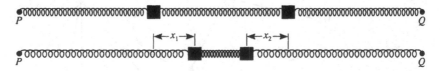

The masses are displaced slightly, and released. What can we say about the subsequent motion of the system?

To investigate this problem, we let x_1 and x_2 be the displacements of the two masses at time t. We measure x_1 from left to right, and x_2 from right to left, as indicated in the figure above. Let k be the stiffness of the spring – in other words, if the extension in the spring is x, then the restoring force is kx.

The spring pulls the left-hand mass towards P with force kx_1 and towards Q with force $-k(x_1 + x_2)$. By dealing with the right-hand mass similarly, we obtain the following equations of motion of the system:

$$m\ddot{x}_1 = -kx_1 - k(x_1 + x_2) = -2kx_1 - kx_2,$$

$$m\ddot{x}_2 = -kx_2 - k(x_1 + x_2) = -kx_1 - 2kx_2,$$

where $\ddot{x}_i$ denotes the second derivative of x_i with respect to t.

These are second order linear differential equations in two unknowns x_1 and x_2, so the general solution involves four arbitrary constants. We shall find the general solution, using a method which can be applied in a much wider context.

Write $x = (x_1, x_2)$, $\ddot{x} = (\ddot{x}_1, \ddot{x}_2)$ and $q = k/m$. Then the equations of motion are equivalent to the matrix equation

(32.5) $\qquad \ddot{x} = xA, \quad \text{where } A = \begin{pmatrix} -2q & -q \\ -q & -2q \end{pmatrix}.$

Notice that A is symmetric. Hence the eigenvalues of A are real, and A has two linearly independent eigenvectors. It is this property which we wish to emphasize and exploit in the present example.

Before we explicitly find the eigenvectors of A, let us explain why they allow us to solve the equation of motion (32.5).

Suppose that u is an eigenvector of A, with eigenvalue $-\omega^2$. For an arbitrary constant β, let

$$x = \sin(\omega t + \beta) u.$$

Then

$$\ddot{x} = -\omega^2 \sin(\omega t + \beta) u$$

$$= \sin(\omega t + \beta) uA \quad (\text{since } uA = -\omega^2 u)$$

$$= xA.$$

Thus x is a solution of the equation of motion. If u_1 and u_2 are linearly independent eigenvectors of A, with eigenvalues $-\omega_1^2$ and $-\omega_2^2$, respectively, then

$$\alpha_1 \sin(\omega_1 t + \beta_1) u_1 + \alpha_2 \sin(\omega_2 t + \beta_2) u_2$$

is a solution of the equation of motion which involves four arbitrary constants α_1, α_2, β_1, β_2, so it is the general solution.

We now adopt this line of attack in the problem to hand. For the matrix given in (32.5), the eigenvalues are $-3q$ and $-q$, with corresponding eigenvectors $(1, 1)$ and $(1, -1)$. Therefore the general solution of the equation of motion (32.5) is

$$\alpha_1 \sin(\sqrt{(3q)}\, t + \beta_1)(1, 1) + \alpha_2 \sin(\sqrt{q} \cdot t + \beta_2)(1, -1).$$

The solutions which involve just one eigenvector of A are called the *normal modes of vibration*. They are as follows.

Mode 1: $\sin(\sqrt{(3q)}\, t + \beta_1)(1, 1)$

Here, $x_1 = x_2 = \sin(\sqrt{(3q)}\, t + \beta_1)$ and the vibration is

Mode 2: $\sin(\sqrt{q} \cdot t + \beta_2)(1, -1)$

Here, $x_1 = -x_2 = \sin(\sqrt{q} \cdot t + \beta_2)$ and the vibration is

The general molecular vibration problem

Suppose we have a molecule consisting of n atoms which vibrate under internal forces. At the equilibrium position of each atom, we assign three coordinate axes, which we use to measure the displacement of the atom. Thus, the state of the molecule at a given time is described by a vector in the $3n$-dimensional vector space $\mathbb{R}^{3n}$. We assume that the internal forces are linear functions of the displacements. It follows that when we apply Newton's Second Law of Motion, we obtain equations which may be expressed in the form

(32.6) $\ddot{x} = xA.$

(Compare (32.5).) Here x is the row vector in $\mathbb{R}^{3n}$ which measures the displacements of all the atoms, and A is a $3n \times 3n$ matrix with real entries which are determined by the internal forces.

Assume, for the moment, that at each atom the three coordinate axes which we have chosen are at right angles to each other. It can be shown, from physical considerations, that in this special case the matrix A is symmetric. In particular, A has real eigenvalues, and A has $3n$ linearly independent eigenvectors. Now, the effect of changing coordinate axes is merely to replace A by a matrix which is conjugate to A. Therefore we have the following proposition, for the general case, where our chosen coordinate axes are not necessarily at right angles to each other.

32.7 Proposition
All the eigenvalues of A are real, and A has $3n$ linearly independent eigenvectors.

To solve the equation of motion (32.6), we look for normal modes of the system, which we define next.

32.8 Definition
A normal mode of vibration for our molecule is a vector in $\mathbb{R}^{3n}$ of one of the following forms:

(1) $\qquad\qquad \sin(\omega t + \beta)\, u \quad (\beta \text{ constant})$

where $-\omega^2$ is a non-zero eigenvalue of A and u is a corresponding eigenvector;

(2) $\qquad\qquad (t + \beta)\, u \quad (\beta \text{ constant})$

where u is an eigenvector of A corresponding to the eigenvalue 0.

32.9 Proposition
Each normal mode of vibration is a solution of the equation of motion (32.6), and for this solution all the atoms vibrate with the same frequency. The general solution of the equation of motion is a linear combination of the normal modes of vibration.

Proof If $uA = -\omega^2 A$ and $x = \sin(\omega t + \beta)\, u$, then

$$\ddot{x} = -\omega^2 \sin(\omega t + \beta)\, u = \sin(\omega t + \beta)\, uA = xA.$$

If $uA = 0$ and $x = (t + \beta)u$, then

$$\ddot{x} = 0 = (t + \beta)uA = xA.$$

This proves that the normal modes of vibration are solutions of the equation of motion (32.6). By construction, all the atoms vibrate with the same frequency (namely, ω or 0) in a normal mode.

Note that A can have no strictly positive eigenvalue; for if λ were such an eigenvalue, with eigenvector u, then $x = e^{\sqrt{\lambda} t} u$ would be a solution to the equation of motion, which is nonsense. Therefore there exist $3n$ linearly independent normal modes, by Proposition 32.7. Since each normal mode involves an arbitrary constant, the general linear combination of normal modes involves $6n$ arbitrary constants, so it is the general solution to the equation of motion (32.6) (as (32.6) consists of second order differential equations in $3n$ unknowns).   ■

Proposition 32.9 reduces the problem of solving the equations of motion to that of finding all the eigenvalues and eigenvectors of the $3n \times 3n$ matrix A. However, this can be a huge and unwieldy task if it is attempted directly – even writing down the matrix A for a given molecule can be a painful operation!

The symmetry group of the molecule and its representation theory can often be used to simplify greatly the calculation of the eigenvectors of A, and we shall describe a method for doing this.

Use of the symmetry group

We continue the discussion of the previous section. Let G be the symmetry group of the molecule in question. Since G permutes the atoms among themselves, each element of G acts as an endomorphism of the space $\mathbb{R}^{3n}$ of displacement vectors. Thus, $\mathbb{R}^{3n}$ is an $\mathbb{R}G$-module.

32.10 Example
Let g be the rotation of order 2 of the H_2O molecule:

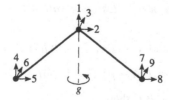

Assign coordinate axes at the initial positions of each atom as shown, and for $1 \leqslant i \leqslant 9$, let v_i denote a unit vector along coordinate axis i. Then g fixes v_1, negates v_2 and v_3, interchanges v_4 and v_7, and interchanges v_5 and v_6 with the negatives of v_8 and v_9. Therefore g acts on $\mathbb{R}^9$ as follows:

$$(x_1, x_2, x_3, x_4, x_5, x_6, x_7, x_8, x_9)g$$
$$= (x_1, -x_2, -x_3, x_7, -x_8, -x_9, x_4, -x_5, -x_6).$$

We return to the general set-up. The equations of motion are $\ddot{x} = xA$, and we are trying to find the eigenvectors of A. The eigenspace for the eigenvalue λ of A is, by definition,

$$\{x \in \mathbb{R}^{3n} : xA = \lambda x\}.$$

We can now present the crucial proposition which allows us to exploit the symmetry group G of our molecule. In effect, it tells us that the function $x \rightarrow xA$ $(x \in \mathbb{R}^{3n})$ is an $\mathbb{R}G$-homomorphism from $\mathbb{R}^{3n}$ to itself.

32.11 Proposition
For all $g \in G$ and $x \in \mathbb{R}^{3n}$,

$$(xg)A = (xA)g,$$

and the eigenspaces of A are $\mathbb{R}G$-submodules of $\mathbb{R}^{3n}$.

Proof Let $-\omega^2$ be a non-zero eigenvalue of A, and let v be a corresponding eigenvector. Then v specifies the directions and relative magnitudes of the displacements of the atoms from the equilibrium position when the molecule is vibrating in a normal mode of frequency ω. For all g in G, vg must also specify the directions and relative magnitudes of displacements in a normal mode of frequency ω, since the relative configuration of the atoms is unaltered by applying g. Therefore, vg is an eigenvector of A, with eigenvalue $-\omega^2$. This shows that the eigenspace for $-\omega^2$ is an $\mathbb{R}G$-submodule of $\mathbb{R}^{3n}$. A similar argument applies to the eigenspace for the eigenvalue 0.

Choose a basis of $\mathbb{R}^{3n}$ which consists of eigenvectors of A (see Proposition 32.7), and let $g \in G$. For all vectors v in the basis, $vA = \lambda v$ for some $\lambda \in \mathbb{R}$, and

$$(vg)A = \lambda(vg) = (\lambda v)g = (vA)g.$$

Hence $(xg)A = (xA)g$ for all $x \in \mathbb{R}^{3n}$. ∎

The idea now is to use representation theory to express the $\mathbb{R}G$-module $\mathbb{R}^{3n}$ as a direct sum of irreducible $\mathbb{R}G$-submodules, and hence to determine the eigenspaces of A, and the normal modes of the molecule.

We can use character theory to see which irreducible $\mathbb{R}G$-modules are contained in $\mathbb{R}^{3n}$; and if χ is the character of an irreducible $\mathbb{R}G$-module which occurs, then the element

$$\sum_{g \in G} \chi(g^{-1})g$$

sends $\mathbb{R}^{3n}$ onto the sum of those irreducible $\mathbb{R}G$-submodules of $\mathbb{R}^{3n}$ which have character χ (see (14.27)). (This procedure sometimes needs to be modified, since the character of the $\mathbb{R}G$-module $\mathbb{R}^{3n}$ might contain an irreducible character which cannot be realized over $\mathbb{R}$ – but in practice, problems like this are uncommon.)

32.12 Definition
Suppose that χ is the character of an irreducible $\mathbb{R}G$-module. Let V_χ denote the sum of those irreducible $\mathbb{R}G$-submodules of $\mathbb{R}^{3n}$ which have character χ. We call V_χ a *homogeneous component* of $\mathbb{R}^{3n}$.

The problem of finding the eigenspaces of A is considerably simplified as a consequence of our next proposition.

32.13 Proposition
Each homogeneous component V_χ of $\mathbb{R}^{3n}$ is A-invariant – that is,

$$xA \in V_\chi \quad \text{for all } x \in V_\chi.$$

Proof By Maschke's Theorem we may write $\mathbb{R}^{3n} = V_\chi \oplus W$ for some $\mathbb{R}G$-module W, and no $\mathbb{R}G$-submodule of W has character χ. The function

$$\varepsilon : v + w \to w \quad (v \in V_\chi, w \in W)$$

is an $\mathbb{R}G$-homomorphism. Therefore, by Proposition 32.11, the function

$$x \to (xA)\varepsilon \quad (x \in V_\chi)$$

is an $\mathbb{R}G$-homomorphism from V_χ into W. By Proposition 11.3, this function is zero, so $xA \in V_\chi$ for all $x \in V_\chi$. (Although Proposition 11.3 is stated in terms of $\mathbb{C}G$-modules, its proof works equally well for $\mathbb{R}G$-modules – compare Exercise 23.8.) ∎

32.14 Corollary
If V_χ is an irreducible $\mathbb{R}G$-module, then all the non-zero vectors in V_χ are eigenvectors of A.

Proof (Compare the proof of Schur's Lemma.) Since V_χ is A-invariant, we may choose $v \in V_\chi$ such that v is an eigenvector of A, with eigenvalue λ, say. Then the intersection of V_χ with the eigenspace for λ is a non-zero $\mathbb{R}G$-submodule of V_χ, so it must equal V_χ. ∎

We now summarize the steps in the procedure for finding the normal modes of vibration of a given molecule.

32.15 Summary
(1) Assign three coordinate axes at each of the n atoms of the molecule, to obtain $\mathbb{R}^{3n}$.

(2) Calculate the symmetry group G of the molecule. Then $\mathbb{R}^{3n}$ is an $\mathbb{R}G$-module.

(3) Calculate the character χ of the $\mathbb{R}G$-module $\mathbb{R}^{3n}$ and express χ as a linear combination of the irreducible characters of G.

(4) Express $\mathbb{R}^{3n}$ as a direct sum of homogeneous components. This can be done by applying the element $\sum_{g \in G} \chi_i(g^{-1})g$ to $\mathbb{R}^{3n}$ for each irreducible character χ_i of G which appears in χ, or by some other method.

(5) Consider, in turn, each homogeneous component V_{χ_i} of $\mathbb{R}^{3n}$, and find the eigenvectors of A in V_{χ_i} (see Proposition 32.13). This involves no extra work if V_{χ_i} is irreducible (see Corollary 32.14). If V_{χ_i} is reducible, then see Remark 32.19 below, or Exercise 32.7, to make further progress.

(6) If v is an eigenvector of A, with eigenvalue $-\omega^2$, then

$$\sin(\omega t + \beta)v \quad (\text{or } (t + \beta)v \text{ if } \omega = 0)$$

is a normal mode, where β is an arbitrary constant. It is usually necessary to know the equations of motion in order to determine the frequency ω.

This programme can often be successfully completed, as we shall illustrate in the examples which make up the rest of this chapter.

32.16 Example

We first return to Example 32.4, with the spring and two vibrating masses:

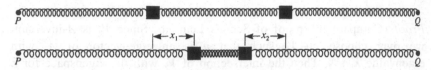

The symmetry group of this system is $G = \langle g\colon g^2 = 1\rangle$, where g is the reflection in the mid-point of PQ. The displacement vectors (x_1, x_2) form an $\mathbb{R}G$-module $\mathbb{R}^2$.

Since $(x_1, x_2)g = (x_2, x_1)$, the $\mathbb{R}G$-submodules of $\mathbb{R}^2$ are $\mathrm{sp}\,(u_1)$ and $\mathrm{sp}\,(u_2)$, where

$$u_1 = (1, 1), \; u_2 = (1, -1).$$

It follows that the normal modes of the system are given by

$$\sin(\omega_1 t + \beta_1)(1, 1), \; \sin(\omega_2 t + \beta_2)(1, -1),$$

where β_1, β_2 are constants and ω_1, ω_2 are the frequencies. This agrees with the conclusion of Example 32.4.

Notice that we have determined the normal modes of vibration (but not their frequencies) using the symmetry group alone.

32.17 Example

Consider a hypothetical triatomic molecule, where the three identical atoms are at the corners of an equilateral triangle. For simplicity, we consider only vibrations of the molecule in the plane, so we assign two displacement coordinates to each atom. We choose to take our axes along the edges of the triangle, as shown, as this eases the calculations:

Thus the position of the molecule is given by a vector $(x_1, \ldots, x_6)$ in $\mathbb{R}^6$, where x_i is the displacement along axis i ($1 \leqslant i \leqslant 6$).

The symmetry group (in two dimensions) of the molecule is the dihedral group D_6, generated by a rotation a of order 3 and a reflection

b (see Example 32.1). It is easy to work out the action of each element of D_6 on $\mathbb{R}^6$. For example, if b is the reflection which fixes the top atom, then

$$(x_1, x_2, x_3, x_4, x_5, x_6)b = (x_2, x_1, x_6, x_5, x_4, x_3).$$

We want to express the $\mathbb{R}D_6$-module $\mathbb{R}^6$ as a direct sum of irreducible $\mathbb{R}D_6$-modules. To do this, we first calculate the character χ of the module. Since the rotation a does not fix any of the atoms, $\chi(a) = 0$. And from the action of b given above, we see that $\chi(b) = 0$. Thus the values of χ are

	1	a	b
χ	6	0	0

By Section 18.3, the character table of D_6 is

	1	a	b
χ_1	1	1	1
χ_2	1	1	−1
χ_3	2	−1	0

Hence $\chi = \chi_1 + \chi_2 + 2\chi_3$.

Thus, we seek to express $\mathbb{R}^6$ as a direct sum of $\mathbb{R}D_6$-submodules with characters χ_1, χ_2, χ_3 and χ_3.

As a matter of notation, if v_1, v_2, v_3 are 2-dimensional displacement vectors for the three atoms, then we represent the displacement vector $(v_1, v_2, v_3) \in \mathbb{R}^6$ pictorially by the diagram

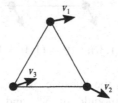

We first calculate the normal modes of the form $(t + \beta)v$, corresponding to the eigenspace of A for eigenvalue 0. These include the *rotation* and *translation* modes, which occur for every molecule.

Rotation mode In this mode, the molecule rotates with constant angular velocity about the centre. The mode is given by $(t + \beta)v$, where $v = (1, -1, 1, -1, 1, -1)$; pictorially,

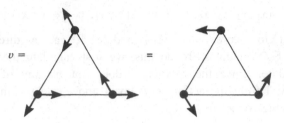

We call sp (v) the *rotation submodule* of $\mathbb{R}^6$. If χ_R is the character of sp (v), then

$$\chi_R(1) = 1, \ \chi_R(a) = 1, \ \chi_R(b) = -1,$$

and so $\chi_R = \chi_2$. Indeed, sp $(v) = \mathbb{R}^6 \varepsilon_2$, where

$$\varepsilon_2 = \sum_{g \in D_6} \chi_2(g^{-1})g = 1 + a + a^2 - b - ab - a^2 b$$

(compare (14.27)).

Translation modes These are modes in which all atoms move in the same constant direction with the same constant speed. The modes are of the form $(t + \beta)v$, where v is a vector in the span of v_1, v_2 and v_3, these vectors being given pictorially by

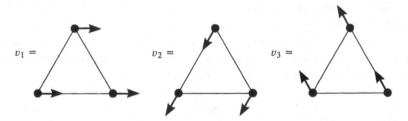

(thus $v_1 = (-1, 1, 0, -1, 1, 0)$, $v_2 = (1, 0, -1, 1, 0, -1)$, $v_3 = (0, -1, 1, 0, -1, 1)$).

Since $v_1 + v_2 + v_3 = 0$, the subspace sp (v_1, v_2, v_3) has dimension 2. It is clearly an $\mathbb{R}D_6$-submodule of $\mathbb{R}^6$, and is called the *translation submodule*; it does not contain the rotation submodule. Thus the character of the translation submodule is part of $\chi - \chi_2 = \chi_1 + 2\chi_3$, so the character must be χ_3.

Vibratory modes The remaining normal modes correspond to eigenspaces of the matrix A with non-zero eigenvalues, and are called *vibratory modes*. The sum of these eigenspaces forms an $\mathbb{R}D_6$-submodule $\mathbb{R}^6_{\text{vib}}$ of $\mathbb{R}^6$ (by Proposition 32.11), with character χ_{vib}, where

$$\chi_{\text{vib}} = \chi - (\chi_2 + \chi_3) = \chi_1 + \chi_3.$$

In particular, $\mathbb{R}^6_{\text{vib}}$ has dimension 3. Since no mode in $\mathbb{R}^6_{\text{vib}}$ can have any translation component, if $w \in \mathbb{R}^6_{\text{vib}}$ then the total component of w in each direction is zero; moreover, $\mathbb{R}^6_{\text{vib}}$ does not contain the rotation submodule, so that total moment of each vector in $\mathbb{R}^6_{\text{vib}}$ about the centre is zero. These constraints imply three independent linear equations in the coordinates of the vectors in $\mathbb{R}^6_{\text{vib}}$, and since $\mathbb{R}^6_{\text{vib}}$ has dimension 3, every vector in $\mathbb{R}^6$ which satisfies these equations lies in $\mathbb{R}^6_{\text{vib}}$. Therefore a basis of $\mathbb{R}^6_{\text{vib}}$ is u_1, u_2, u_3, where

$u_3 = (0,1,1,0,0,0) =$

$u_2 = (0,0,0,1,1,0) =$

$u_3 = (1,0,0,0,0,1) =$

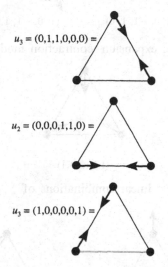

Clearly $\mathrm{sp}\,(u_1 + u_2 + u_3)$ is an $\mathbb{R}D_6$-submodule of $\mathbb{R}^6_{\text{vib}}$ with character χ_1. The vibratory mode given by $u_1 + u_2 + u_3$ is sometimes called the *expansion–contraction* mode (you will see the reason for this name in the picture in (32.18(3)) below).

Finally, since D_6 permutes the vectors u_1, u_2, u_3 among themselves, it is easy to see that $\mathrm{sp}\,(u_1 - u_2, u_1 - u_3)$ is an $\mathbb{R}D_6$-submodule of $\mathbb{R}^6_{\text{vib}}$. Its character is χ_3 and it gives us the last eigenspace for the matrix A.

Our calculation of the normal modes is now complete, and we summarize our findings below.

(32.18)   (1) Rotation mode:

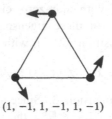

$$(1, -1, 1, -1, 1, -1)$$

(2) Translation modes: linear combinations of

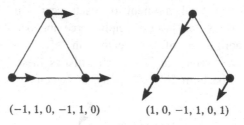

$$(-1, 1, 0, -1, 1, 0) \qquad (1, 0, -1, 1, 0, 1)$$

(3) Vibratory mode: expansion–contraction mode

$$(1,1,1,1,1,1)$$

(4) Vibratory modes: linear combinations of

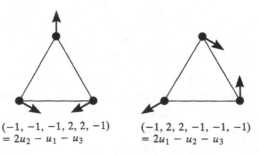

$$(-1, -1, -1, 2, 2, -1) \qquad (-1, 2, 2, -1, -1, -1)$$
$$= 2u_2 - u_1 - u_3 \qquad\qquad = 2u_1 - u_2 - u_3$$

(We have chosen $2u_2 - u_1 - u_3$, $2u_1 - u_2 - u_3$ as the basis for the vibratory modes in (4) merely because these modes look simpler than $u_1 - u_2$, $u_1 - u_3$ pictorially.)

We emphasize that we have found the normal modes of vibration without explicit knowledge of the equations of motion. In order to check our results, we now calculate the equations of motion.

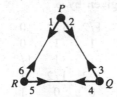

Let m be the mass of each atom, and assume that the magnitude of the force between two atoms is k times the decrease in distance between them.

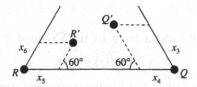

For a general displacement $(x_1, x_2, x_3, x_4, x_5, x_6)$, denote the new positions of the atoms by P', Q', R'. From the diagram, the difference in length between QR and $Q'R'$ is

$$(x_4 + x_5) + \tfrac{1}{2}(x_3 + x_6).$$

(We always assume that $x_1, \ldots, x_6$ are small compared with the distance between the atoms, so that we may ignore second order terms.)

Similarly,

$$PR - P'R' = (x_1 + x_6) + \tfrac{1}{2}(x_2 + x_5),$$
$$PQ - P'Q' = (x_2 + x_3) + \tfrac{1}{2}(x_1 + x_4).$$

Hence the force on the molecule at P in the direction of the first coordinate axis is

$$-k(PR - P'R') = -k(x_1 + x_6) - \tfrac{1}{2}k(x_2 + x_5).$$

Therefore,

$$\frac{m}{k}\ddot{x}_1 = -(x_1 + x_6) - \tfrac{1}{2}(x_2 + x_5).$$

In the same way,

$$\frac{m}{k}\ddot{x}_2 = -(x_2 + x_3) - \tfrac{1}{2}(x_1 + x_4),$$

and we obtain similar equations for $\ddot{x}_3, \ldots, \ddot{x}_6$. The matrix A for which $\ddot{x} = xA$ is therefore given by

$$A = \frac{-k}{m} \begin{pmatrix} 1 & 1/2 & 1/2 & 0 & 0 & 1 \\ 1/2 & 1 & 1 & 0 & 0 & 1/2 \\ 0 & 1 & 1 & 1/2 & 1/2 & 0 \\ 0 & 1/2 & 1/2 & 1 & 1 & 0 \\ 1/2 & 0 & 0 & 1 & 1 & 1/2 \\ 1 & 0 & 0 & 1/2 & 1/2 & 1 \end{pmatrix}.$$

You should check that the vectors which we gave in (32.18) are indeed eigenvectors of A.

32.19 Remark

In Example 32.17, the character χ of the $\mathbb{R}G$-module $\mathbb{R}^6$ was given by

$$\chi = \chi_1 + \chi_2 + 2\chi_3.$$

All the non-zero vectors in the homogeneous components for χ_1 and χ_2 gave normal modes, since these homogenous components were irreducible (see Corollary 32.14). The homogeneous component V_{χ_3} for χ_3 was reducible, but we were able to write it as a sum of two subspaces of eigenvectors (those appearing in (32.18)(2) and (4)) because $V_{\chi_3} \cap \mathbb{R}^6_{\text{vib}}$ was an A-invariant $\mathbb{R}G$-submodule of V_{χ_3} different from $\{0\}$ and V_{χ_3}. This illustrates a method which sometimes helps to deal with reducible homogeneous components. In our next example, the situation is more complicated.

32.20 Example

We analyse the normal modes for the methane molecule CH_4.

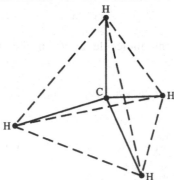

We determined the symmetry group G of this molecule in Example 32.2, where we found that G is isomorphic to S_4. Label the corners of the tetrahedron 1, 2, 3, 4, as shown below, and identify G with S_4; thus, for example, the rotations about the vertical axis through 1 are written as

$$1, (2\ 3\ 4), (2\ 4\ 3).$$

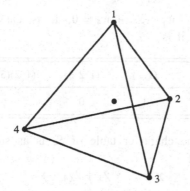

In order fully to exploit the symmetry of the methane molecule, at each hydrogen atom we choose displacement axes along the edges of the tetrahedron. Let v_{12}, v_{13}, v_{14} be unit vectors at corner 1 in the directions of the edges 12, 13, 14, respectively; similarly, let v_{21}, v_{23}, v_{24} be unit vectors at corner 2 in the directions of the edges 21, 23, 24, and so on, giving twelve vectors v_{ij}, in all.

We now introduce a new idea, by taking *four* unit vectors w_1, w_2, w_3 and w_4 at the carbon atom, with w_i pointing towards corner i ($1 \leqslant i \leqslant 4$). Since $w_1 + w_2 + w_3 + w_4 = 0$, these four vectors span a 3-dimensional space.

Let V be the vector space over $\mathbb{R}$ with basis

$$v_{12}, v_{13}, v_{14}, v_{21}, v_{23}, v_{24}, v_{31}, v_{32}, v_{34}, v_{41}, v_{42}, v_{43},$$

and let W be the vector space over $\mathbb{R}$ spanned by w_1, w_2, w_3, w_4. Then $V \cong \mathbb{R}^{12}$, $W \cong \mathbb{R}^3$ and V and W are $\mathbb{R}G$-modules. Our main task is to find $\mathbb{R}G$-submodules of $\mathbb{R}^{15} = V \oplus W$.

The action of G on V is easy to describe; for g in G, we have

$$v_{ij}g = v_{ig,jg} \quad \text{for all } i, j.$$

Thus, G permutes our twelve basis vectors of V, and we can quickly calculate the character χ of V:

	1	(1 2)	(1 2 3)	(1 2)(3 4)	(1 2 3 4)
χ	12	2	0	0	0

For example, (1 2) fixes the basis vectors v_{34} and v_{43} only; all the basis vectors are moved by (1 2 3); and so on.

The group G acts on W as follows; for g in G, we have

$$w_i g = w_{ig} \quad (1 \leqslant i \leqslant 4).$$

After recalling that $w_1 + \ldots + w_4 = 0$, it is easy to calculate the character ϕ of W – it is

	1	(1 2)	(1 2 3)	(1 2)(3 4)	(1 2 3 4)
ϕ	3	1	0	-1	-1

By Section 18.1, the character table of S_4 is as shown at the top of p. 387. We find that

$$\chi = \chi_1 + \chi_3 + 2\chi_4 + \chi_5,$$

$$\phi = \chi_4.$$

By applying the elements

$$\sum_{g \in G} \chi_i(g^{-1})g \quad (i = 1, 3, 5, 4)$$

to $\mathbb{R}^{15}$, we can find $\mathbb{R}G$-submodules with characters χ_1, χ_3, χ_5 and $3\chi_4$ (see (14.27)).

The $\mathbb{R}G$-submodule W_1 with character χ_1 is spanned by

$$\sum_{i,j} v_{ij}$$

and this gives the expansion–contraction normal mode:

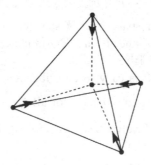

We next describe the $\mathbb{R}G$-submodule W_5 with character χ_5. Let

$$p_1 = (v_{23} - v_{32}) + (v_{34} - v_{43}) + (v_{42} - v_{24}),$$

$$p_2 = (v_{31} - v_{13}) + (v_{14} - v_{41}) + (v_{43} - v_{34}),$$

Character table of S_4

	1	(1 2)	(1 2 3)	(1 2)(3 4)	(1 2 3 4)
χ_1	1	1	1	1	1
χ_2	1	−1	1	1	−1
χ_3	2	0	−1	2	0
χ_4	3	1	0	−1	−1
χ_5	3	−1	0	−1	1

$$p_3 = (v_{12} - v_{21}) + (v_{41} - v_{14}) + (v_{24} - v_{42}),$$

$$p_4 = (v_{21} - v_{12}) + (v_{13} - v_{31}) + (v_{32} - v_{23}).$$

The vector p_i gives a rotation about the axis through the corner i and the centroid of the tetrahedron.

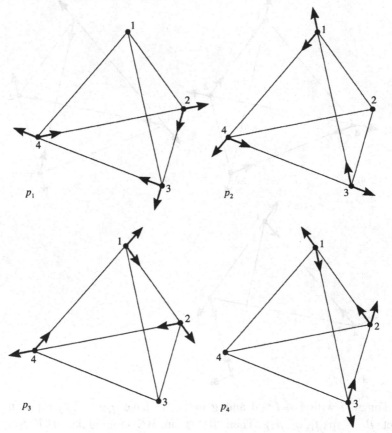

It should be clear from the pictures that for all i with $1 \leqslant i \leqslant 4$ and all g in G, we have $p_i g = \pm p_j$ for some j. Therefore, if we let $W_5 = \mathrm{sp}\,(p_1, p_2, p_3, p_4)$, then W_5 is an $\mathbb{R}G$-submodule of V. Now $p_1 + p_2 + p_3 + p_4 = 0$, so dim $W_5 = 3$. Check that the character of W_5 is χ_5. The $\mathbb{R}G$-module W_5 is the rotation submodule. (Compare, for example, the picture for p_4 with the picture for the rotation vector v in Example 32.17.)

Now we construct the $\mathbb{R}G$-submodule W_3 of V with character χ_3. Let

$$q_1 = (v_{12} + v_{21}) + (v_{34} + v_{43}) - (v_{13} + v_{31}) - (v_{24} + v_{42}),$$

$$q_2 = (v_{13} + v_{31}) + (v_{24} + v_{42}) - (v_{14} + v_{41}) - (v_{23} + v_{32}),$$

$$q_3 = (v_{14} + v_{41}) + (v_{23} + v_{32}) - (v_{12} + v_{21}) - (v_{34} + v_{43}).$$

(Each q_i is associated with an 'opposite pair of edges'.)

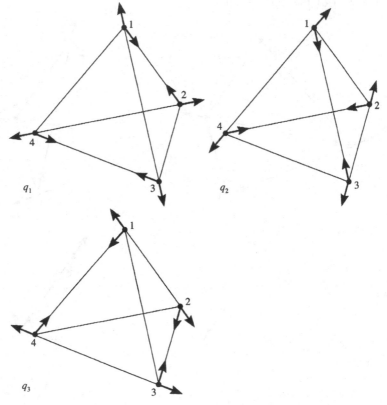

For all i with $1 \leqslant i \leqslant 4$ and g in G, we have $q_i g = \pm q_j$ for some j. Let $W_3 = \mathrm{sp}\,(q_1, q_2, q_3)$. Then W_3 is an $\mathbb{R}G$-submodule of V. Since $q_1 + q_2 + q_3 = 0$, the dimension of W_3 is 2; its character is χ_3.

In the $\mathbb{R}G$-submodules W_1, W_5 and W_3 which we have found so far, all the non-zero vectors are eigenvectors of A, by Corollary 32.14.

We now come to the homogeneous component $(V \oplus W)_{\chi_4}$ of $\mathbb{R}^{15}$. Define the vectors r_1, r_2, r_3, r_4 by

$$r_1 = (v_{12} + v_{21}) + (v_{13} + v_{31}) + (v_{14} + v_{41})$$
$$- (v_{23} + v_{32}) - (v_{24} + v_{42}) - (v_{34} + v_{43}),$$

$$r_2 = (v_{12} + v_{21}) + (v_{23} + v_{32}) + (v_{24} + v_{42})$$
$$- (v_{13} + v_{31}) - (v_{14} + v_{41}) - (v_{34} + v_{43}),$$

$$r_3 = (v_{13} + v_{31}) + (v_{23} + v_{32}) + (v_{34} + v_{43})$$
$$- (v_{12} + v_{21}) - (v_{14} + v_{41}) - (v_{24} + v_{42}),$$

$$r_4 = (v_{14} + v_{41}) + (v_{24} + v_{42}) + (v_{34} + v_{43})$$
$$- (v_{13} + v_{31}) - (v_{12} + v_{21}) - (v_{23} + v_{32}).$$

(The vector r_i is associated with corner i.)

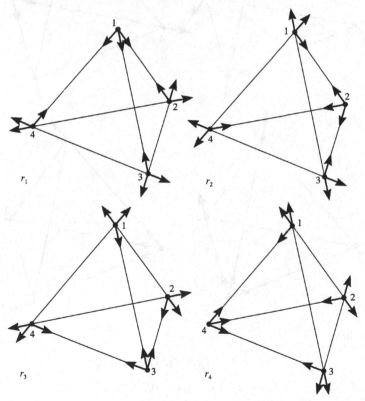

For each g in G and i with $1 \leqslant i \leqslant 4$, we have $r_i g = r_{ig}$. Thus G permutes the vectors r_1, r_2, r_3, r_4 among themselves. Note that $r_1 + r_2 + r_3 + r_4 = 0$, so r_1, r_2, r_3, r_4 span a 3-dimensional $\mathbb{R}G$-submodule W_4 of V. The character of W_4 is χ_4 (see Proposition 13.24).

Next, define the vectors s_1, s_2, s_3, s_4 by

$$s_1 = (v_{12} + v_{13} + v_{14}) - (v_{21} + v_{31} + v_{41}),$$

$$s_2 = (v_{21} + v_{23} + v_{24}) - (v_{12} + v_{32} + v_{42}),$$

$$s_3 = (v_{31} + v_{32} + v_{34}) - (v_{13} + v_{23} + v_{43}),$$

$$s_4 = (v_{41} + v_{42} + v_{43}) - (v_{14} + v_{24} + v_{34}).$$

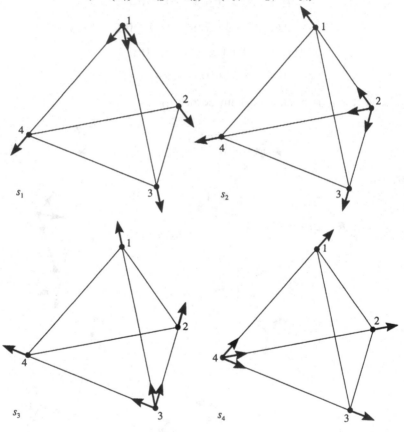

We have

$$s_i g = s_{ig} \quad (g \in G, 1 \leqslant i \leqslant 4),$$

$$s_1 + s_2 + s_3 + s_4 = 0,$$

and s_1, s_2, s_3, s_4 span a 3-dimensional $\mathbb{R}G$-submodule W_4' of V with character χ_4.

Now recall that w_1, w_2, w_3, w_4 span W; we have

$$w_i g = w_{ig} \quad (g \in G, 1 \leqslant i \leqslant 4),$$

$$w_1 + w_2 + w_3 + w_4 = 0,$$

and the character of W is χ_4. The sum of W_4, W_4' and W is direct, so

$$(V \oplus W)_{\chi_4} = W_4 \oplus W_4' \oplus W.$$

We now break off temporarily from studying the methane molecule, in order to deal with the easier case of a molecule with identical atoms at the corners of the tetrahedron, and no central atom. In this case, the space W does not enter our calculations, and we can decompose $V_{\chi_4} = W_4 \oplus W_4'$ in the following way.

(32.21) (1) The vectors $r_1 - 2s_1$, $r_2 - 2s_2$, $r_3 - 2s_3$, $r_4 - 2s_4$ span the 3-dimensional space of translation modes.

(2) The vectors r_1, r_2, r_3, r_4 span the subspace $V_{\chi_4} \cap \mathbb{R}_{\text{vib}}^{12}$ of V_{χ_4}, and so they give the final 3-dimensional space of eigenvectors (see Remark 32.19). The normal mode $(\sin \omega t) r_1$ is sometimes called an 'umbrella mode'. To see why, look at the picture of the vector r_1!

We return now to the methane molecule.

The task in front of us is to find the eigenvectors of A which lie in

$$(V \oplus W)_{\chi_4} = W_4 \oplus W_4' \oplus W.$$

The solution of this problem depends, in fact, upon the constants which appear in the equations of motion, so we cannot complete the work using only representation theory. Since $\dim (V \oplus W)_{\chi_4} = 9$, it is very difficult to calculate the eigenvectors of A in $(V \oplus W)_{\chi_4}$ directly. We shall therefore press on and explain how to reduce the difficulty to that of calculating the eigenvectors of a 3×3 matrix.

Let H be the subgroup of S_4 consisting of those permutations which fix the number 1, and let

$$U_1 = \{v \in (V \oplus W)_{\chi_4} : vh = v \text{ for all } h \in H\}.$$

Since $(vh)A = (vA)h$ for all $v \in V_{\chi_4}$ and all $h \in H$, it follows that U_1 is A-invariant.

We find that

$$\langle 3\chi_4 \downarrow H, 1_H \rangle_H = 3,$$

and so $\dim U_1 = 3$. But for all $h \in H$, $r_1 h = r_1$, $s_1 h = s_1$ and $w_1 h = w_1$. Therefore

$$U_1 = \text{sp}\,(r_1, s_1, w_1).$$

Once the equations of motion, and hence the matrix A, have been calculated, it is possible to calculate the 3×3 matrix B of the action of A on r_1, s_1, w_1 (see Exercise 32.5). The eigenvectors of B then give three eigenvectors of A.

Better still,

$$r_1(1\ 2) = r_2,\ s_1(1\ 2) = s_2,\ w_1(1\ 2) = w_2,$$

and since A commutes with the action of G, the space U_2, defined by

$$U_2 = \text{sp}\,(r_2, s_2, w_2)$$

is A-invariant, and the matrix of A acting on r_2, s_2, w_2 is again B. A similar remark applies to U_3, where

$$U_3 = \text{sp}\,(r_3, s_3, w_3).$$

Therefore, the process of calculating the eigenvectors of the 3×3 matrix B gives nine eigenvectors of A which form a basis of $(V \oplus W)_{\chi_4}$.

One eigenvector of A acting on r_1, s_1, w_1 is easy to find, namely the translation vector

$$r_1 - 2s_1 + 3\cos\vartheta w_1,$$

where ϑ is the angle between an edge of the tetrahedron and the line joining a corner on the edge to the centroid. Thus we obtain the translation submodule $\text{sp}\,(r_1 - 2s_1 + 3\cos\vartheta w_1,\ r_2 - 2s_2 + 3\cos\vartheta w_2,\ r_3 - 2s_3 + 3\cos\vartheta w_3)$.

By means of representation theory, we have therefore reduced the initial problem of finding the eigenvectors of a 15×15 matrix A to that of calculating two of the eigenvectors of a 3×3 matrix. It is hard to imagine a more spectacular application of representation theory with which to conclude our text.

Summary of Chapter 32

1. The symmetry group G of a molecule with n atoms consists of those distance-preserving endomorphisms of $\mathbb{R}^3$ which send each atom to an atom of the same kind.

2. The equations of motion of the molecule have the form

$$\ddot{x} = xA,$$

where $x \in \mathbb{R}^{3n}$ and the $3n \times 3n$ matrix A has $3n$ linearly independent eigenvectors.

3. If u is an eigenvector of A, with eigenvalue $-\omega^2$, then $x = \sin(\omega t + \beta)u$ (or $x = (t + \beta)u$ if $\omega = 0$) is a solution of the equations of motion, and is called a normal mode. All solutions are linear combinations of normal modes.

4. The space $\mathbb{R}^{3n}$ of displacement vectors is an $\mathbb{R}G$-module. The action of any $g \in G$ on $\mathbb{R}^{3n}$ commutes with A.

5. To determine the eigenvectors of A (and hence all solutions of the equations of motion), it is sufficient to find the eigenvectors of A restricted to each homogeneous component V_{χ_i} of the $\mathbb{R}G$-module $\mathbb{R}^{3n}$. If V_{χ_i} is irreducible, then all non-zero vectors in V_{χ_i} are eigenvectors of A.

Exercises for Chapter 32

1. Suppose that $b \in O(\mathbb{R}^3)$, and let $e_1 = (1, 0, 0)$, $e_2 = (0, 1, 0)$, $e_3 = (0, 0, 1)$.
   (a) Show that the matrix B of b with respect to the basis e_1, e_2, e_3 of $\mathbb{R}^3$ satisfies $BB^t = I$ (where B^t denotes the transpose of B). Deduce that $\det B = \pm 1$.
   (b) Let $C = (\det B)B$. Prove successively that
      (i) C has a real eigenvalue,
      (ii) C has a positive eigenvalue,
      (iii) 1 is an eigenvalue of C.
   (c) Deduce from (a) and (b) that b is a rotation if $\det B = 1$, and $-b$ is a rotation otherwise.
   (d) Show that if b is a rotation through an angle ϕ about some axis, then $\operatorname{tr} B = 1 + 2 \cos \phi$.

2. Suppose that G is the symmetry group of some molecule in $\mathbb{R}^3$.

Show that the sum of the character χ_T of the translation submodule and the character χ_R of the rotation submodule has value at $g \in G$ given by

$$(\chi_T + \chi_R)(g) = \begin{cases} 2(1 + 2\cos\phi), & \text{if } g \text{ is a rotation} \\ & \text{through angle } \phi \\ & \text{about some axis,} \\ 0, & \text{if } g \text{ is not a rotation.} \end{cases}$$

3. Consider the triatomic molecule studied in Example 32.17. Take the axes for the displacement coordinates as shown below:

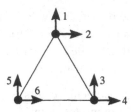

Calculate the equations of motion $\ddot{x} = xA$ with respect to these axes, and verify that A is symmetric. (See the paragraph before Proposition 32.7.)

4. Consider the space spanned by the vectors r_1, r_2, r_3, r_4 given in Example 32.20. Find a basis for this space which is simpler than the one which we have used. What property of r_1, r_2, r_3, r_4 prompted us to use these vectors?

5. The purpose of this exercise is to derive the equations of motion of the methane molecule, and so find explicitly the 3×3 matrix B which appears at the end of Example 32.20.

   Label the corners of the tetrahedron 1, 2, 3, 4 and let 0 denote the centroid of the tetrahedron. Work with 15 unit displacement vectors

$$v_{12}, v_{13}, \ldots, v_{43}, w_1, w_2, w_3$$

as described in Example 32.20, and let the position vector of the molecule be

$$\sum_{i \neq j} x_{ij} v_{ij} + \sum_{i=1}^{3} y_i w_i.$$

(a) Prove that $\cos(\angle\,012) = \sqrt{(2/3)}$ and $\cos(\angle\,102) = -1/3$.

(b) Show that the decrease in the length of the edge 12 from its original length is

$$x_{12} + x_{21} + \tfrac{1}{2}(x_{13} + x_{14} + x_{23} + x_{24}),$$

with similar expressions for the edges 13, 14, 23, 24, 34.

Also, show that the length of the edge 01 has decreased by

$$\sqrt{(2/3)}(x_{12} + x_{13} + x_{14}) + y_1 - \tfrac{1}{3}(y_2 + y_3),$$

with similar expressions for the edges 02, 03.

Finally, show that the length of the edge 04 has decreased by

$$\sqrt{(2/3)}(x_{41} + x_{42} + x_{43}) - \tfrac{1}{3}(y_1 + y_2 + y_3).$$

(c) Let m_1 denote the mass of a hydrogen atom, and m_2 the mass of a carbon atom. Assume that the magnitude of the force between hydrogen atoms is k_1 times the decrease in distance between them, and the magnitude of the force between a hydrogen atom and a carbon atom is k_2 times the decrease in distance between them.

Prove that

$$m_1 \ddot{x}_{12} = -k_1[x_{12} + x_{21} + \tfrac{1}{2}(x_{13} + x_{14} + x_{23} + x_{24})]$$
$$- \tfrac{1}{3}k_2[x_{12} + x_{13} + x_{14} + \sqrt{(3/2)}(y_1 - \tfrac{1}{3}y_2 - \tfrac{1}{3}y_3)],$$

with similar expressions for $\ddot{x}_{13}, \ddot{x}_{14}, \ddot{x}_{21}, \ddot{x}_{23}, \ddot{x}_{24}, \ddot{x}_{31}, \ddot{x}_{32}, \ddot{x}_{34}$.

Also, show

$$m_1 \ddot{x}_{41} = -k_1[x_{14} + x_{41} + \tfrac{1}{2}(x_{42} + x_{43} + x_{12} + x_{13})]$$
$$- \tfrac{1}{3}k_2[x_{41} + x_{42} + x_{43} - \tfrac{1}{3}\sqrt{(3/2)}(y_1 + y_2 + y_3)],$$

with similar expressions for $\ddot{x}_{42}$ and $\ddot{x}_{43}$.

Finally, show

$$m_2 \ddot{y}_1 = -k_2[\sqrt{(2/3)}(x_{12} + x_{13} + x_{14} - x_{41}$$
$$- x_{42} - x_{43}) + \tfrac{4}{3}y_1],$$

with similar expressions for $\ddot{y}_2$ and $\ddot{y}_3$.

(d) The equations in part (c) determine the 15×15 matrix A in the equations of motion $\ddot{x} = xA$. Verify that the vectors

$$\sum_{i,j} v_{ij}, \; p_1, \; p_2, \; p_3, \; q_1, \; q_2,$$

which appear in Example 32.20, are eigenvectors of A.
(e) Find the entries b_{ij} in the 3×3 matrix B which are given by

$$r_1 A = b_{11} r_1 + b_{12} s_1 + b_{13} w_1,$$

$$s_1 A = b_{21} r_1 + b_{22} s_1 + b_{23} w_1,$$

$$w_1 A = b_{31} r_1 + b_{32} s_1 + b_{33} w_1,$$

where the vectors r_1, s_1, w_1 are as in Example 32.20.
(f) Verify that

$$(1, -2, \sqrt{6})$$

is an eigenvector of B.

6. Consider a hypothetical molecule in which there are four identical atoms at the corners of a square. Assume that the only internal forces are along the sides of the square.
   (a) Find the normal modes of the molecule.
   (b) Calculate the equations of motion, $\ddot{x} = xA$, and check that the vectors you found in part (a) are, indeed, eigenvectors of A.

7. In this exercise, we derive a method for simplifying the problem of finding the eigenvectors of A when the homogeneous component V_{χ_i} is reducible. (See 32.15(5).) We assume that χ_i is the character of an irreducible $\mathbb{R}G$-module which remains irreducible as a $\mathbb{C}G$-module.

   Suppose that $V_{\chi_i} = U_1 \oplus \ldots \oplus U_m$, a sum of m isomorphic irreducible $\mathbb{R}G$-modules. We reduce our problem to that of finding the eigenvectors of an $m \times m$ matrix.

   For $1 \le i \le m$, let ϑ_i be an $\mathbb{R}G$-isomorphism from U_1 to U_i.
   (a) Prove that for all non-zero u in U_1,

   $$\mathrm{sp}\,(u\vartheta_1, \ldots, u\vartheta_m)$$

   is an A-invariant vector space of dimension m.
   (Hint: compose the function $w \rightarrow wA$ with a projection, and use Exercise 23.8.)
   (b) Let A_u denote the matrix of the endomorphism $w \rightarrow wA$ of $\mathrm{sp}\,(u\vartheta_1, \ldots, u\vartheta_m)$ with respect to the basis $u\vartheta_1, \ldots, u\vartheta_m$. Prove that if u and v are non-zero elements of U_1, then $A_u = A_v$.
   (c) Assume that the eigenvectors of the $m \times m$ matrix A_u are known. Show how to find the eigenvectors of A in V_{χ_i}.

Solutions to exercises

Chapter 1

1. Note that all subgroups of G are normal, since G is abelian; and $G \neq \{1\}$ since G is simple. Let g be a non-identity element of G. Then $\langle g \rangle$ is a normal subgroup of G, so $\langle g \rangle = G$. If G were infinite, then $\langle g^2 \rangle$ would be a normal subgroup different from G and $\{1\}$; hence G is finite. Let p be a prime number which divides $|G|$. Then $\langle g^p \rangle$ is a normal subgroup of G which is not equal to G. Therefore $g^p = 1$, and so G is cyclic of prime order.

2. Since G is simple and $\operatorname{Ker} \vartheta \lhd G$, either $\operatorname{Ker} \vartheta = \{1\}$ or $\operatorname{Ker} \vartheta = G$. If $\operatorname{Ker} \vartheta = \{1\}$ then ϑ is an isomorphism; and if $\operatorname{Ker} \vartheta = G$ then $H = \{1\}$.

3. First, $G \cap A_n = \{g \in G: g \text{ is even}\}$, so $G \cap A_n \lhd G$. Since $G \cap A_n \neq G$, we may choose $h \in G$ with $h \notin A_n$. For all odd g in G, we have $g = (gh^{-1})h \in (G \cap A_n)h$. Therefore $G \cap A_n$ and $(G \cap A_n)h$ are the only right cosets of $G \cap A_n$ in G, and $G/(G \cap A_n) \cong C_2$.

4. (a) Using the method of Example 1.4, it is routine to verify that ϕ and ψ are homomorphisms. $\operatorname{Ker} \phi = \{1, a^2\}$ and $\operatorname{Ker} \psi = \{1, c^2\}$.
   (b) Since $b^2 \lambda = I$ but $(b\lambda)^2 = Y^2 = -I$, it follows that λ is not a homomorphism. Check using the method of Example 1.4 again that μ is a homomorphism. Also $\operatorname{Ker} \mu = \{1\}$ and $\operatorname{Im} \mu = L$, so μ is an isomorphism.

5. Let

$$D_{4m} = \langle a, b: a^{2m} = b^2 = 1, b^{-1}ab = a^{-1} \rangle, \text{ and}$$

$$D_{2m} = \langle c, d: c^m = d^2 = 1, d^{-1}cd = c^{-1} \rangle,$$

where m is odd. The elements of $D_{2m} \times C_2$ are

$$(c^i d^j, (-1)^k)$$

for $0 \leqslant i \leqslant m - 1$, $0 \leqslant j \leqslant 1$, $0 \leqslant k \leqslant 1$. Let $x = (c^{(m+1)/2}, -1)$ and $y = (d, 1)$. Check that

$$x^{2m} = y^2 = 1, \quad y^{-1}xy = x^{-1}.$$

By Example 1.4, the function $\vartheta: D_{4m} \to D_{2m} \times C_2$ defined by

$$\vartheta: \ a^i b^j \rightarrow x^i y^j \quad (0 \leqslant i \leqslant 2m - 1, 0 \leqslant j \leqslant 1)$$

is a homomorphism. Since $\text{Im} \, \vartheta = \langle x, y \rangle$, it contains

$$x^2 = (c, 1) \text{ and } x^m = (1, -1)$$

and hence $\text{Im} \, \vartheta = D_{2m} \times C_2$. As $|D_{4m}| = |D_{2m} \times C_2|$, we conclude that ϑ is an isomorphism.

6. (a) Let $G = \langle a \rangle$ and suppose that $1 \neq H \leqslant G$. First observe that there exists $i > 0$ such that $a^i \in H$. Choose k as small as possible such that $k > 0$ and $a^k \in H$. If $1 \neq a^j \in H$ then $j = qk + r$ for some integers q, r with $0 \leqslant r < k$. Hence $a^r = a^j a^{-qk} = a^j (a^k)^{-q} \in H$. Since $r < k$, we have $r = 0$. Therefore $a^j = a^{kq}$ and so $H = \langle a^k \rangle$; thus H is cyclic.

(b) Assume that $G = \langle a \rangle$ and $|G| = dn$. If $g \in G$ and $g^n = 1$, then $g = a^j$ for some integer j and $dn | jn$, so $d | j$; hence $g \in \langle a^d \rangle$. It follows that

$$\{g \in G: \ g^n = 1\} = \langle a^d \rangle,$$

which is a cyclic group of order n.

(c) If x and y are elements of order n in the finite cyclic group G, then x, $y \in H$, where $H = \{g \in G: g^n = 1\}$. Now $\langle x \rangle$ and $\langle y \rangle$ have order n; also H has order n, by part (b). We deduce that

$$\langle x \rangle = H = \langle y \rangle.$$

Thus $x \in \langle y \rangle$, and so x is a power of y.

7. Let G be the set of non-zero complex numbers. If g, $h \in G$ then $gh \neq 0$, so $gh \in G$. If g, h, $k \in G$ then $(gh)k = g(hk)$; also $1 \in G$ and $1g = g1 = g$ for all $g \in G$. Finally, if $g \in G$ then $g^{-1} = 1/g \in G$, and $g^{-1}g = gg^{-1} = 1$. Thus G is a group under multiplication.

If H is a subgroup of G of order n, then $h^n = 1$ for all $h \in H$ (since the order of h divides n, by Lagrange's Theorem). Therefore

$$H \leqslant \{g \in G: \ g^n = 1\} = \langle e^{2\pi i/n} \rangle.$$

Since $|H| = n = |\langle e^{2\pi i/n} \rangle|$, it follows that $H = \langle e^{2\pi i/n} \rangle$.

8. Partition G into subsets $\{g, g^{-1}\}$ $(g \in G)$. Each such subset has size 1 or 2, and the identity element is in a subset of size 1. Hence, if $|G|$ is even then there exists $g \in G$ such that $g \neq 1$ and the subset $\{g, g^{-1}\}$ has size 1; so $g = g^{-1}$ and g has order 2.

9. Define matrices A, B as follows:

$$A = \begin{pmatrix} e^{i\pi/4} & 0 \\ 0 & e^{-i\pi/4} \end{pmatrix}, \quad B = \begin{pmatrix} 0 & 1 \\ -1 & 0 \end{pmatrix}.$$

Check that $A^8 = I$, $B^2 = A^4$ and $B^{-1}AB = A^{-1}$. These relations show that every element of the group $\langle A, B \rangle$ has the form $A^j B^k$ with $0 \leqslant j \leqslant 7$, $0 \leqslant k \leqslant 1$. Moreover,

$$A^j = \begin{pmatrix} e^{ij\pi/4} & 0 \\ 0 & e^{-ij\pi/4} \end{pmatrix}, \quad A^j B = \begin{pmatrix} 0 & e^{ij\pi/4} \\ -e^{-ij\pi/4} & 0 \end{pmatrix}.$$

Since these matrices, with $0 \leqslant j \leqslant 7$, are all distinct, $\langle A, B \rangle$ has order 16.

10. Suppose $|G: H| = 2$ and let $g \in G$. If $g \in H$ then $g^{-1}Hg = H$. And if $g \notin H$ then H, Hg are the two right cosets of H in G, while H, gH are the two left cosets. Therefore $Hg = gH$, and so $g^{-1}Hg = H$ again. Hence $H \lhd G$.

Chapter 2

1. Let $u, w \in W$ and $\lambda \in F$. Since ϑ is a linear transformation, we have

$$(u\vartheta^{-1} + w\vartheta^{-1})\vartheta = (u\vartheta^{-1})\vartheta + (w\vartheta^{-1})\vartheta = u + w,$$

$$(\lambda(w\vartheta^{-1}))\vartheta = \lambda(w\vartheta^{-1})\vartheta = \lambda w.$$

Hence $(u + w)\vartheta^{-1} = u\vartheta^{-1} + w\vartheta^{-1}$ and $(\lambda w)\vartheta^{-1} = \lambda(w\vartheta^{-1})$, so ϑ^{-1} is a linear transformation.

2. (1) $\Rightarrow$ (2): If ϑ is invertible then ϑ is injective, so $\operatorname{Ker}\vartheta = \{0\}$.
   (2) $\Rightarrow$ (3): If $\operatorname{Ker}\vartheta = \{0\}$ then $\dim(\operatorname{Im}\vartheta) = \dim V$ (by (2.12)), so $\operatorname{Im}\vartheta = V$ (by (2.7)).
   (3) $\Rightarrow$ (1): Assume that $\operatorname{Im}\vartheta = V$, so ϑ is surjective. By (2.12), $\operatorname{Ker}\vartheta = \{0\}$. If $u, v \in V$ and $u\vartheta = v\vartheta$ then $(u - v)\vartheta = 0$, so $u - v \in \operatorname{Ker}\vartheta = \{0\}$, and so $u = v$. Thus ϑ is injective. As ϑ is surjective and injective, ϑ is invertible.

3. First suppose that $V = U \oplus W$. Then $V = U + W$. Let $v \in U \cap W$. Then $v = v + 0 = 0 + v$ and this gives us two expressions for v as the sum of an element in U and an element in W. Since such expressions are unique, $v = 0$. Thus $U \cap W = \{0\}$.
   Now suppose that $V = U + W$ and $U \cap W = \{0\}$. If $u_1 + w_1 = u_2 + w_2$ with $u_1, u_2 \in U$ and $w_1, w_2 \in W$, then $u_1 - u_2 = w_2 - w_1 \in U \cap W = \{0\}$; hence $u_1 = u_2$ and $w_1 = w_2$. This shows that $V = U \oplus W$.

4. Assume first that $V = U \oplus W$. If $v \in V$ then $v = u + w$ for some $u \in U$ and $w \in W$; since u is a linear combination of $u_1, \ldots, u_r$ and w is a linear combination of $w_1, \ldots, w_s$, it follows that v is a linear combination of $u_1, \ldots, u_r, w_1, \ldots, w_s$. Therefore $u_1, \ldots, u_r, w_1, \ldots, w_s$ span V. Suppose that

$$\lambda_1 u_1 + \ldots + \lambda_r u_r + \mu_1 w_1 + \ldots + \mu_s w_s = 0$$

with all λ_i, μ_j in F. Since $V = U \oplus W$, the expression $0 = 0 + 0$ is the unique expression for 0 as the sum of vectors in U and W, and so

$$\lambda_1 u_1 + \ldots + \lambda_r u_r = \mu_1 w_1 + \ldots + \mu_s w_s = 0.$$

As $u_1, \ldots, u_r$ are linearly independent, this forces $\lambda_i = 0$ for all i; similarly $\mu_i = 0$ for all i. Therefore $u_1, \ldots, u_r, w_1, \ldots, w_s$ are linearly independent; hence they form a basis of V.
   Conversely, suppose that $u_1, \ldots, u_r, w_1, \ldots, w_s$ is a basis of V. If $v \in U \cap W$ then $v = \lambda_1 u_1 + \ldots + \lambda_r u_r = \mu_1 w_1 + \ldots + \mu_s w_s$ for some λ_i, $\mu_j \in F$; this gives $\lambda_i = \mu_j = 0$ for all i, j, since $u_1, \ldots, u_r, w_1, \ldots, w_s$ are linearly independent. Thus $v = 0$ and so $U \cap W = \{0\}$. It is easy to see that $V = U + W$, so by Exercise 3, $V = U \oplus W$.

5. (a) Assume first that $V = U_1 \oplus U_2 \oplus U_3$. Let $u \in U_1 \cap (U_2 + U_3)$. Then $u = u_1 = u_2 + u_3$ for some $u_i \in U_i$ $(1 \leqslant i \leqslant 3)$. Since $u_1 + 0 + 0 = 0 + u_2 + u_3$ and the sum $U_1 + U_2 + U_3$ is direct, we have $u_1 =$

$u_2 = u_3 = 0$. Therefore $U_1 \cap (U_2 + U_3) = \{0\}$. Similarly,
$U_2 \cap (U_1 + U_3) = U_3 \cap (U_1 + U_2) = \{0\}$.
Now suppose that $U_1 \cap (U_2 + U_3) = U_2 \cap (U_1 + U_3) = U_3 \cap (U_1 + U_2) = \{0\}$. Assume that $u_i, u_i' \in U_i \ (1 \le i \le 3)$ and $u_1 + u_2 + u_3 = u_1' + u_2' + u_3'$. Then
$u_1 - u_1' = (u_2' - u_2) + (u_3' - u_3) \in U_1 \cap (U_2 + U_3) = \{0\}$, so $u_1 = u_1'$.
Similarly, $u_2 = u_2'$ and $u_3 = u_3'$. Therefore $V = U_1 \oplus U_2 \oplus U_3$.
(b) Let $V = \mathbb{R}^2$, and $U_1 = \mathrm{sp}\,((1, 0))$, $U_2 = \mathrm{sp}\,((0, 1))$, $U_3 = \mathrm{sp}\,((1, 1))$.

6. By Exercise 4, if $V = U \oplus W$ then $\dim V = \dim U + \dim W$. More generally, if $V = U_1 \oplus \ldots \oplus U_r$ then $V = U_1 \oplus (U_2 \oplus \ldots \oplus U_r)$ (see (2.10)); by induction on r, $\dim (U_2 \oplus \ldots \oplus U_r) = \dim U_2 + \ldots + \dim U_r$, so $\dim V = \dim U_1 + \ldots + \dim U_r$.

7. Let $V = \mathbb{R}^2$. Define $\vartheta, \phi \colon V \to V$ by

$$\vartheta \colon (x, y) \to (x, 0) \text{ and } \phi \colon (x, y) \to (y, 0).$$

Then $\mathrm{Im}\,\vartheta = \mathrm{sp}\,((1, 0))$, $\mathrm{Ker}\,\vartheta = \mathrm{sp}\,((0, 1))$, so $V = \mathrm{Im}\,\vartheta \oplus \mathrm{Ker}\,\vartheta$; and $\mathrm{Im}\,\phi = \mathrm{Ker}\,\phi = \mathrm{sp}\,((1, 0))$, so $V \neq \mathrm{Im}\,\phi \oplus \mathrm{Ker}\,\phi$.

8. Suppose first that ϑ is a projection. Then $V = \mathrm{Im}\,\vartheta \oplus \mathrm{Ker}\,\vartheta$ by Proposition 2.32. Take a basis $u_1, \ldots, u_r$ for $\mathrm{Im}\,\vartheta$ and a basis $w_1, \ldots, w_s$ for $\mathrm{Ker}\,\vartheta$. Then $u_1, \ldots, u_r, w_1, \ldots, w_s$ is a basis, say $\mathscr{B}$, of V, by Exercise 4. Since $u_i\vartheta = u_i$ for all i and $w_j\vartheta = 0$ for all j, the matrix $[\vartheta]_{\mathscr{B}}$ is diagonal, the diagonal entries being r 1's followed by s 0's.
   Conversely, if $[\vartheta]_{\mathscr{B}}$ has the given form, then clearly $\vartheta^2 = \vartheta$, so ϑ is a projection.

9. Let $v \in V$. Then

$$v = \tfrac{1}{2}(v + v\vartheta) + \tfrac{1}{2}(v - v\vartheta).$$

Observe that $\tfrac{1}{2}(v + v\vartheta)\vartheta = \tfrac{1}{2}(v\vartheta + v)$, so $\tfrac{1}{2}(v + v\vartheta) \in U$. Similarly, $\tfrac{1}{2}(v - v\vartheta) \in W$. Thus $V = U + W$. If $v \in U \cap W$ then $v = v\vartheta = -v$, so $v = 0$. Therefore $V = U \oplus W$, by Exercise 3. The construction of the basis $\mathscr{B}$ is similar to that in Solution 8.

Chapter 3

1. First, suppose that ρ is a representation of G. Then

$$I = 1\rho = (a^m)\rho = (a\rho)^m = A^m.$$

   Conversely, assume that $A^m = I$. Then $(a^i)\rho = A^i$ for all integers i (including $i > m - 1$ and $i < 0$). Therefore for all integers i, j,

$$(a^i a^j)\rho = (a^{i+j})\rho = A^{i+j} = A^i A^j = (a^i\rho)(a^j\rho),$$

   and so ρ is a representation.

2. Check that $A^3 = B^3 = C^3 = I$. Hence by Exercise 1, each ρ_j is a representation. The representations ρ_2 and ρ_3 are faithful, but ρ_1 is not.

3. Define ρ by

$$(a^i b^j)\rho = (-1)^j \quad (0 \leqslant i \leqslant n-1, 0 \leqslant j \leqslant 1).$$

It is easy to check that ρ is a representation of G.

4. (1) For all $g \in G$, $I^{-1}(g\rho)I = g\rho$; hence ρ is equivalent to ρ.
(2) If ρ is equivalent to σ then there is an invertible matrix T such that $g\sigma = T^{-1}(g\rho)T$ for all $g \in G$; then $g\rho = (T^{-1})^{-1}(g\sigma)T^{-1}$, so σ is equivalent to ρ.
(3) If ρ is equivalent to σ and σ is equivalent to τ, then there are invertible matrices S and T such that $g\sigma = S^{-1}(g\rho)S$ and $g\tau = T^{-1}(g\sigma)T$ for all $g \in G$; then $g\tau = (ST)^{-1}(g\rho)(ST)$, so ρ is equivalent to τ.

5. Check that in each of the cases (1) $S = A$, $T = B$, (2) $S = A^3$, $T = -B$, (3) $S = -A$, $T = B$, (4) $S = C$, $T = D$, we have

$$S^6 = T^2 = I, \ T^{-1}ST = S^{-1}.$$

It follows that each ρ_k is a representation (see Example 1.4).

The matrices $A^r B^s$ ($0 \leqslant r \leqslant 5, 0 \leqslant s \leqslant 1$) are all different, so ρ_1 is faithful. Similarly ρ_4 is faithful. But ρ_2 and ρ_3 are not faithful, since $a^2\rho_2 = I$ and $a^3\rho_3 = I$.

The representations ρ_1 and ρ_4 are equivalent: to see this, let

$$T = \begin{pmatrix} 1 & 1 \\ -i & i \end{pmatrix}.$$

Then $T^{-1}(g\rho_4)T = g\rho_1$ for all $g \in G$. (To find T, first work out the eigenvectors of C.)

If $j \neq 2$, then $a^2\rho_j \neq I$; hence ρ_2 is not equivalent to any of the others. And if $j \neq 3$, then $a^3\rho_j \neq I$; hence ρ_3 is not equivalent to any of the others.

6. Define the matrices A and B by

$$A = \begin{pmatrix} 0 & 1 & 0 \\ -1 & 0 & 0 \\ 0 & 0 & 1 \end{pmatrix}, \quad B = \begin{pmatrix} 1 & 0 & 0 \\ 0 & -1 & 0 \\ 0 & 0 & 1 \end{pmatrix}.$$

Then the function $\rho: a^r b^s \to A^r B^s$ ($0 \leqslant r \leqslant 3, 0 \leqslant s \leqslant 1$) is a faithful representation of $D_8 = \langle a, b: a^4 = b^2 = 1, b^{-1}ab = a^{-1} \rangle$. Compare Example 3.2(1).

7. By Theorem 1.10, $G/\mathrm{Ker}\,\rho \cong \mathrm{Im}\,\rho$. But $\mathrm{Im}\,\rho \leqslant \mathrm{GL}(1, F)$ and $\mathrm{GL}(1, F)$ is abelian. Therefore $G/\mathrm{Ker}\,\rho$ is abelian.

8. No: let G be any non-abelian group and let ρ be the trivial representation.

Chapter 4

1.

g	1	(1 2)	(1 3)
$[g]_{\mathscr{B}_1}$	$\begin{pmatrix} 1 & 0 & 0 \\ 0 & 1 & 0 \\ 0 & 0 & 1 \end{pmatrix}$	$\begin{pmatrix} 0 & 1 & 0 \\ 1 & 0 & 0 \\ 0 & 0 & 1 \end{pmatrix}$	$\begin{pmatrix} 0 & 0 & 1 \\ 0 & 1 & 0 \\ 1 & 0 & 0 \end{pmatrix}$
$[g]_{\mathscr{B}_2}$	$\begin{pmatrix} 1 & 0 & 0 \\ 0 & 1 & 0 \\ 0 & 0 & 1 \end{pmatrix}$	$\begin{pmatrix} 1 & 0 & 0 \\ 0 & -1 & 0 \\ 0 & -1 & 1 \end{pmatrix}$	$\begin{pmatrix} 1 & 0 & 0 \\ 0 & 1 & -1 \\ 0 & 0 & -1 \end{pmatrix}$

g	(2 3)	(1 2 3)	(1 3 2)
$[g]_{\mathscr{B}_1}$	$\begin{pmatrix} 1 & 0 & 0 \\ 0 & 0 & 1 \\ 0 & 1 & 0 \end{pmatrix}$	$\begin{pmatrix} 0 & 1 & 0 \\ 0 & 0 & 1 \\ 1 & 0 & 0 \end{pmatrix}$	$\begin{pmatrix} 0 & 0 & 1 \\ 1 & 0 & 0 \\ 0 & 1 & 0 \end{pmatrix}$
$[g]_{\mathscr{B}_2}$	$\begin{pmatrix} 1 & 0 & 0 \\ 0 & 0 & 1 \\ 0 & 1 & 0 \end{pmatrix}$	$\begin{pmatrix} 1 & 0 & 0 \\ 0 & -1 & 1 \\ 0 & -1 & 0 \end{pmatrix}$	$\begin{pmatrix} 1 & 0 & 0 \\ 0 & 0 & -1 \\ 0 & 1 & -1 \end{pmatrix}$

All the matrices $[g]_{\mathscr{B}_2}$ have the form

$$\begin{pmatrix} 1 & 0 & 0 \\ 0 & \blacksquare & \blacksquare \\ 0 & \blacksquare & \blacksquare \end{pmatrix}.$$

2. Let $g \in S_n$. For all u, v in V and λ in F, we have

$$vg \in V, \quad v1 = v, \quad (\lambda v)g = \lambda(vg), \quad (u + v)g = ug + vg.$$

It remains to check (2) of Definition 4.2. Let $v \in V$ and g, $h \in S_n$. Assume first that $gh \in A_n$. Then $v(gh) = v$; and $(vg)h = v$, since either $vg = v = vh$ (if g, $h \in A_n$) or $vg = -v = vh$ (if g, $h \notin A_n$). Next, assume that $gh \notin A_n$. Then $v(gh) = -v$; and $(vg)h = -v$, since one of g, h is in A_n and the other is not. We have now checked all the conditions in Definition 4.2, so V is an FG-module.

3. Let $A = \begin{pmatrix} 0 & 1 & 0 & 0 \\ -1 & 0 & 0 & 0 \\ 0 & 0 & 0 & -1 \\ 0 & 0 & 1 & 0 \end{pmatrix}$ and $B = \begin{pmatrix} 0 & 0 & 1 & 0 \\ 0 & 0 & 0 & 1 \\ -1 & 0 & 0 & 0 \\ 0 & -1 & 0 & 0 \end{pmatrix}$.

Check that

$$A^4 = I, \quad B^2 = A^2, \quad B^{-1}AB = A^{-1}.$$

Hence $\rho: a^i b^j \to A^i B^j$ $(0 \leqslant i \leqslant 3, 0 \leqslant j \leqslant 1)$ is a representation of Q_8 over $\mathbb{R}$. Let $V = \mathbb{R}^4$. By Theorem 4.4(1), V becomes an $\mathbb{R}Q_8$-module if we define $vg = v(g\rho)$ for all $v \in V$, $g \in Q_8$. If we put

$v_1 = (1, \ 0, \ 0, \ 0), v_2 = (0, \ 1, \ 0, \ 0), v_3 = (0, \ 0, \ 1, \ 0), v_4 = (0, \ 0, \ 0, \ 1),$
then for all i, $v_i a$ and $v_i b$ are as required in the question.

4. You may find it helpful first to check that if

$$M = \begin{pmatrix} 0 & 1 & & & 0 \\ 1 & 0 & & & \\ & & 1 & & \\ & & & \ddots & \\ 0 & & & & 1 \end{pmatrix}$$

then MA is obtained from A by swapping the first two rows, and AM is obtained from A by swapping the first two columns. To solve the exercise, let g be the permutation in S_n which has the property that for all i,

$$\text{row } i \text{ of } B = \text{row } ig \text{ of } A.$$

Let P be the $n \times n$ matrix (p_{ij}) defined by

$$p_{ij} = \begin{cases} 1, & \text{if } j = ig, \\ 0, & \text{if } j \neq ig. \end{cases}$$

Then P is a permutation matrix, and the ij-entry of PA is

$$\sum_{k=1}^{n} p_{ik} a_{kj} = a_{ig,j}.$$

Hence $PA = B$.

If C is a matrix obtained from A by permuting the columns, then $C = AQ$ for some permutation matrix Q; the proof is similar to that for the rows.

Chapter 5

1. It is easy to verify that V is an FG-module. Let U be a non-zero FG-submodule of V, and let $(\alpha, \beta) \in U$ with $(\alpha, \beta) \neq (0, 0)$. Then $(\alpha, \beta) + (\alpha, \beta)a = (\alpha + \beta, \alpha + \beta) \in U$, and $(\alpha, \beta) - (\alpha, \beta)a = (\alpha - \beta, \beta - \alpha) \in U$. Since at least one of $\alpha + \beta$ and $\alpha - \beta$ is non-zero, we deduce that $(1, 1)$ or $(1, -1)$ belongs to U. Hence the FG-submodules of V are $\{0\}$, $\text{sp}((1, 1))$, $\text{sp}((1, -1))$ and V.

2. Suppose that ρ has degree n and ρ is reducible. Then ρ is equivalent to a representation τ of the form

$$\tau: g \rightarrow \left(\begin{array}{c|c} X_g & 0 \\ \hline Y_g & Z_g \end{array} \right) \quad (g \in G)$$

where X_g is a $k \times k$ matrix and $0 < k < n$. Then σ is equivalent to τ, since σ is equivalent to ρ. Therefore σ is reducible.

3. Let $G = D_{12}$ and let $\rho_1, \ldots, \rho_4$ be the representations of G defined in Exercise 3.5. First consider the FG-module $V = F^2$, where $vg = v(g\rho_1)$ for

$v \in V$, $g \in G$. Suppose that U is a non-zero FG-submodule of V. Then U is an FH-module, where H is the subgroup $\{1, b\}$. By the solution to Exercise 1, either $(1, 1)$ or $(1, -1)$ lies in U. Since $(1, 1)$ and $(1, 1)a$ are linearly independent, and also $(1, -1)$ and $(1, -1)a$ are linearly independent, it follows that dim $U \geqslant 2$. Consequently $U = V$ and so V is irreducible. (See Example 5.5(2) for an alternative argument.) Therefore ρ_1 is irreducible; also ρ_4 is irreducible, since ρ_1 and ρ_4 are equivalent.

Now let $V = F^2$ with $vg = v(g\rho_2)$ for $v \in V$, $g \in G$. Then $(1, 1)a = -(1, 1) = (1, 1)b$. Hence sp$((1, 1))$ is an FG-submodule of V and ρ_2 is reducible.

Finally, ρ_3 is irreducible, by an argument similar to that for ρ_1.

4. (a) It is easy to check the given relations. Using the relations, we may write every element of G in the form

$$a^i b^j c^k \quad (0 \leqslant i \leqslant 2, \, 0 \leqslant j \leqslant 2, \, 0 \leqslant k \leqslant 1).$$

Thus $|G| \leqslant 18$. However, it is clear that $|\langle a, b \rangle| = 9$ and $\langle a, b \rangle \neq G$. Hence, by Lagrange's Theorem, $|G|$ is a multiple of 9 and $|G| > 9$. Therefore $|G| = 18$.

(b) Let

$$A = \begin{pmatrix} \varepsilon & 0 \\ 0 & \varepsilon^{-1} \end{pmatrix}, \, B = \begin{pmatrix} \eta & 0 \\ 0 & \eta^{-1} \end{pmatrix}, \, C = \begin{pmatrix} 0 & 1 \\ 1 & 0 \end{pmatrix}.$$

Check that $A^3 = B^3 = C^2 = I$, $AB = BA$, $C^{-1}AC = A^{-1}$ and $C^{-1}BC = B^{-1}$. Hence ρ is a representation (compare Example 1.4).

(c) For every element g of $\langle a, b \rangle$, there exists a cube root ξ of unity such that

$$g\rho = \begin{pmatrix} \xi & 0 \\ 0 & \xi^{-1} \end{pmatrix}.$$

But there are only three distinct cube roots of unity, so there exist distinct g_1, $g_2 \in \langle a, b \rangle$ with $g_1\rho = g_2\rho$. Therefore ρ is never faithful.

(d) Let $V = \mathbb{C}^2$ be the $\mathbb{C}G$-module obtained by defining $vg = v(g\rho)$ for all $v \in \mathbb{C}^2$, $g \in G$. If U is a non-zero $\mathbb{C}G$-submodule of V, then U is a $\mathbb{C}H$-submodule, where H is the subgroup $\{1, c\}$. Hence either $(1, 1)$ or $(1, -1)$ lies in U, by the solution to Exercise 1; accordingly, let u be $(1, 1)$ or $(1, -1)$ (so that $u \in U$). Now u and ua are linearly independent unless $\varepsilon = 1$, and u and ub are linearly independent unless $\eta = 1$. Hence, if either $\varepsilon \neq 1$ or $\eta \neq 1$ then dim $U = 2$ and so ρ is irreducible. On the other hand, if $\varepsilon = \eta = 1$ then sp$((1, 1))$ is a $\mathbb{C}G$-submodule of V, so ρ is reducible.

5. Let $V = \{0\}$ and let $0g = 0$ for all $g \in G$. Then V is neither reducible nor irreducible.

Chapter 6

1. (a) $\quad xy = -2 \cdot 1 - a^3 + ab + 3a^2 b + 2a^3 b,$

$\quad\quad yx = -2 \cdot 1 - a^3 + b + 2a^2 b + 3a^3 b,$

$\quad\quad x^2 = 4 \cdot 1 + a^2 + 4a^3.$

(b) $az = ab + a^3b = a^2ba + ba = za$, and $bz = 1 + a^2 = zb$. Hence $a^ib^jz = za^ib^j$ for all i, j and so $gz = zg$ for all $g \in G$. If $r \in CG$ then $r = \sum_g \lambda_g g$ with $\lambda_g \in C$, so $rz = \sum \lambda_g gz = \sum \lambda_g zg = zr$.

2. Let $C_2 \times C_2 = \langle a, b: a^2 = b^2 = 1, ab = ba \rangle$. Relative to the basis 1, a, b, ab of $F(C_2 \times C_2)$, the regular representation ρ is given by

$$1\rho = \begin{pmatrix} 1 & 0 & 0 & 0 \\ 0 & 1 & 0 & 0 \\ 0 & 0 & 1 & 0 \\ 0 & 0 & 0 & 1 \end{pmatrix}, \quad a\rho = \begin{pmatrix} 0 & 1 & 0 & 0 \\ 1 & 0 & 0 & 0 \\ 0 & 0 & 0 & 1 \\ 0 & 0 & 1 & 0 \end{pmatrix},$$

$$b\rho = \begin{pmatrix} 0 & 0 & 1 & 0 \\ 0 & 0 & 0 & 1 \\ 1 & 0 & 0 & 0 \\ 0 & 1 & 0 & 0 \end{pmatrix}, \quad (ab)\rho = \begin{pmatrix} 0 & 0 & 0 & 1 \\ 0 & 0 & 1 & 0 \\ 0 & 1 & 0 & 0 \\ 1 & 0 & 0 & 0 \end{pmatrix}.$$

3. No: let $G = \langle a: a^2 = 1 \rangle$, and take $r = 1 + a$, $s = 1 - a$.

4. (a) As g runs through G, so do gh and hg. Hence $ch = hc = c$.
   (b) $c^2 = c\sum_{h \in G} h = \sum_{h \in G} ch = |G|c$.
   (c) All the entries in $[\vartheta]_{\mathscr{B}}$ are 1 (compare the solution to Exercise 2). The reason is that for all i, j, there exists a unique h in G such that $g_i h = g_j$.

5. Note first that if u is an element of a vector space, and $u + u = u$, then $u = 0$. Now $0r = (0 + 0)r = 0r + 0r$, and $v0 = v(0 + 0) = v0 + v0$; hence $0r = v0 = 0$.
   Let V be the trivial FG-module and let $0 \neq v \in V$ and $1 \neq g \in G$. If $r = 1 - g$, then $vr = 0$ and neither v nor r is 0.

6. Let $v_1 = 1 + \omega^2 a + \omega a^2$ and $v_2 = b + \omega^2 ab + \omega a^2 b$. Check that $v_1 a = \omega v_1$, $v_2 a = \omega^2 v_2$, $v_1 b = v_2$ and $v_2 b = v_1$. Hence W is a CG-submodule of CG. Use the argument of either Example 5.5(2) or the solution to Exercise 5.3 to prove that W is irreducible.

Chapter 7

1. For all u_1, $u_2 \in U$, $\lambda \in F$ and $g \in G$, we have

$$(u_1 + u_2)\vartheta\phi = (u_1\vartheta + u_2\vartheta)\phi = u_1(\vartheta\phi) + u_2(\vartheta\phi),$$

$$(\lambda u_1)\vartheta\phi = (\lambda(u_1\vartheta))\phi = \lambda(u_1(\vartheta\phi)),$$

$$(u_1 g)\vartheta\phi = ((u_1\vartheta)g)\phi = ((u_1\vartheta)\phi)g = (u_1(\vartheta\phi))g.$$

2. Let $a = (1\,2\,3\,4\,5)$ and let $v_1, \ldots, v_5$ be the natural basis for the permutation module for G over F. Then

$$\vartheta: \lambda_1 v_1 + \ldots + \lambda_5 v_5 \rightarrow \lambda_1 1 + \lambda_2 a + \lambda_3 a^2 + \lambda_4 a^3 + \lambda_5 a^4$$

is the required FG-isomorphism. (Note that $v_i\vartheta = a^i$, so $(v_i a)\vartheta = v_{i+1}\vartheta = a^{i+1} = (v_i\vartheta)a$; hence ϑ is an FG-homomorphism.)

3. It is easy to show that V_0 is an FG-submodule of V. Let $x \in G$. Then

$$\sum_{g \in G} vxg = \sum_{g \in G} vg = \sum_{g \in G} vgx.$$

Hence $(vx)\vartheta = v\vartheta = (v\vartheta)x$; noting that $V\vartheta \subseteq V_0$, we see that ϑ is an FG-homomorphism from V to V_0.

If $v \in V_0$ then $(v/|G|)\vartheta = v$; hence ϑ is surjective.

4. Suppose that $\phi: V \to W$ is an FG-isomorphism. Let $g \in G$. For all $v \in V_0$, $(v\phi)g = (vg)\phi = v\phi$, and so $V_0\phi \subseteq W_0$. For all $w \in W_0$, $(w\phi^{-1})g = (wg)\phi^{-1} = w\phi^{-1}$, so $W_0\phi^{-1} \subseteq V_0$. Hence the function ϕ, restricted to V_0, is an FG-isomorphism from V_0 to W_0.

5. No: let $v_1, \ldots, v_4$ be the natural basis of the permutation module V for G over F. In the notation of Exercise 3,

$$V_0 = \mathrm{sp}\,(v_1 + v_2, v_3 + v_4) \quad \text{and} \quad (FG)_0 = \mathrm{sp}\left(\sum_{g \in G} g\right).$$

Since V_0 and $(FG)_0$ have different dimensions, it follows from Exercise 4 that V and FG are not isomorphic FG-modules.

6. (a) ϑ is easily seen to be a linear transformation. Also

$$(\alpha 1 + \beta x)x\vartheta = (\beta 1 + \alpha x)\vartheta = (\beta - \alpha)(1 - x) = (\alpha - \beta)(1 - x)x$$
$$= (\alpha 1 + \beta x)\vartheta x.$$

Hence ϑ is an FG-homomorphism.

(b) $(\alpha - \beta)(1 - x)\vartheta = ((\alpha - \beta) - (\beta - \alpha))(1 - x) = 2(\alpha - \beta)(1 - x)$. Hence $\vartheta^2 = 2\vartheta$.

(c) Let $\mathscr{B}$ be the basis $1 - x$, $1 + x$.

Chapter 8

1. $V = \mathrm{sp}\,(-\omega v_1 + v_2) \oplus \mathrm{sp}\,(-\omega^2 v_1 + v_2)$, where $\omega = e^{2\pi i/3}$. (Find eigenvectors for x.)

2. Let $G = \{1, a, b, ab\} \cong C_2 \times C_2$ (so $a^2 = b^2 = 1$, $ab = ba$). Then

$$\mathbb{R}G = \mathrm{sp}\,(1 + a + b + ab) \oplus \mathrm{sp}\,(1 + a - b - ab) \oplus \mathrm{sp}\,(1 - a + b - ab)$$
$$\oplus \mathrm{sp}\,(1 - a - b + ab).$$

3. Let G be any group, and let V be a 2-dimensional vector space over $\mathbb{C}$ with basis v_1, v_2. Define $vg = v$ for all $v \in V$, $g \in G$; this makes V into a $\mathbb{C}G$-module. If we let

$$\vartheta: \lambda v_1 + \mu v_2 \to \lambda v_2 \quad (\lambda, \mu \in \mathbb{C})$$

then ϑ is a $\mathbb{C}G$-homomorphism from V to V, and $\mathrm{Ker}\,\vartheta = \mathrm{Im}\,\vartheta = \mathrm{sp}\,(v_2)$.

4. Suppose ρ is reducible. Then by Maschke's Theorem, ρ is equivalent to a representation σ of the form

$$g\sigma = \begin{pmatrix} \lambda_g & 0 \\ 0 & \mu_g \end{pmatrix} \quad (\lambda_g, \mu_g \in \mathbb{C}).$$

Then $(g\sigma)(h\sigma) = (h\sigma)(g\sigma)$ for all $g, h \in G$, since all diagonal matrices commute with each other; hence also $(g\rho)(h\rho) = (h\rho)(g\rho)$ for all g, $h \in G$. This is a contradiction. Therefore ρ is irreducible.

5. $U = \mathrm{sp}\,((1, 0))$ is the only 1-dimensional $\mathbb{C}G$-submodule of V, so there is no $\mathbb{C}G$-submodule W of V with $V = U \oplus W$.

6. (1) It is straightforward to verify for $[\ ,\ ]$ the axioms of a complex inner product. For example, if $u \neq 0$ then $(ux, ux) > 0$ for all $x \in G$, so $[u, u] > 0$. Also

$$[ug, vg] = \sum_{x \in G}(ugx, vgx) = \sum_{x \in G}(ux, vx) = [u, v].$$

(2) It is easy to prove that $U^{\perp}$ is a subspace of V. Let $v \in U^{\perp}$ and $g \in G$. Then for all $u \in U$,

$$[u, vg] = [ug^{-1}, vgg^{-1}] \quad \text{by part (1)}$$

$$= [ug^{-1}, v] = 0 \quad \text{since } ug^{-1} \in U.$$

Therefore $vg \in U^{\perp}$, and so $U^{\perp}$ is a $\mathbb{C}G$-submodule of V.

(3) Let $W = U^{\perp}$. Then $V = U \oplus W$, and W is a $\mathbb{C}G$-submodule of V by part (2).

7. We know that the regular $\mathbb{C}G$-module $\mathbb{C}G$ is faithful (Proposition 6.6). Let $\mathbb{C}G = U_1 \oplus \ldots \oplus U_r$, where $U_1, \ldots, U_r$ are irreducible $\mathbb{C}G$-submodules of $\mathbb{C}G$. Then there exist $i \in \{1, \ldots, r\}$ and $g \in G$ such that $ug \neq u$ for some $u \in U_i$ (otherwise $vg = v$ for all $v \in \mathbb{C}G$). Define

$$K = \{x \in G : vx = v \text{ for all } v \in U_i\}.$$

Check that K is a normal subgroup of G; also $K \neq G$ since $g \notin K$. Since G is simple, we must therefore have $K = \{1\}$. This means that U_i is a faithful irreducible $\mathbb{C}G$-module.

Chapter 9

1. Let $C_2 = \langle a : a^2 = 1 \rangle$. Irreducible representations ρ_1, ρ_2:

$$1\rho_1 = a\rho_1 = (1); \ 1\rho_2 = (1), \ a\rho_2 = (-1).$$

Let $C_3 = \langle b : b^3 = 1 \rangle$ and let $\omega = e^{2\pi i/3}$. Irreducible representations ρ_1, ρ_2, ρ_3:

$$1\rho_1 = b\rho_1 = b^2\rho_1 = (1);$$

$$b^i\rho_2 = (\omega^i);$$

$$b^i\rho_3 = (\omega^{2i}).$$

Let $C_2 \times C_2 = \{(1, 1), (x, 1), (1, y), (x, y)\}$, where $x^2 = y^2 = 1$. Irreducible representations ρ_1, ρ_2, ρ_3, ρ_4:

$$gp_1 = (1) \text{ for all } g \in C_2 \times C_2;$$

$$(x^i, y^j)\rho_2 = (-1)^j;$$

$$(x^i, y^j)\rho_3 = (-1)^i;$$

$$(x^i, y^j)\rho_4 = (-1)^{i+j}.$$

2. Let $C_4 \times C_4 = \langle (x, 1), (1, y): x^4 = y^4 = 1 \rangle$.
   (a) $\rho: (x^i, y^j) \to (-1)^i$.
   (b) If $g_1 = (x^2, 1)$ and $g_2 = (1, y^2)$ then g_1, g_2 and $g_1 g_2$ all have order 2. Since $(g_1 g_2)\sigma = (g_1\sigma)(g_2\sigma)$ for all representations σ, we cannot have $(g_1 g_2)\sigma = g_1\sigma = g_2\sigma = (-1)$.

3. For $1 \leqslant j \leqslant r$, let g_j generate C_{n_j}, and let $\varepsilon_j = e^{2\pi i/n_j}$. Then

$$\rho: (g_1^{i_1}, \ldots, g_r^{i_r}) \to \begin{pmatrix} \varepsilon_1^{i_1} & & 0 \\ & \ddots & \\ 0 & & \varepsilon_r^{i_r} \end{pmatrix}$$

is a faithful representation of $C_{n_1} \times \ldots \times C_{n_r}$ of degree r.
Yes: if $r = 2$, $n_1 = 2$, $n_2 = 3$, then

$$\sigma: (g_1^{i_1}, g_2^{i_2}) \to (\varepsilon_1^{i_1} \varepsilon_2^{i_2})$$

is a faithful representation of degree $1 < r$.

4. Check that $A^4 = B^2 = I$, $B^{-1}AB = A^{-1}$ when $A = a\rho$ and $B = b\rho$. Hence ρ gives a representation; similarly for σ.
   If $M(g\rho) = (g\rho)M$ for $g = a$ and for $g = b$, then $M = \lambda I$ for some $\lambda \in \mathbb{C}$. Hence ρ is irreducible (Corollary 9.3).
   Notice that the matrix

$$\begin{pmatrix} 5 & -6 \\ 4 & -5 \end{pmatrix}$$

commutes with $g\sigma$ for all $g \in G$; hence σ is reducible (Corollary 9.3).

5. Let $z = \sum_{g \in G} g$. Then $xz = z = zx$ for all $x \in G$. Hence $z \in Z(\mathbb{C}G)$, and the result follows from Proposition 9.14.

6. (a) Clearly a commutes with $a + a^{-1}$. Also $b^{-1}(a + a^{-1})b = a^{-1} + a$, so b commutes with $a + a^{-1}$.
   (b) Check that $w(a + a^{-1}) = -w$ for all $w \in W$.

7. (a) Let $C_n = \langle x: x^n = 1 \rangle$. Then $\rho: x^j \to (e^{2\pi i j/n})$ is a faithful irreducible representation of C_n.
   (b) $\rho: a \to \begin{pmatrix} 0 & 1 \\ -1 & 0 \end{pmatrix}$, $b \to \begin{pmatrix} 1 & 0 \\ 0 & -1 \end{pmatrix}$

   gives a faithful irreducible representation of D_8 (see Example 5.5(2)).
   (c) The centre of $C_2 \times D_8$ is isomorphic to $C_2 \times C_2$, so is not cyclic. Therefore Proposition 9.16 shows that $C_2 \times D_8$ has no faithful irreducible representation.
   (d) Let $C_3 = \langle x: x^3 = 1 \rangle$ and let $\omega = e^{2\pi i/3}$. Check that

$$\rho: (x, a) \to \begin{pmatrix} 0 & \omega \\ -\omega & 0 \end{pmatrix}, (x, b) \to \begin{pmatrix} \omega & 0 \\ 0 & -\omega \end{pmatrix}$$

gives a representation of $C_3 \times D_8$. It is irreducible (see for example Exercise 8.4) and faithful.

Chapter 10

1. Let $V = \mathrm{sp}(\sum_{g \in G} g)$. Then V is a trivial $\mathbb{C}G$-submodule of $\mathbb{C}G$. Now suppose that U is an arbitrary trivial $\mathbb{C}G$-submodule of $\mathbb{C}G$, so $U = \mathrm{sp}(u)$ for some u. Then $ug = u$ for all $g \in G$, so $|G|u = u(\sum_{g \in G} g) = (\sum_{g \in G} g)u \in V$. Thus $U = V$, and so $\mathbb{C}G$ has exactly one trivial $\mathbb{C}G$-submodule, namely V.

2. Let $G = \langle x : x^4 = 1 \rangle$. Then

$$\mathbb{C}G = \mathrm{sp}(1 + x + x^2 + x^3) \oplus \mathrm{sp}(1 + ix - x^2 - ix^3)$$
$$\oplus \mathrm{sp}(1 - x + x^2 - x^3) \oplus \mathrm{sp}(1 - ix - x^2 + ix^3).$$

3. Let

$$u_1 = 1 + a + a^2 + a^3 - b - ab - a^2b - a^3b,$$
$$u_2 = 1 - a + a^2 - a^3 + b - ab + a^2b - a^3b,$$
$$u_3 = 1 - a + a^2 - a^3 - b + ab - a^2b + a^3b.$$

4. We decompose $\mathbb{C}G$ as a direct sum of irreducible $\mathbb{C}G$-submodules. Let

$$v_0 = 1 + a + a^2 + a^3, \qquad v_1 = 1 + ia - a^2 - ia^3,$$
$$v_2 = 1 - a + a^2 - a^3, \qquad v_3 = 1 - ia - a^2 + ia^3$$

(compare the solution to Exercise 2). For $0 \le j \le 3$, let $w_j = bv_j$. Then, as in Example 10.8(2), the subspaces $\mathrm{sp}(v_0, w_0)$, $\mathrm{sp}(v_1, w_3)$, $\mathrm{sp}(v_2, w_2)$ and $\mathrm{sp}(v_3, w_1)$ are $\mathbb{C}G$-submodules of $\mathbb{C}G$. We have

$$\mathrm{sp}(v_0, w_0) = U_0 \oplus U_1, \quad \mathrm{sp}(v_2, w_2) = U_2 \oplus U_3,$$

where $U_i = \mathrm{sp}(u_i)$ ($0 \le i \le 3$) and u_1, u_2, u_3 are as in the solution to Exercise 3, while $u_0 = \sum_{g \in G} g$.

Let $U_4 = \mathrm{sp}(v_1, w_3)$, $U_5 = \mathrm{sp}(v_3, w_1)$. As in Example 5.5(2) (or see Exercise 8.4), U_4 and U_5 are irreducible $\mathbb{C}G$-modules. Moreover $U_4 \cong U_5$, since there is a $\mathbb{C}G$-isomorphism sending $v_1 \to w_1$, $w_3 \to v_3$.

Theorem 10.5 now shows that there are exactly five non-isomorphic irreducible $\mathbb{C}G$-modules, namely U_0, U_1, U_2, U_3 and U_4. Therefore every irreducible representation of D_8 over $\mathbb{C}$ is equivalent to precisely one of the following:

$$\rho_0 : a \to (1), \, b \to (1)$$
$$\rho_1 : a \to (1), \, b \to (-1)$$
$$\rho_2 : a \to (-1), \, b \to (1)$$
$$\rho_3 : a \to (-1), \, b \to (-1)$$
$$\rho_4 : a \to \begin{pmatrix} i & 0 \\ 0 & -i \end{pmatrix}, \, b \to \begin{pmatrix} 0 & 1 \\ 1 & 0 \end{pmatrix}.$$

5. Let $\vartheta\colon U_1 \to U_2$ be a $\mathbb{C}G$-isomorphism. For $\lambda \in \mathbb{C}$, define the function $\phi_\lambda\colon U_1 \to V$ by

$$\phi_\lambda\colon u \to u + \lambda u\vartheta \quad (u \in U_1).$$

Then ϕ_λ is easily seen to be a $\mathbb{C}G$-homomorphism; moreover,

$$u \in \operatorname{Ker}\phi_\lambda \Leftrightarrow u + \lambda u\vartheta = 0 \Leftrightarrow u = 0,$$

since the sum $U_1 + U_2$ is direct. Thus $U_1 \cong \operatorname{Im}\phi_\lambda$. It is easy to check that if $\lambda \neq \mu$ then $\operatorname{Im}\phi_\lambda \neq \operatorname{Im}\phi_\mu$. Therefore we have constructed infinitely many $\mathbb{C}G$-submodules $\operatorname{Im}\phi_\lambda$ of the required form.

6. V is irreducible, either by the method of Example 5.5(2) or by Exercise 8.4.
   Let $u_1 = 1 - ia - a^2 + ia^3$, $u_2 = b - iab - a^2b + ia^3b$. Then $\operatorname{sp}(u_1, u_2)$ is a $\mathbb{C}G$-submodule of $\mathbb{C}G$ which is isomorphic to V. A $\mathbb{C}G$-isomorphism is given by $v_1 \to u_1$, $v_2 \to u_2$.

Chapter 11

1. Since G is non-abelian, not all the dimensions are 1 (see Proposition 9.18). Hence, by Theorem 11.12, the dimensions are 1, 1, 2.

2. Compare Example 11.13 to see that the possible answers are 1^{12}, 1^82, 1^42^2 and 1^33 (where 1^{12} means twelve 1s, 1^82 means eight 1s and one 2, and so on). It will be shown later (Exercises 15.4, 17.3) that 1^82 cannot occur.
   By Exercise 5.3, D_{12} has at least two inequivalent irreducible representations of degree 2. Hence the answer for D_{12} is 1^42^2.

3. For each $g \in G$, define $\phi_g\colon \mathbb{C}G \to \mathbb{C}G$ by $r\phi_g = gr$ ($r \in \mathbb{C}G$). Then $\{\phi_g\colon g \in G\}$ gives a basis of $\operatorname{Hom}_{\mathbb{C}G}(\mathbb{C}G, \mathbb{C}G)$ (compare the proof of Proposition 11.8).

4. Let $v_1, \ldots, v_n$ be the natural basis of V. Then $\operatorname{sp}(v_1 + \ldots + v_n)$ is the unique trivial $\mathbb{C}G$-submodule of V (compare Exercise 10.1). Hence by Corollary 11.6, $\dim(\operatorname{Hom}_{\mathbb{C}G}(V, U)) = 1$.

5. Let v_1, w_2 be the basis of U_3 described in Example 10.8(2). Define ϑ_1 and ϑ_2 by $r\vartheta_1 = v_1r$, $r\vartheta_2 = w_2r$ ($r \in \mathbb{C}G$). Then ϑ_1, ϑ_2 is a basis of $\operatorname{Hom}_{\mathbb{C}G}(\mathbb{C}G, U_3)$, by the proof of Proposition 11.8. Also, define ϕ_1, ϕ_2 by $u\phi_1 = u$, $u\phi_2 = bu$ ($u \in U_3$). Then ϕ_1, ϕ_2 is a basis of $\operatorname{Hom}_{\mathbb{C}G}(U_3, \mathbb{C}G)$.

6. Let $V = X_1 \oplus \ldots \oplus X_r$ and $W = Y_1 \oplus \ldots \oplus Y_s$, where each X_a and each Y_b is an irreducible $\mathbb{C}G$-module. Then by (11.5)(3) and Proposition 11.2, $\dim(\operatorname{Hom}_{\mathbb{C}G}(V, W))$ is equal to the number of ordered pairs (a, b) such that $X_a \cong Y_b$. This, in turn, equals

$$\sum_{i=1}^{k} |\{(a, b)\colon X_a \cong Y_b \cong V_i\}|.$$

Now the number of integers a with $X_a \cong V_i$ is $\dim(\operatorname{Hom}_{\mathbb{C}G}(V, V_i)) = d_i$, by Corollary 11.6, and similarly the number of integers b with $Y_b \cong V_i$ is e_i. Therefore, $\dim(\operatorname{Hom}_{\mathbb{C}G}(V, W)) = \sum_{i=1}^{k} d_i e_i$.

Chapter 12

1. Assume that g, $h \in C_G(x)$. Then $gx = xg$ and $hx = xh$, so $h^{-1}x = xh^{-1}$ and $gh^{-1}x = gxh^{-1} = xgh^{-1}$; thus $gh^{-1} \in C_G(x)$. Also $1x = x1$, so $1 \in C_G(x)$. Therefore $C_G(x)$ is a subgroup of G.

   If $z \in Z(G)$ then $zg = gz$ for all $g \in G$, so $zx = xz$ and $z \in C_G(x)$.

2. Note that $x \in C_G(g) \Leftrightarrow x \in C_G(gz)$. Now the required result follows from Theorem 12.8.

3. (a) $(1\,2)^G = \{(i\,j): 1 \leqslant i < j \leqslant n\}$ and this set has size $\binom{n}{2}$. The centralizer $C_G((1\,2))$ consists of all elements x and $(1\,2)x$, where x is a permutation fixing 1 and 2. Thus $|C_G((1\,2))| = 2 \cdot (n - 2)!$, in agreement with Theorem 12.8 (since $\binom{n}{2} = n!/(2 \cdot (n - 2)!)$).

   (b) $(1\,2\,3)^G$ consists of all 3-cycles $(i\,j\,k)$. There are $\binom{n}{3}$ choices for the numbers i, j, k (unordered). Each choice gives exactly two 3-cycles, namely $(i\,j\,k)$ and $(i\,k\,j)$.

   $(1\,2)(3\,4)^G$ consists of all permutations of the form $(i\,j)(k\,l)$ with i, j, k, l distinct. There are $\binom{n}{4}$ choices for the numbers i, j, k, l (unordered), and three permutations for each choice, namely $(i\,j)(k\,l)$, $(i\,k)(j\,l)$ and $(i\,l)(j\,k)$.

   (c) Every element of $(1\,2\,3)(4\,5\,6)^G$ has the form $(1\,i\,j)(k\,l\,m)$, with i, j, k, l, m distinct. There are five choices for i; then four choices for j; then we can make two different 3-cycles $(k\,l\,m)$ and $(k\,m\,l)$ from the remaining numbers. This gives $5 \cdot 4 \cdot 2 = 40$ elements in all.

   The elements of $(1\,2)(3\,4)(5\,6)^G$ have the form $(1\,i)(j\,k)(l\,m)$. There are five choices for i; then we can make three permutations $(j\,k)(l\,m)$, $(j\,l)(k\,m)$ and $(j\,m)(k\,l)$ of cycle shape $(2, 2)$ from the remaining numbers. Hence $|(1\,2)(3\,4)(5\,6)^G| = 5 \cdot 3 = 15$.

   The sizes of the conjugacy classes of S_6 are given in the following table:

Cycle-shape	(1)	(2)	(3)	(2^2)	(4)	(3, 2)	(5)	(2^3)	(3^2)	(4, 2)	(6)
Class size	1	15	40	45	90	120	144	15	40	90	120

4. An element x of cycle-shape (5) has $C_{S_6}(x) = \langle x \rangle$ (note that $|x^{S_6}| = 144$ and use Theorem 12.8). Hence by Proposition 12.17, $x^{A_6} \neq x^{S_6}$. For elements g of other cycle-shapes, $g^{A_6} = g^{S_6}$.

5. By Example 12.18(2), the conjugacy classes of A_5 have sizes 1, 12, 12, 15, 20. If H is a normal subgroup of A_5 then $|H|$ divides 60, and $1 \in H$, and H is a union of conjugacy classes of A_5. Hence $|H| = 1$ or 60; therefore A_5 is simple.

6. We have $Q_8 = \langle a, b: a^4 = 1, b^2 = a^2, b^{-1}ab = a^{-1} \rangle$. The conjugacy classes of Q_8 are

$$\{1\}, \{a^2\}, \{a, a^3\}, \{b, a^2b\}, \{ab, a^3b\},$$

and a basis of $Z(\mathbb{C}Q_8)$ is

$$1, a^2, a + a^3, b + a^2b, ab + a^3b.$$

7. The class equation gives

$$|G| = |Z(G)| + \sum_{x_i \notin Z(G)} |x_i^G|.$$

(a) For $x_i \notin Z(G)$, $|x_i^G|$ divides p^n and $|x_i^G| \neq 1$ by Theorem 12.8 and (12.9). Therefore p divides $|x_i^G|$. Hence p divides $|Z(G)|$, so $Z(G) \neq \{1\}$.

(b) If no conjugacy class of G has size p, then p^2 divides $|x_i^G|$ for all $x_i \notin Z(G)$. If, in addition, $|G| \geqslant p^3$, then by the class equation, p^2 divides $|Z(G)|$. This is a contradiction.

Chapter 13

1. The characters χ_i of ρ_i ($i = 1, 2$) are as follows:

	1	a^3	a, a^5	a^2, a^4	b, a^2b, a^4b	ab, a^3b, a^5b
χ_1	2	2	-1	-1	0	0
χ_2	2	0	0	2	0	-2

Also $\operatorname{Ker} \rho_1 = \{1, a^3\}$ and $\operatorname{Ker} \rho_2 = \{1, a^2, a^4\}$.

2. Let $C_4 = \langle x : x^4 = 1 \rangle$. The irreducible characters $\chi_1, \ldots, \chi_4$ of C_4 are as follows:

	1	x	x^2	x^3
χ_1	1	1	1	1
χ_2	1	i	-1	$-i$
χ_3	1	-1	1	-1
χ_4	1	$-i$	-1	i

We have $\chi_{\text{reg}} = \chi_1 + \chi_2 + \chi_3 + \chi_4$.

3. Since $\chi(g) = |\text{fix}(g)|$, we have $\chi((1\,2)) = 5$ and $\chi((1\,6)(2\,3\,5)) = 2$.

4. If χ is a non-zero character which is a homomorphism, then $\chi(1) = \chi(1^2) = (\chi(1))^2$, so $\chi(1) = 1$.

5. Let ρ be a representation with character χ. Then $z\rho = \lambda I$ for some $\lambda \in \mathbb{C}$, by Proposition 9.14. Thus, for all g in G, $(zg)\rho = (z\rho)(g\rho) = \lambda(g\rho)$, and hence $\chi(zg) = \lambda\chi(g)$. Moreover, $I = 1\rho = z^m\rho = (z\rho)^m = \lambda^m I$, so $\lambda^m = 1$.

6. Let ρ be a representation with character χ. If $g \in Z(G)$ then $g\rho = \lambda I$ for some $\lambda \in \mathbb{C}$, by Proposition 9.14. Conversely, if $g\rho = \lambda I$ for some $\lambda \in \mathbb{C}$, then $(g\rho)(h\rho) = (h\rho)(g\rho)$ for all $h \in G$, and hence $g \in Z(G)$ since ρ is faithful.

   We have now proved that $g\rho = \lambda I$ for some $\lambda \in \mathbb{C}$ if and only if $g \in Z(G)$. The required result now follows from Theorem 13.11(1).

7. (a) For all $g, h \in G$, $\det((gh)\rho) = \det((g\rho)(h\rho)) = \det(g\rho)\det(h\rho)$. Hence $g \to (\det(g\rho))$ is a representation of G over $\mathbb{C}$ of degree 1, and so δ is a linear character of G.

   (b) $G/\operatorname{Ker} \delta \cong \operatorname{Im} \delta$ by Theorem 1.10, and $\operatorname{Im} \delta$ is a subgroup of the multiplicative group $\mathbb{C}^*$ of non-zero complex numbers, which is abelian. Therefore $G/\mathrm{Ker}\,\delta$ is abelian.

(c) $\mathrm{Im}\,\delta$ is a finite subgroup of $\mathbb{C}^*$, hence is cyclic, by Exercise 1.7. Also $-1 \in \mathrm{Im}\,\delta$, so $\mathrm{Im}\,\delta$ has even order. Hence $\mathrm{Im}\,\delta$ contains a subgroup H of index 2. It is easy to check that $\{g \in G\colon \delta(g) \in H\}$ is a normal subgroup of G of index 2.

8. Let ρ be the regular representation of G, and define the character δ as in Exercise 7. By Exercise 1.8, G has an element x of order 2. Order the natural basis $g_1, \ldots, g_{2k}$ of $\mathbb{C}G$ so as to obtain a basis $\mathscr{B}$ in which g and gx are adjacent for all $g \in G$. Then

$$[x]_{\mathscr{B}} = \begin{pmatrix} 0 & 1 & & & & \text{\Large 0} \\ 1 & 0 & & & & \\ & & 0 & 1 & & \\ & & 1 & 0 & & \\ \text{\Large 0} & & & & \ddots & \end{pmatrix}.$$

There are k blocks $\binom{0\,1}{1\,0}$, and since k is odd, $\det([x]_{\mathscr{B}}) = (-1)^k = -1$. Thus $\delta(x) = -1$. The required result now follows from Exercise 7.

9. Let V be a $\mathbb{C}G$-module with character χ. We may choose a basis $\mathscr{B}$ of V so that $[g]_{\mathscr{B}}$ is diagonal with all diagonal entries ± 1 (see (9.10)). Say there are r entries 1 and s entries -1. Define δ as in Exercise 7. If s is odd then $\delta(g) = -1$, and G has a normal subgroup of index 2 by Exercise 7. And if s is even then $-s \equiv s \bmod 4$, so $\chi(g) = r - s \equiv r + s = \chi(1) \bmod 4$.

10. As $x \neq 1$, we have $\chi_{\mathrm{reg}}(x) \neq \chi_{\mathrm{reg}}(1)$ (see Proposition 13.20), so $\chi_i(x) \neq \chi_i(1)$ for some irreducible character χ_i of G, by Theorem 13.19.

Chapter 14

1. Using Proposition 14.5(2), we obtain

$$\langle \chi, \chi \rangle = \frac{3 \cdot 3}{24} + \frac{(-1)(-1)}{4} + 0 + \frac{3 \cdot 3}{8} + \frac{(-1)(-1)}{4} = 2,$$

$$\langle \chi, \psi \rangle = \frac{3 \cdot 3}{24} + \frac{(-1) \cdot 1}{4} + 0 + \frac{3 \cdot (-1)}{8} + \frac{(-1)(-1)}{4} = 0,$$

$$\langle \psi, \psi \rangle = \frac{3 \cdot 3}{24} + \frac{1 \cdot 1}{4} + 0 + \frac{(-1)(-1)}{8} + \frac{(-1)(-1)}{4} = 1.$$

Hence ψ is irreducible by Theorem 14.20 (but χ is not).

2. Let χ_i be the character of ρ_i ($i = 1, 2, 3$). The values of these characters are as follows:

Conjugacy class	1	a^2	a, a^3	$b, a^2 b$	$ab, a^3 b$
χ_1	2	-2	0	0	0
χ_2	2	-2	0	0	0
χ_3	2	2	0	0	-2

By Theorem 14.21, ρ_1 and ρ_2 are equivalent, but ρ_3 is not equivalent to ρ_1 or ρ_2.

3. The representations ρ and σ have the same character, by Proposition 13.2; hence ρ and σ are equivalent, by Theorem 14.21, and this gives the required matrix T.

4. Let χ_1 be the trivial character of G. Then

$$\langle \chi, \chi_1 \rangle = \frac{1}{|G|} \sum_{g \in G} \chi(g).$$

Since $\chi(1) > 0$ and by hypothesis $\chi(g) \geqslant 0$ for all $g \in G$, we have $\langle \chi, \chi_1 \rangle \neq 0$. As $\chi \neq \chi_1$, Theorem 14.17 shows that χ is reducible.

5. We have

$$\langle \chi_{\mathrm{reg}}, \chi \rangle = \frac{1}{|G|} \sum_{g \in G} \chi_{\mathrm{reg}}(g) \overline{\chi(g)}.$$

But $\chi_{\mathrm{reg}}(g)$ is $|G|$ if $g = 1$ and is 0 if $g \neq 1$. Hence $\langle \chi_{\mathrm{reg}}, \chi \rangle = \chi(1)$.

6. This follows at once from Exercise 11.4 and Theorem 14.24.

7. Recall that $\langle \psi, \psi \rangle = \sum_{i=1}^{k} d_i^2$. Hence if $\langle \psi, \psi \rangle = a$ where $a = 1, 2$ or 3, then exactly a of the integers d_i are 1 and the rest are 0. If $\langle \psi, \psi \rangle = 4$, then either exactly four of the d_i are 1, or exactly one of the d_i is 2; the rest are 0.

8. No: let $G = C_2$ and $\chi = \chi_{\mathrm{reg}}$, the regular character of C_2.

Chapter 15

1.
$$\langle \chi, \chi_1 \rangle = \tfrac{1}{6}(19 \cdot 1 + 3 \cdot (-1) \cdot 1 + 2 \cdot (-2) \cdot 1) = 2,$$
$$\langle \chi, \chi_2 \rangle = \tfrac{1}{6}(19 \cdot 1 + 3 \cdot (-1)(-1) + 2 \cdot (-2) \cdot 1) = 3,$$
$$\langle \chi, \chi_3 \rangle = \tfrac{1}{6}(19 \cdot 2 + 0 + 2 \cdot (-2)(-1)) = 7.$$

Hence $\chi = 2\chi_1 + 3\chi_2 + 7\chi_3$. Since all the coefficients here are non-negative integers, it follows that χ is a character of S_3.

2. By the method used in the solution to Exercise 1, we obtain
$$\psi_1 = \tfrac{1}{6}\chi_1 + \tfrac{1}{6}\chi_2 + \tfrac{1}{3}\chi_3,$$
$$\psi_2 = \tfrac{1}{2}\chi_1 - \tfrac{1}{2}\chi_2,$$
$$\psi_3 = \tfrac{1}{3}\chi_1 + \tfrac{1}{3}\chi_2 - \tfrac{1}{3}\chi_3.$$

3. We find that $\psi = -\chi_2 + \chi_3 + \chi_5 + 2\chi_6$. Since the coefficient of χ_2 is a negative integer, ψ is not a character of G.

4. (a) For all groups G, if $x \in G$ then the subgroup generated by x and the elements of $Z(G)$ is abelian (since the elements of $Z(G)$ commute with powers of x). Hence, if $G = Z(G) \cup Z(G)x$ then $G = Z(G)$. Therefore the centre of a group G never has index 2 in G.
   Every abelian group of order 12 has 12 conjugacy classes. If $|G| = 12$

and G is non-abelian, then $|Z(G)|$ divides 12 and $|Z(G)| \neq 6$ or 12, so $|Z(G)| \leq 4$. Therefore, at most 4 conjugacy classes of G have size 1 (see (12.9)); the remaining conjugacy classes have size at least 2, so there cannot be as many as 9 conjugacy classes in total.

(b) Since the number of irreducible representations is equal to the number of conjugacy classes, it follows from the solution to Exercise 11.2 and part (a) that G has 4, 6 or 12 conjugacy classes. If G is abelian (e.g. $G = C_4 \times C_3$) then G has 12 conjugacy classes; if $G = D_{12}$ then G has 6 conjugacy classes (see (12.12)); and if $G = A_4$ then G has 4 conjugacy classes (see Example 12.18(1)).

Chapter 16

1. Let $C_2 \times C_2 = \{(1, 1), (x, 1), (1, y), (x, y): x^2 = y^2 = 1\}$. The character table of $C_2 \times C_2$ is (cf. Exercise 9.1)

	(1, 1)	(x, 1)	(1, y)	(x, y)
χ_1	1	1	1	1
χ_2	1	1	-1	-1
χ_3	1	-1	1	-1
χ_4	1	-1	-1	1

2. The last row of the character table is (cf. Example 16.5(2))

	g_1	g_2	g_3	g_4	g_5
χ_5	2	-2	0	0	0

3. The complete character table of G is

g_i $|C_G(g_i)|$	g_1 10	g_2 5	g_3 5	g_4 2
χ_1	1	1	1	1
χ_2	2	$(-1+\sqrt{5})/2$	$(-1-\sqrt{5})/2$	0
χ_3	1	1	1	-1
χ_4	2	$(-1-\sqrt{5})/2$	$(-1+\sqrt{5})/2$	0

The two unknown degrees $\chi_3(1)$, $\chi_4(1)$ are 1, 2 since $\sum_{i=1}^{4}(\chi_i(1))^2 = 10$. Because g_4 has order 2, Corollary 13.10, together with the relation $\sum_{i=1}^{4}\chi_i(1)\overline{\chi_i(g_4)} = 0$, gives the values on g_4. Then $\sum_{i=1}^{4}\chi_i(g_2)\overline{\chi_i(g_4)} = 0$ gives $\chi_3(g_2) = 1$; similarly $\chi_3(g_3) = 1$. Finally, $\sum_{i=1}^{4}\chi_i(1)\overline{\chi_i(g_2)} = 0$ gives $\chi_4(g_2) = (-1 - \sqrt{5})/2$; similarly $\chi_4(g_3) = (-1 + \sqrt{5})/2$.

4. (a) $\sum_{i=1}^{5}\chi_i(g_1)\overline{\chi_i(g_2)} = 0$ gives $3 + 3\zeta + 3\bar{\zeta} = 0$, and $\sum_{i=1}^{5}\chi_i(g_2)\overline{\chi_i(g_2)} = 7$
   gives $3 + 2\zeta\bar{\zeta} = 7$.
   Hence $\zeta = (-1 \pm i\sqrt{7})/2$.

   (b) The column which corresponds to the conjugacy class containing g_2^{-1}
   has values which are the complex conjugates of those in the column of
   g_2 (see Proposition 13.9(3)); since ζ is non-real, this is a different
   column of the character table of G.

5. Let $g \in G$. By the column orthogonality relations applied to the column
   corresponding to g, we have $\sum_{i=1}^{k}\chi_i(g)\overline{\chi_i(g)} = |C_G(g)|$. This number is
   equal to $|G|$ if and only if $C_G(g) = G$, which occurs if and only if
   $g \in Z(G)$.

6. The matrix $\bar{C}$ is obtained from C by rearranging the columns (see
   Proposition 13.9(3)). Therefore $\det \bar{C} = \pm \det C$; if $\det \bar{C} = \det C$ then $\det C$
   is real, and if $\det \bar{C} = -\det C$ then $\det C$ is purely imaginary.
      By the column orthogonality relations, $\bar{C}^t C$ is the $k \times k$ diagonal
   matrix whose diagonal entries are $|C_G(g_i)|$ $(1 \le i \le k)$. Hence
   $|\det C|^2 = \prod |C_G(g_i)|$.
      If $G = C_3$ then $\det C = \pm i 3\sqrt{3}$. (The sign depends upon the ordering of
   rows and columns.)

Chapter 17

1. (a) The conjugacy classes of Q_8 are $\{1\}$, $\{a^2\}$, $\{a, a^3\}$, $\{b, a^2b\}$ and
   $\{ab, a^3b\}$.
   (b) $G' = \{1, a^2\}$ and $G/G' = \{G', G'a, G'b, G'ab\} \cong C_2 \times C_2$. The
   character table of $C_2 \times C_2$ is given in the solution to Exercise 16.1.
   Hence the linear characters of G are

g_i $|C_G(g_i)|$	1 8	a^2 8	a 4	b 4	ab 4
χ_1	1	1	1	1	1
χ_2	1	1	1	-1	-1
χ_3	1	1	-1	1	-1
χ_4	1	1	-1	-1	1

   (c) Using the column orthogonality relations, the last irreducible character of
   G is

	1	a^2	a	b	ab
χ_5	2	-2	0	0	0

   The character table of Q_8 is the same as that of D_8.

2. It is easy to see that $a^7 = b^3 = 1$. Use Proposition 12.13 to see quickly that
   $b^{-1}ab = a^2$.
   (a) Using the relations, every element of G has the form $a^m b^n$ with

$0 \leqslant m \leqslant 6$, $0 \leqslant n \leqslant 2$; hence $|G| \leqslant 21$. But a has order 7 and b has order 3, so 21 divides $|G|$ by Lagrange's Theorem. Therefore $|G| = 21$.

(b) The conjugacy classes of G are $\{1\}$, $\{a, a^2, a^4\}$, $\{a^3, a^5, a^6\}$, $\{a^m b: 0 \leqslant m \leqslant 6\}$ and $\{a^m b^2: 0 \leqslant m \leqslant 6\}$.

(c) First, $G' = \langle a \rangle$, so we get three linear characters of G:

g_i	1	a	a^3	b	b^2
$|C_G(g_i)|$	21	7	7	3	3
χ_1	1	1	1	1	1
χ_2	1	1	1	ω	ω^2
χ_3	1	1	1	ω^2	ω

where $\omega = e^{2\pi i/3}$. To find the two remaining irreducible characters χ_4 and χ_5, we note that a is not conjugate to a^{-1}; therefore for some irreducible character χ, we have $\chi(a) \neq \bar{\chi}(a)$ (see Corollary 15.6). Hence χ_4 and χ_5 must be complex conjugates of each other. Applying the column orthogonality relations, we obtain

	1	a	a^3	b	b^2
χ_4	3	α	$\bar{\alpha}$	0	0
χ_5	3	$\bar{\alpha}$	α	0	0

where $\alpha = (-1 + i\sqrt{7})/2$.

3. The number of linear characters of G divides $|G|$ by Theorem 17.11. Now consult the solution to Exercise 11.2 to see that there are 3, 4 or 12 linear characters. If there are 12, then G is abelian (see Proposition 9.18), so G is certainly not simple. If G has 3 or 4 linear characters then $|G/G'| = 3$ or 4 and again G is not simple as $G' \lhd G$.

4. In the character table below, we have $\chi_1 = 1_G$, $\chi_2 = \chi$, $\chi_3 = \chi^2$, $\chi_4 = \chi_2\chi_3$, $\chi_5 = \phi$, $\chi_6 = \phi\chi$; all of these are irreducible characters by Proposition 17.14. The centralizer orders are obtained by using the orthogonality relations, and the class sizes $|g_i^G|$ come from the equations $|G| = |C_G(g_i)||g_i^G|$ (Theorem 12.8).

g_i	g_1	g_2	g_3	g_4	g_5	g_6
$|C_G(g_i)|$	12	4	4	6	6	12
$|g_i^G|$	1	3	3	2	2	1
χ_1	1	1	1	1	1	1
χ_2	1	$-i$	i	1	-1	-1
χ_3	1	-1	-1	1	1	1
χ_4	1	i	$-i$	1	-1	-1
χ_5	2	0	0	-1	-1	2
χ_6	2	0	0	-1	1	-2

5. The normal subgroups of D_8 are

$$D_8 = \operatorname{Ker}\chi_1, \langle a \rangle = \operatorname{Ker}\chi_2,$$

$$\langle a^2, b \rangle = \operatorname{Ker}\chi_3, \langle a^2, ab \rangle = \operatorname{Ker}\chi_4,$$

$$\langle a^2 \rangle = \operatorname{Ker}\chi_2 \cap \operatorname{Ker}\chi_3, \{1\} = \operatorname{Ker}\chi_5.$$

6. (a) Check that the given matrices satisfy the relevant relations:

$$\begin{pmatrix} \varepsilon & 0 \\ 0 & \varepsilon^{-1} \end{pmatrix}^{2n} = \begin{pmatrix} 1 & 0 \\ 0 & 1 \end{pmatrix}, \begin{pmatrix} \varepsilon & 0 \\ 0 & \varepsilon^{-1} \end{pmatrix}^{n} = \begin{pmatrix} 0 & 1 \\ \varepsilon^{n} & 0 \end{pmatrix}^{2},$$

$$\begin{pmatrix} 0 & 1 \\ \varepsilon^{n} & 0 \end{pmatrix}^{-1} \begin{pmatrix} \varepsilon & 0 \\ 0 & \varepsilon^{-1} \end{pmatrix} \begin{pmatrix} 0 & 1 \\ \varepsilon^{n} & 0 \end{pmatrix} = \begin{pmatrix} \varepsilon & 0 \\ 0 & \varepsilon^{-1} \end{pmatrix}^{-1}.$$

Hence we have representations of T_{4n} (cf. Example 1.4).

(b) The representations in part (a) are irreducible unless $\varepsilon = \pm 1$, by Exercise 8.4. For $\varepsilon = e^{2\pi i r/2n}$, with $r = 1, 2, \ldots, n-1$, we get $n-1$ irreducible representations, no two of which are equivalent, since they have distinct characters. Moreover $G' = \langle a^2 \rangle$, so $|G/G'| = 4$ and there are four representations of degree 1 (see Theorem 17.11). Now the sum of the squares of the degrees of the irreducible representations we have found so far is

$$(n-1) \cdot 2^2 + 4 \cdot 1^2 = 4n.$$

Hence we have found all the irreducible representations, by Theorem 11.12. (For further details on the representations of degree 1, see the solution to Exercise 18.3; note that the structure of G/G' depends upon whether n is even or odd.)

7. (a) Check that the given matrices satisfy the relevant relations.

(b) The given representations, for $\varepsilon = e^{2\pi i k/2n}$ with $0 \leqslant k \leqslant n-1$, are irreducible (by Exercise 8.4), and inequivalent (consider the character values on a^2). Also $G' = \langle b \rangle$, so $|G/G'| = 2n$ and there are $2n$ representations of degree 1. The sum of the squares of the degrees of the irreducible representations we have found is

$$n \cdot 2^2 + 2n \cdot 1^2 = 6n,$$

so we have obtained all the irreducible representations.

8. (a) Check that the given matrices satisfy the relevant relations.

(b) The given representations, for $\varepsilon = e^{2\pi i j/n}$ with $0 \leqslant j \leqslant n-1$, are irreducible (by Exercise 8.4) and inequivalent (their characters are distinct). Note that b^2 does not belong to the kernel of any of these representations.

We get further representations by

$$a \to \begin{pmatrix} \eta & 0 \\ 0 & \eta^{-1} \end{pmatrix}, b \to \begin{pmatrix} 0 & 1 \\ 1 & 0 \end{pmatrix},$$

where η is any $(2n)$th root of unity in $\mathbb{C}$. For $\eta = e^{2\pi i j/2n}$ with $1 \leqslant j \leqslant n-1$, these representations are irreducible and inequivalent. Moreover, they are not equivalent to any representation found earlier, since b^2 is in the kernel of each of these representations.

Finally, $G' = \langle a^2, b^2 \rangle$ and $G/G' \cong C_2 \times C_2$, so we get four representations of degree 1.

We have now found all the irreducible representations, since the sum of the squares of the irreducible representations given above is

$$n \cdot 2^2 + (n-1) \cdot 2^2 + 4 \cdot 1^2 = 8n.$$

Chapter 18

1. The character table of D_8 is as shown.

Character table of D_8

| g_i | 1 | a^2 | a | b | ab |
$\|C_G(g_i)\|$	8	8	4	4	4
χ_1	1	1	1	1	1
χ_2	1	1	1	-1	-1
χ_3	1	1	-1	1	-1
χ_4	1	1	-1	-1	1
χ_5	2	-2	0	0	0

(See Example 16.3(3) or Section 18.3.) Regarding D_8 as the symmetry group of a square, take b to be a reflection in a diagonal of the square. Then π takes the following values:

	1	a^2	a	b	ab
π	4	0	0	2	0

Hence $\pi = \chi_1 + \chi_3 + \chi_5$. (Compare Example 14.28(2), where we took b to be a different reflection.)

2. Let $\omega = e^{2\pi i/6}$. Then $\omega + \omega^{-1} = 1$, $\omega^2 + \omega^{-2} = \omega^4 + \omega^{-4} = -1$. Hence, using Section 18.3, the character table of D_{12} is as shown.

Character table of D_{12}

| g_i | 1 | a^3 | a | a^2 | b | ab |
$\|C_G(g_i)\|$	12	12	6	6	4	4
χ_1	1	1	1	1	1	1
χ_2	1	1	1	1	-1	-1
χ_3	1	-1	-1	1	1	-1
χ_4	1	-1	-1	1	-1	1
χ_5	2	-2	1	-1	0	0
χ_6	2	2	-1	-1	0	0

Seven normal subgroups of D_{12} are $G = \mathrm{Ker}\,\chi_1$, $\langle a \rangle = \mathrm{Ker}\,\chi_2$, $\langle a^2, b \rangle = \mathrm{Ker}\,\chi_3$, $\langle a^2, ab \rangle = \mathrm{Ker}\,\chi_4$, $\langle a^2 \rangle = \mathrm{Ker}\,\chi_3 \cap \mathrm{Ker}\,\chi_4$, $\langle a^3 \rangle = \mathrm{Ker}\,\chi_6$ and $\{1\} = \mathrm{Ker}\,\chi_5$.

3. The $n + 3$ conjugacy classes of G are

$$\{1\}, \{a^n\}, \{a^r, a^{-r}\}(1 \leqslant r \leqslant n - 1), \{a^{2j}b: 0 \leqslant j \leqslant n - 1\},$$
$$\{a^{2j+1}b: 0 \leqslant j \leqslant n - 1\}.$$

We get $n - 1$ irreducible characters ψ_j $(1 \leqslant j \leqslant n - 1)$ of G from Exercise 17.6 as follows:

g_i	1	a^n	a^r $(1 \leqslant r \leqslant n - 1)$	b	ab
$\lvert C_G(g_i) \rvert$	$4n$	$4n$	$2n$	4	4
ψ_j	2	$2(-1)^j$	$\omega^{rj} + \omega^{-rj}$	0	0

where $\omega = e^{2\pi i/2n}$.

The remaining four irreducible characters of G are linear. If n is odd, then $G/G' = \langle G'b \rangle \cong C_4$ and the linear characters are

g_i	1	a^n	a^r $(1 \leqslant r \leqslant n - 1)$	b	ab
χ_1	1	1	1	1	1
χ_2	1	-1	$(-1)^r$	i	$-i$
χ_3	1	1	1	-1	-1
χ_4	1	-1	$(-1)^r$	$-i$	i

If n is even, then $G/G' \cong C_2 \times C_2$ and the linear characters are

g_i	1	a^n	a^r $(1 \leqslant r \leqslant n - 1)$	b	ab
χ_1	1	1	1	1	1
χ_2	1	1	1	-1	-1
χ_3	1	1	$(-1)^r$	1	-1
χ_4	1	1	$(-1)^r$	-1	1

Note that $T_4 \cong C_4$, $T_8 \cong Q_8$ and T_{12} is the Example in Section 18.4.

4. The $3n$ conjugacy classes of G are, for $0 \leqslant r \leqslant n - 1$,

$$\{a^{2r}\}, \{a^{2r}b, a^{2r}b^2\}, \{a^{2r+1}, a^{2r+1}b, a^{2r+1}b^2\}.$$

We have $G' = \langle b \rangle$ and $G/G' = \langle G'a \rangle \cong C_{2n}$. Hence we get $2n$ linear characters χ_j $(0 \leqslant j \leqslant 2n - 1)$, as shown. Exercise 17.7 gives us n irreducible characters ψ_k $(0 \leqslant k \leqslant n - 1)$.

Character table of U_{6n}

g_i	a^{2r}	$a^{2r}b$	a^{2r+1}
$\|C_G(g_i)\|$	$6n$	$3n$	$2n$
χ_j $(0 \leqslant j \leqslant 2n-1)$	ω^{2jr}	ω^{2jr}	$\omega^{j(2r+1)}$
ψ_k $(0 \leqslant k \leqslant n-1)$	$2\omega^{2kr}$	$-\omega^{2kr}$	0

Note: $\omega = e^{2\pi i/2n}$.

Observe that $U_6 \cong D_6$, $U_{12} \cong T_{12}$ and $U_{18} \cong D_6 \times C_3$.

5. The $2n+3$ conjugacy classes of G are

$$\{1\}, \{b^2\}, \{a^{2r+1}, a^{-2r-1}b^2\}(0 \leqslant r \leqslant n-1),$$

$$\{a^{2s}, a^{-2s}\}, \{a^{2s}b^2, a^{-2s}b^2\}(1 \leqslant s \leqslant (n-1)/2),$$

$$\{a^jb^k: j \text{ even}, k = 1 \text{ or } 3\}, \text{ and}$$

$$\{a^jb^k: j \text{ odd}, k = 1 \text{ or } 3\}.$$

Using Exercise 17.8, we get four linear characters $\chi_1, \ldots, \chi_4$, n characters ψ_j $(0 \leqslant j \leqslant n-1)$ of degree 2, and a further $n-1$ characters ϕ_j $(1 \leqslant j \leqslant n-1)$ of degree 2, as shown below. For example, the character table of V_{24} is given at the top of p. 422.

Character table of V_{8n}

g_i	1	b^2	a^{2r+1} $(0 \leqslant r \leqslant n-1)$	a^{2s} $(1 \leqslant s \leqslant (n-1)/2)$	$a^{2s}b^2$	b	ab
$\|C_G(g_i)\|$	$8n$	$8n$	$4n$	$4n$	$4n$	4	4
χ_1	1	1	1	1	1	1	1
χ_2	1	1	1	1	1	-1	-1
χ_3	1	1	-1	1	1	1	-1
χ_4	1	1	-1	1	1	-1	1
ψ_j $(0 \leqslant j \leqslant n-1)$	2	-2	$\omega^{2j(2r+1)}$ $-\omega^{-2j(2r+1)}$	ω^{4js} $+\omega^{-4js}$	$-\omega^{4js}$ $-\omega^{-4js}$	0	0
ϕ_j $(1 \leqslant j \leqslant n-1)$	2	2	$\omega^{j(2r+1)}$ $+\omega^{-j(2r+1)}$	ω^{2js} $+\omega^{-2js}$	ω^{2js} $+\omega^{-2js}$	0	0

Note: $\omega = e^{2\pi i/2n}$

Character table of V_{24}

g_i	1	b^2	a	a^3	a^5	a^2	a^2b^2	b	ab
$\|C_G(g_i)\|$	24	24	12	12	12	12	12	4	4
χ_1	1	1	1	1	1	1	1	1	1
χ_2	1	1	1	1	1	1	1	-1	-1
χ_3	1	1	-1	-1	-1	1	1	1	-1
χ_4	1	1	-1	-1	-1	1	1	-1	1
ψ_0	2	-2	0	0	0	2	-2	0	0
ψ_1	2	-2	$i\sqrt{3}$	0	$-i\sqrt{3}$	-1	1	0	0
ψ_2	2	-2	$-i\sqrt{3}$	0	$i\sqrt{3}$	-1	1	0	0
ϕ_1	2	2	1	-2	1	-1	-1	0	0
ϕ_2	2	2	-1	2	-1	-1	-1	0	0

Chapter 19

1. $$\langle \chi\psi, \phi \rangle = \frac{1}{|G|}\sum_{g\in G}\chi(g)\psi(g)\overline{\phi(g)} = \frac{1}{|G|}\sum_{g\in G}\chi(g)\overline{\bar\psi(g)\phi(g)} = \langle \chi, \bar\psi\phi \rangle.$$

Similarly, $\langle \chi\psi, \phi \rangle = \langle \psi, \bar\chi\phi \rangle$.

2. $\langle \chi\psi, 1_G \rangle = \langle \chi, \bar\psi \rangle$, by Exercise 1. The result now follows from Proposition 13.15 and (14.13).

3. Let V be a $\mathbb{C}G$-module with character χ. Since χ is not faithful, there exists $1 \neq g \in G$ with $vg = v$ for all $v \in V$. By Proposition 15.5 there is an irreducible character ψ of G such that $\psi(g) \neq \psi(1)$. Let n be an integer with $n \geq 0$. Then $wg = w$ for all $w \in V \otimes \ldots \otimes V$ (n factors). Hence $\phi(g) = \phi(1)$ for all irreducible characters ϕ for which $\langle \chi^n, \phi \rangle \neq 0$. Therefore $\langle \chi^n, \psi \rangle = 0$.

4. Note that $(1\,2\,3\,4\,5)^2$ is conjugate to $(1\,3\,4\,5\,2)$ in A_5. Using Proposition 19.14 we obtain

	1	$(1\,2\,3)$	$(1\,2)(3\,4)$	$(1\,2\,3\,4\,5)$	$(1\,3\,4\,5\,2)$
χ_S	15	0	3	0	0
χ_A	10	1	-2	0	0
ϕ_S	6	0	2	1	1
ϕ_A	3	0	-1	$(1 + \sqrt{5})/2$	$(1 - \sqrt{5})/2$

Then

$$\chi_S = \psi_1 + \psi_2 + 2\psi_3,$$

$$\chi_A = \psi_2 + \psi_4 + \psi_5,$$

$$\phi_S = \psi_1 + \psi_3,$$

$$\phi_A = \psi_4.$$

5. We have recorded χ as χ_5, χ_S as χ_4 and χ_A as χ_2, below. Since $\langle \chi_i, \chi_i \rangle = 1$ for $i = 2, 4, 5$, these characters are irreducible. The table also records the trivial character χ_1, $\chi_3 = \bar{\chi}_2$, $\chi_6 = \bar{\chi}_5$ and $\chi_7 = \chi_2\chi_5$; these are irreducible by Propositions 13.15 and 17.14. Since G has seven conjugacy classes, the character table is complete.

Character table of G (cf. Exercise 27.2)

g_i $|C_G(g_i)|$	g_1 24	g_2 24	g_3 4	g_4 6	g_5 6	g_6 6	g_7 6
χ_1	1	1	1	1	1	1	1
χ_2	1	1	1	ω	ω^2	ω^2	ω
χ_3	1	1	1	ω^2	ω	ω	ω^2
χ_4	3	3	-1	0	0	0	0
χ_5	2	-2	0	$-\omega^2$	$-\omega$	ω	ω^2
χ_6	2	-2	0	$-\omega$	$-\omega^2$	ω^2	ω
χ_7	2	-2	0	-1	-1	1	1

Note: $\omega = e^{2\pi i/3}$

6. Taking $D_6 = \langle a, b: a^3 = b^2 = 1, b^{-1}ab = a^{-1} \rangle$, the character table of $D_6 \times D_6$ is as shown.

Character table of $D_6 \times D_6$

(g_i, h_j) $|C_G(g_i, h_j)|$	$(1, 1)$ 36	$(a, 1)$ 18	$(b, 1)$ 12	$(1, a)$ 18	(a, a) 9	(b, a) 6	$(1, b)$ 12	(a, b) 6	(b, b) 4
$\chi_1 \times \chi_1$	1	1	1	1	1	1	1	1	1
$\chi_2 \times \chi_1$	1	1	-1	1	1	-1	1	1	-1
$\chi_3 \times \chi_1$	2	-1	0	2	-1	0	2	-1	0
$\chi_1 \times \chi_2$	1	1	1	1	1	1	-1	-1	-1
$\chi_2 \times \chi_2$	1	1	-1	1	1	-1	-1	-1	1
$\chi_3 \times \chi_2$	2	-1	0	2	-1	0	-2	1	0
$\chi_1 \times \chi_3$	2	2	2	-1	-1	-1	0	0	0
$\chi_2 \times \chi_3$	2	2	-2	-1	-1	1	0	0	0
$\chi_3 \times \chi_3$	4	-2	0	-2	1	0	0	0	0

Chapter 20

1. (a) Regard D_8 as the subgroup of S_4 which permutes the four corners of a square, as in Example 1.1(3). Take b to be the reflection in the axis shown:

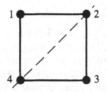

Then $a \to (1\,2\,3\,4)$, $b \to (1\,3)$ gives the required isomorphism.

(b) Take the irreducible characters $\chi_1, \ldots, \chi_5$ of S_4 as in Section 18.1, and take the character table of H to be

g_i $|C_H(g_i)|$	1 8	(1 3)(2 4) 8	(1 2 3 4) 4	(1 3) 4	(1 2)(3 4) 4
ψ_1	1	1	1	1	1
ψ_2	1	1	1	-1	-1
ψ_3	1	1	-1	1	-1
ψ_4	1	1	-1	-1	1
ψ_5	2	-2	0	0	0

(see Example 16.3(3) or Section 18.3). We obtain

$$\chi_1 \downarrow H = \psi_1, \chi_2 \downarrow H = \psi_4, \chi_3 \downarrow H = \psi_1 + \psi_4, \chi_4 \downarrow H = \psi_3 + \psi_5,$$

$$\chi_5 \downarrow H = \psi_2 + \psi_5.$$

2. Let $\chi_1, \ldots, \chi_{11}$ be the irreducible characters of S_6, as in Example 19.17. Either by direct calculation, or using (20.13), we find that the characters $\chi_i \downarrow A_6$ ($i = 1, 3, 5, 7, 9$) are distinct irreducible characters of A_6; these give the characters $\psi_1, \ldots, \psi_5$ in our character table below. Also, $\langle \chi_{11} \downarrow A_6, \chi_{11} \downarrow A_6 \rangle = 2$. Arguing as in Example 20.14, we obtain from $\chi_{11} \downarrow A_6$ the two irreducible characters which we have called ψ_6 and ψ_7.

Character table of A_6

g_i $|C_{A_6}(g_i)|$	1 360	(1 2 3) 9	(1 2)(3 4) 8	(1 2 3 4 5) 5	(1 3 4 5 2) 5	(1 2 3)(4 5 6) 9	(1 2 3 4)(5 6) 4
ψ_1	1	1	1	1	1	1	1
ψ_2	5	2	1	0	0	-1	-1
ψ_3	10	1	-2	0	0	1	0
ψ_4	9	0	1	-1	-1	0	1
ψ_5	5	-1	1	0	0	2	-1
ψ_6	8	-1	0	α	β	-1	0
ψ_7	8	-1	0	β	α	-1	0

Note: $\alpha = (1 + \sqrt{5})/2$, $\beta = (1 - \sqrt{5})/2$

3. Let $\psi_1, \ldots, \psi_r$ be the irreducible characters of H. Then $\chi \downarrow H = d_1\psi_1 + \ldots + d_r\psi_r$ for some non-negative integers d_i. Since each ψ_i has degree 1, the inequality (20.6) gives

$$\chi(1) = d_1 + \ldots + d_r \leq d_1^2 + \ldots + d_r^2 \leq n.$$

4. The inequality $\langle \chi \downarrow H, \chi \downarrow H \rangle_H \leq 3$ follows at once from Proposition 20.5.

Write $d = \langle \chi \downarrow H, \chi \downarrow H \rangle_H$. For examples with $d = 1$ or 2, take $G = S_3$ and H a subgroup of order 2, and χ an irreducible character of G of degree d. For an example with $d = 3$, take $G = A_4$, $H = V_4$ and χ an irreducible character of G of degree 3 (see Section 18.2).

5. See (20.13). Since 20 occurs only once in the list of degrees for S_7, the restriction of the irreducible character of degree 20 to A_7 must be the sum of two different irreducible characters of degree 10. From the remaining fourteen irreducible characters of S_7, upon restriction to A_7 we get at least seven irreducible characters of A_7; and we get precisely seven if and only if the restriction of each of the fourteen characters is irreducible. We are told that A_7 has exactly nine conjugacy classes. Hence the irreducible characters of A_7 have degrees

$$1, 6, 14, 14, 15, 10, 10, 21, 35.$$

Chapter 21

1. (a) Let $u = 1 - a^2 + b - a^2 b$. Then $ua^2 = -u$ and $ub = u$. Hence $\mathrm{sp}(u)$ is a $\mathbb{C}H$-submodule of $\mathbb{C}H$.
   (b) Since every element of G belongs to H or to Ha, the elements u and ua form a basis of $U \uparrow G$.
   (c) The character ψ of U and the character $\psi \uparrow G$ of $U \uparrow G$ are given by

	1	a^2	b	a^2b
ψ	1	-1	1	-1

	1	a^2	a	b	ab
$\psi \uparrow G$	2	-2	0	0	0

Since $\langle \psi \uparrow G, \psi \uparrow G \rangle = 1$, the induced module $U \uparrow G$ is irreducible.

2. Label the characters of H as follows:

	1	$(1\,2\,3)$	$(1\,3\,2)$
ψ_1	1	1	1
ψ_2	1	ω	ω^2
ψ_3	1	ω^2	ω

where $\omega = e^{2\pi i/3}$.
   (a) $\chi_1 \downarrow H = \chi_2 \downarrow H = \psi_1$, $\chi_3 \downarrow H = \psi_2 + \psi_3$,
      $\chi_4 \downarrow H = \chi_5 \downarrow H = \psi_1 + \psi_2 + \psi_3$.
   (b) Using the Frobenius Reciprocity Theorem, we obtain

   $$\psi_1 \uparrow G = \chi_1 + \chi_2 + \chi_4 + \chi_5, \ \psi_2 \uparrow G = \psi_3 \uparrow G = \chi_3 + \chi_4 + \chi_5.$$

3. It is sufficient to prove that if U is a $\mathbb{C}H$-submodule of $\mathbb{C}H$ then $\dim(U \uparrow G) = |G:H| \dim U$. Let Hg_j $(1 \leq j \leq m)$ be the distinct right

cosets of H in G. Then $U(\mathbb{C}G) = Ug_1 + \ldots + Ug_m$, where $Ug_j = \{ug_j : u \in U\}$. The sum $Ug_1 + \ldots + Ug_m$ is direct (since the elements in Ug_j are linear combinations of elements in the right coset Hg_j). Also, $\dim(Ug_j) = \dim U$ (since $u \rightarrow ug_j$ ($u \in U$) is a vector space isomorphism). Hence $\dim(U \uparrow G) = \dim(U(\mathbb{C}G)) = m \dim U$.

4. Let ϕ be an irreducible character of G. By using the Frobenius Reciprocity Theorem twice, together with the result of Exercise 19.1 (also twice), we obtain

$$\langle (\psi(\chi \downarrow H)) \uparrow G, \phi \rangle_G = \langle \psi(\chi \downarrow H), \phi \downarrow H \rangle_H = \langle \psi, (\bar{\chi}\phi) \downarrow H \rangle_H$$
$$= \langle \psi \uparrow G, \bar{\chi}\phi \rangle_G = \langle (\psi \uparrow G)\chi, \phi \rangle_G.$$

Since this holds for all irreducible characters ϕ of G, we deduce from Theorem 14.17 that $(\psi(\chi \downarrow H)) \uparrow G = (\psi \uparrow G)\chi$.

5. The values of $\phi \uparrow G$ and $\psi \uparrow G$ are given by Proposition 21.23. On elements of cycle-shapes (1), (7) and (3, 3), the values are as follows, and on all other elements the values are zero.

Cycle-shape	(1)	(7)	(3, 3)
$\phi \uparrow G$	240	2	12
$\psi \uparrow G$	720	-1	0

6. We have $|G : H|\psi(1) = d_1\chi_1(1) + \ldots + d_k\chi_k(1)$, where

$$d_i = \langle \psi \uparrow G, \chi_i \rangle_G = \langle \psi, \chi_i \downarrow H \rangle_H,$$

by the Frobenius Reciprocity Theorem. Hence, since ψ is irreducible, $\chi_i \downarrow H = d_i\psi + \beta$ where either β is a character of H or $\beta = 0$. Thus $\chi_i(1) \geqslant d_i\psi(1)$, and therefore

$$|G : H|\psi(1) \geqslant (d_1^2 + \ldots + d_k^2)\psi(1).$$

The required result follows.

7. By applying the result of Exercise 6, we deduce, as in the proof of Proposition 20.9, that either
   (1) $\psi \uparrow G$ is irreducible, or
   (2) $\psi \uparrow G$ is the sum of two different irreducible characters of the same degree.

   Suppose first that $\psi \uparrow G$ is irreducible, say $\psi \uparrow G = \chi$. Then $\chi(1) = 2\psi(1)$ and $\langle \chi \downarrow H, \psi \rangle_H = 1$, by the Frobenius Reciprocity Theorem; hence $\chi \downarrow H$ is reducible, say $\chi \downarrow H = \psi + \phi$. Now suppose that ψ' is an irreducible character of H. We have

$$\langle \psi' \uparrow G, \chi \rangle_G \neq 0 \Leftrightarrow \langle \psi', \chi \downarrow H \rangle_H \neq 0 \Leftrightarrow \psi' = \psi \text{ or } \phi.$$

Thus

   If $\psi \uparrow G$ is irreducible, then for precisely one other irreducible character ϕ of H we have $\psi \uparrow G = \phi \uparrow G$. (Compare Proposition 20.11.)

Suppose next that $\psi \uparrow G$ is reducible, say $\psi \uparrow G = \chi_1 + \chi_2$. Then $\chi_1(1) = \psi(1)$ and $\langle \chi_1 \downarrow H, \psi \rangle_H = 1$; hence $\chi_1 \downarrow H = \psi$. Now suppose that ψ' is an irreducible character of H. We have

$$\langle \psi' \uparrow G, \chi_1 \rangle_G \neq 0 \Leftrightarrow \langle \psi', \chi_1 \downarrow H \rangle_H \neq 0 \Leftrightarrow \psi' = \psi.$$

Thus

If $\psi \uparrow G = \chi_1 + \chi_2$ where χ_1 and χ_2 are irreducible characters of G, and ψ' is an irreducible character of H such that $\psi' \uparrow G$ has χ_1 or χ_2 as a constituent, then $\psi' = \psi$. (Compare Proposition 20.12.)

Chapter 22

1. The number of linear characters divides 15 (Theorem 17.11), and the sum of the squares of the degrees of the irreducible characters is 15 (Theorem 11.12); moreover, each degree divides 15 (Theorem 22.11). Hence every irreducible character has degree 1, and so G is abelian by Proposition 9.18.

2. Use Theorems 11.12, 17.11 and 22.11 again. The degree of each irreducible character is 1 or 2, and if there are r characters of degree 1 and s of degree 2, then

$$r \text{ divides } 16, \text{ and } r \cdot 1^2 + s \cdot 2^2 = 16.$$

Hence $r = 4$ or 8 or 16, and $r + s = 7$ or 10 or 16.

3. (a) Since G is non-abelian, not every irreducible character has degree 1 (Proposition 9.18). This time, Theorems 11.12, 17.11 and 22.11 show that there are r irreducible characters of degree 1 and s irreducible characters of degree q, where

$$r \text{ divides } pq, \ 1 \leq s \text{ and } r + sq^2 = pq.$$

Hence $r = q$ and $s = (p - 1)/q$.
(b) $|G'| = p$ by Theorem 17.11.
(c) The number of conjugacy classes of G is $r + s$.
(For more information on groups of order pq, see Chapter 25.)

4. (a) By hypothesis, there exist $a, b \in \mathbb{C}$ such that $\phi(g) = a$ for all $g \neq 1$ and $\phi(1) = a + b|G|$. Then $\phi = a1_G + b\chi_{\text{reg}}$, since both sides of this equation take the same values on all elements of G.
(b) We have

$$\langle 1_G, \phi \rangle = \frac{1}{|G|}(a + b|G| + (|G| - 1)a) = a + b, \text{ and}$$

$$\langle \chi_{\text{reg}}, \phi \rangle = \frac{1}{|G|}|G|(a + b|G|) = a + b|G|.$$

Since ϕ is a character, both $\langle 1_G, \phi \rangle$ and $\langle \chi_{\text{reg}}, \phi \rangle$ are integers.
(c) If χ is a non-trivial irreducible character of G, then $\langle \phi, \chi \rangle \in \mathbb{Z}$ and $\langle 1_G, \chi \rangle = 0$. Hence $\langle \phi - a1_G, \chi \rangle \in \mathbb{Z}$. But $\langle \phi - a1_G, \chi \rangle = \langle b\chi_{\text{reg}}, \chi \rangle = b|G|\chi(1)/|G| = b\chi(1)$.

(d) Since $\chi(1)$ divides $|G|$, part (c) implies that $b|G|$ is an integer. Therefore, by part (b), a, and hence also b, is an integer.

5. (a) If $g \in G$ then g has odd order, by Lagrange's Theorem. Therefore, if $g^2 = 1$ then $g = 1$.

(b) For all $g \in G$, we have $\chi(g) + \chi(g^{-1}) = \chi(g) + \bar{\chi}(g) = 2\chi(g)$, and $\chi(g)$ is an algebraic integer. Partition $G \backslash \{1\}$ into subsets by putting each element with its inverse. Each such subset has size 2, by part (a). Hence

$$\sum_{g \in G} \chi(g) = \chi(1) + 2\alpha$$

for some algebraic integer α. The stated result follows, since

$$\langle \chi, 1_G \rangle = \frac{1}{|G|} \sum_{g \in G} \chi(g).$$

(c) If $\chi \neq 1_G$ in part (b), then $\langle \chi, 1_G \rangle = 0$, and hence $\alpha = -\chi(1)/2$. But $-\chi(1)/2$ is a rational number which is not an integer (since $\chi(1)$ divides $|G|$, hence is odd). This contradicts Proposition 22.5. Thus $\chi = 1_G$.

(Further information about the number of characters χ such that $\chi = \bar{\chi}$ appears in Theorem 23.1 and Corollary 23.2.)

6. (a) By Theorem 22.16, $\chi(g)$ is an integer for all characters χ. Let $\chi_1, \ldots, \chi_7$ be the irreducible characters of G, with $\chi_1 = 1_G$. By the column orthogonality relations, we have

$$\text{(I)} \quad 1 + \sum_{i=2}^{7} (\chi_i(g))^2 = 5, \text{ and} \quad \text{(II)} \quad 1 + \sum_{i=2}^{7} \chi_i(1)\chi_i(g) = 0.$$

From equation (I) we deduce that either $\chi_i(g) = 0, \pm 1$ for all i, or $\chi_i(g) = \pm 2$ for exactly one i and $\chi_i(g) = 0$ for all other $i > 1$; the second possibility is ruled out by equation (II).

(b) We deduce from part (a) that $\chi_i(g) = 0$ for two values of i, say $i = 2, 3$, and $\chi_i(g) = \pm 1$ for $4 \leqslant i \leqslant 7$. By Corollary 22.27, $\chi_2(1) \equiv \chi_3(1) \equiv 0 \bmod 5$. Also

$$\text{(III)} \quad \sum_{i=1}^{7} (\chi_i(1))^2 = 120.$$

Since $5^2 + 10^2 > 120$, we deduce that $\chi_2(1) = \chi_3(1) = 5$.

(c) By Corollary 22.27, $\chi_i(1) \equiv \chi_i(g) \bmod 5$ for all i, and from equation (III) we have

$$\text{(IV)} \quad \sum_{i=4}^{7} (\chi_i(1))^2 = 69.$$

Hence the possibilities for the pairs of integers $(\chi_i(1), \chi_i(g))$ with $4 \leqslant i \leqslant 7$ are $(1, 1), (4, -1), (6, 1)$. The only possibility which is consistent with equation (IV) is that the values of $\chi_i(1)$ for $4 \leqslant i \leqslant 7$ are 1, 4, 4, 6 in some order.

(d) We now have the following part of the character table of G:

g_i	g_1	g_2	g_3	g_4	g_5	g_6	g_7
Order of g_i	1	2	2	3	4	6	5
$\lvert C_G(g_i)\rvert$	120	12	8	6	4	6	5
χ_1	1	1	1	1	1	1	1
χ_2	5						0
χ_3	5						0
χ_4	1						1
χ_5	4						−1
χ_6	4						−1
χ_7	6						1

We successively calculate the five remaining columns of the character table. We are told that $\chi_i(g_j)$ is an integer for all i, j.

(1) First, $\chi_i(g_4) \equiv \chi_i(1) \bmod 3$ and $\sum_{i=1}^{7}(\chi_i(g_4))^2 = 6$. Hence the values of $\chi_i(g_4)$ for $1 \leq i \leq 7$ are 1, −1, −1, 1, 1, 1, 0, respectively.

(2) Next, $\chi_i(g_5) \equiv \chi_i(1) \bmod 2$ and $\sum_{i=1}^{7}(\chi_i(g_5))^2 = 4$. Hence $\chi_i(g_5) = \pm 1$ for $1 \leq i \leq 4$ and $\chi_i(g_5) = 0$ for $5 \leq i \leq 7$. Since $\sum_{i=1}^{7}\chi_i(1)\chi_i(g_5) = 0$, we deduce that $\chi_4(g_5) = -1$ and (without loss of generality) $\chi_2(g_5) = -\chi_3(g_5) = 1$.

(3) Since $\chi_i(g_3) \equiv \chi_i(1) \bmod 2$ and $\sum_{i=1}^{7}(\chi_i(g_3))^2 = 8$, we deduce that $\chi_i(g_3) = \pm 1$ for $1 \leq i \leq 4$ and that the values of $\chi_i(g_3)$ for $5 \leq i \leq 7$ are 0, 0, ± 2 in some order. Also $\sum_{i=1}^{7}\chi_i(g_3)\chi_i(g_r) = 0$ for $r = 4, 7$, from which we see that $\chi_i(g_3) = 1$ for $1 \leq i \leq 4$. From the relation $\sum_{i=1}^{7}\chi_i(1)\chi_i(g_3) = 0$ we now deduce that the entries in column 3 are 1, 1, 1, 1, 0, 0, −2 in order from the top.

(4) We have $\chi_i(g_6) \equiv \chi_i(g_4) \bmod 2$ and $\sum_{i=1}^{7}(\chi_i(g_6))^2 = 6$. Therefore $\chi_i(g_6) = \pm 1$ for $1 \leq i \leq 6$ and $\chi_7(g_6) = 0$. By applying the column orthogonality relations involving column 6 and columns 3, 4, 5 and 7 we obtain $\chi_2(g_6) = -\chi_3(g_6) = \chi_4(g_6) = -1$ and (without loss of generality) $\chi_5(g_6) = -\chi_6(g_6) = 1$.

(5) Only the entries in column 2 remain to be calculated. These can be obtained from the column orthogonality relations.

The character table of G is as shown.

g_i	g_1	g_2	g_3	g_4	g_5	g_6	g_7
Order of g_i	1	2	2	3	4	6	5
$\lvert C_G(g_i)\rvert$	120	12	8	6	4	6	5
χ_1	1	1	1	1	1	1	1
χ_2	5	−1	1	−1	1	−1	0
χ_3	5	1	1	−1	−1	1	0
χ_4	1	−1	1	1	−1	−1	1
χ_5	4	−2	0	1	0	1	−1
χ_6	4	2	0	1	0	−1	−1
χ_7	6	0	−2	0	0	0	1

7. Suppose first that λ is an eigenvalue of an $n \times n$ matrix A, all of whose entries are integers. Then $\det(A - \lambda I_n) = 0$, and hence λ is a root of the polynomial $\det(xI_n - A)$, which is of the form

$$x^n + a_{n-1}x^{n-1} + \ldots + a_1 x + a_0 \quad (a_r \in \mathbb{Z}).$$

Conversely, assume that λ is a root of the polynomial $p(x) = a_0 + a_1 x + \ldots + a_{n-1}x^{n-1} + x^n$ $(a_r \in \mathbb{Z})$. Let

$$A = \begin{pmatrix} 0 & 1 & 0 & \cdots & 0 \\ 0 & 0 & 1 & & 0 \\ 0 & 0 & 0 & & 0 \\ \vdots & & & & \vdots \\ 0 & 0 & 0 & & 1 \\ -a_0 & -a_1 & -a_2 & \cdots & -a_{n-1} \end{pmatrix}.$$

Check that $\det(xI_n - A) = p(x)$. As $p(\lambda) = 0$, it follows that λ is an eigenvalue of A. Since A has integer entries, λ is therefore an algebraic integer.

Chapter 23

1. Assume that $x \in G$ and x is real. Then $g^{-1}xg = x^{-1}$ for some $g \in G$. Hence $g^{-2}xg^2 = x$, so $g^2 \in C_G(x)$. Let m be the order of g. Since $|G|$ is odd, $m = 2n + 1$ for some integer n, by Lagrange's Theorem. Then $g = g^{2(n+1)} \in C_G(x)$. Therefore $x^{-1} = g^{-1}xg = x$. Since $x^2 = 1$ and x has odd order, it follows that $x = 1$.

2. Adopt the notation of Theorem 9.8. The character χ of $G = C_{n_1} \times \ldots \times C_{n_r}$ which is given by

$$\chi(g_1^{i_1} \ldots g_r^{i_r}) = \lambda_1^{i_1} \ldots \lambda_r^{i_r}$$

is real if and only if $\lambda_i = \pm 1$ for all i with $1 \leqslant i \leqslant r$. Now the n_ith root of unity λ_i can be -1 if and only if n_i is even. Hence the number of real irreducible characters is 2^m, where m is the number of the integers $n_1, \ldots, n_r$ which are even. However, the elements g of G which satisfy $g^2 = 1$ are precisely those elements $g_1^{i_1} \ldots g_r^{i_r}$ where for each j, either $i_j = 0$ or n_j is even and $i_j = n_j/2$. The number of such elements is also 2^m.

3. The elements g of D_{2n} for which $g^2 = 1$ are

$$1, \quad a^i b \ (0 \leqslant i \leqslant n - 1) \quad \text{(and also } a^{n/2} \text{ if } n \text{ is even)}.$$

This gives $n + 1$ elements if n is odd, and $n + 2$ elements if n is even. These numbers coincide with $\sum \chi(1)$, summing over all the irreducible characters χ. Since $\iota\chi \leqslant 1$ for all χ, it follows from the Frobenius–Schur Count of Involutions that we must have $\iota\chi = 1$ for all χ.

4. Let λ_1 and λ_2 be the eigenvalues of $g\rho$. Then $\chi_A(g) = \frac{1}{2}((\lambda_1 + \lambda_2)^2 - (\lambda_1^2 + \lambda_2^2)) = \lambda_1\lambda_2 = \det(g\rho)$ (see Proposition 19.14). Since $\chi(1) = 2$ we have $\chi_A(1) = 1$. It now follows from the Definition 23.13 of $\iota\chi$ that $\iota\chi = -1$ if and only if $\chi_A = 1_G$. The result follows.

5. (a) First, it is easy to check that $\chi(g)$ is real for all $g \in G$, so $\iota\chi = \pm 1$. Let ρ be the representation obtained by using the basis v_1, v_2 of V. Then $\det(a\rho) = 1$ and $\det(b\rho) = -\varepsilon^n$; hence $\det(g\rho) = 1$ for all $g \in G$ if and only if $\varepsilon^n = -1$. The result now follows from Exercise 4.

(b) It is easy to check that if $g = a$ or b and i, $j \in \{1, 2\}$ then

$$\beta(v_i g, v_j) = \beta(v_i, v_j g^{-1}).$$

For example, $\beta(v_1 b, v_1) = \beta(v_2, v_1) = \varepsilon^n = \beta(v_1, \varepsilon^n v_2) = \beta(v_1, v_1 b^{-1})$. Hence β is G-invariant. The definition of β shows that β is symmetric if $\varepsilon^n = 1$ and β is skew-symmetric if $\varepsilon^n = -1$. The result now follows from Theorem 23.16.

(c) The elements of T_{4n} are a^i and $a^i b$ ($0 \leqslant i \leqslant 2n - 1$); a has order $2n$ and $a^i b$ has order 4. Hence a^n is the only element of order 2.

(d) Refer the Exercise 18.3 for the characters ψ_j ($1 \leqslant j \leqslant n - 1$) and χ_j ($1 \leqslant j \leqslant 4$) of T_{4n}. Clearly $\iota\chi_1 = \iota\chi_3 = 1$, and $\iota\chi_2 = \iota\chi_4 = 0$ or 1, according to whether n is odd or even, respectively. By part (a) (or part (b)) and the construction of the characters ψ_j in Exercise 17.6, we get $\iota\psi_j = -1$ or 1, according to whether j is odd or even, respectively. Therefore $\sum_{j=1}^{n-1} (\iota\psi_j)\psi_j(1) = 0$ or -2, according to whether n is odd or even, respectively. Hence $\sum_\chi (\iota\chi)\chi(1) = 2$.

6. Let V be a $\mathbb{C}G$-module with character χ. Since $\iota\chi = -1$, there exists a non-zero G-invariant skew-symmetric bilinear form β on V. As β is G-invariant, the subspace $\{u \in V : \beta(u, v) = 0 \text{ for all } v \in V\}$ is a $\mathbb{C}G$-submodule of V; since V is irreducible it follows that

$$\{u \in V : \beta(u, v) = 0 \text{ for all } v \in V\} = \{0\}. \tag{$*$}$$

Pick a basis $v_1, \ldots, v_n$ of V and let A be the $n \times n$ matrix with ij-entry $\beta(v_i, v_j)$. Since β is skew-symmetric, we have $A^t = -A$. Therefore $\det(A^t) = (-1)^n \det A$, so $\det A = (-1)^n \det A$. Also A is invertible by $(*)$, so $\det A \neq 0$. It follows that n is even; as $n = \chi(1)$ the result is proved.

7. Choose a basis $f_1, \ldots, f_n$ of V and define the symmetric $n \times n$ matrices $A = (a_{ij})$ and $B = (b_{ij})$ by

$$a_{ij} = \beta_1(f_i, f_j), \quad b_{ij} = \beta(f_i, f_j).$$

By applying the Gram–Schmidt orthogonalization process, we may construct a basis $f'_1, \ldots, f'_n$ of V such that $\beta_1(f'_i, f'_j) = \delta_{ij}$ for all i, j. Let $P = (p_{ij})$ be the $n \times n$ matrix which is given by

$$f'_i = \sum_j p_{ij} f_j.$$

Then $PAP^t = I_n$ and PBP^t is symmetric. By a well known property of symmetric matrices, there is an orthogonal matrix Q (i.e. $QQ^t = I$) such that $Q(PBP^t)Q^{-1}$ is diagonal. Write $Q = (q_{ij})$, and define the basis $e_1, \ldots, e_n$ of V by

$$e_i = \sum_j q_{ij} f'_j.$$

Then

$$\beta_1(e_i, e_j) = \delta_{ij}, \quad \text{since } QPAP^tQ^t = I_n, \text{ and}$$

$$\beta(e_i, e_j) = 0 \quad \text{if } i \neq j, \quad \text{since } QPBP^tQ^t \text{ is diagonal.}$$

8. (a) The proof is similar to that of part (1) of Schur's Lemma 9.1.
   (b) Let $v_1, \ldots, v_n$ be a basis of the $\mathbb{R}G$-module V, and consider the vector space V' over $\mathbb{C}$ with basis $v_1, \ldots, v_n$. Then V' is a $\mathbb{C}G$-module, which we are assuming to be an irreducible $\mathbb{C}G$-module. By Schur's Lemma there exists $\lambda \in \mathbb{C}$ such that $v\vartheta = \lambda v$ for all $v \in V'$. But $v_1\vartheta = \lambda v_1 \in V$, so $\lambda \in \mathbb{R}$.
   (c) Let $G = C_3 = \langle a: a^3 = 1 \rangle$, and let V be the $\mathbb{R}G$-submodule of the regular $\mathbb{R}G$-module which is spanned by $1 - a$ and $1 - a^2$. Then V is an irreducible $\mathbb{R}G$-module. Define $\vartheta: V \to V$ by $v\vartheta = av$ $(v \in V)$.

9. ρ_g is a permutation, as $Hxg = Hyg \Rightarrow Hx = Hy$, and ρ is a homomorphism as $(Hx)(\rho_{gh}) = Hxgh = (Hx)(\rho_g)(\rho_h)$. We have

$$g \in \ker \rho \Leftrightarrow Hxg = Hx, \quad \forall x \in G \Leftrightarrow$$

$$xgx^{-1} \in H, \forall x \in G \Leftrightarrow g \in \bigcap_{x \in G} x^{-1}Hx.$$

Finally, ρ is a homomorphism $G \to \mathrm{Sym}(\Omega) \cong S_n$ with kernel $\bigcap_{x \in G} x^{-1}Hx$, which is contained in H.

10. Let c_1, c_2 be the columns of the character table of G corresponding to the classes $\{1\}$ and t^G. By the orthogonality relation for c_2 we have $\sum \chi_i(t)^2 = |C_G(t)| = 2$ (the sum over all irreducible characters χ_i, with $\chi_1 = 1_G$), so we may take $\chi_1(t) = 1$, $\chi_2(t) = \pm 1$ and $\chi_i(t) = 0$ for $i \geq 3$. Now the orthogonality of c_1 and c_2 gives $\chi_1(1) = \chi_2(1) = 1$ and $\chi_2(t) = -1$. Further, χ_1 and χ_2 are the only linear characters, since a linear character must take the value ± 1 on t. Hence $|G: G'| = 2$ by Theorem 17.11. For the last part, if G is simple then since $G' \lhd G$, we have $G' = 1$, i.e. G is abelian. Hence $G \cong C_2$.

Chapter 25

1. It is clear that the given set of matrices has size $p(p - 1)$. Call it G. For closure, note that

$$\begin{pmatrix} 1 & y \\ 0 & x \end{pmatrix}\begin{pmatrix} 1 & y' \\ 0 & x' \end{pmatrix} = \begin{pmatrix} 1 & y' + yx \\ 0 & xx' \end{pmatrix};$$

associativity is a property of matrix multiplication, identity is

$$\begin{pmatrix} 1 & 0 \\ 0 & 1 \end{pmatrix}; \text{ inverse of } \begin{pmatrix} 1 & y \\ 0 & x \end{pmatrix} \text{ is } \begin{pmatrix} 1 & -yx^{-1} \\ 0 & x^{-1} \end{pmatrix}.$$

Therefore G is a group.

2. Let $\eta = e^{2\pi i/5}$ and $\varepsilon = e^{2\pi i/11}$, and write

$$\alpha = \varepsilon + \varepsilon^3 + \varepsilon^4 + \varepsilon^5 + \varepsilon^9, \beta = \varepsilon^2 + \varepsilon^6 + \varepsilon^7 + \varepsilon^8 + \varepsilon^{10}.$$

Character table of $F_{11,5}$

g_i	1	a	a^2	b	b^2	b^3	b^4
$|C_G(g_i)|$	55	11	11	5	5	5	5
χ_1	1	1	1	1	1	1	1
χ_2	1	1	1	η	η^2	η^3	η^4
χ_3	1	1	1	η^2	η^4	η	η^3
χ_4	1	1	1	η^3	η	η^4	η^2
χ_5	1	1	1	η^4	η^3	η^2	η
χ_6	5	α	β	0	0	0	0
χ_7	5	β	α	0	0	0	0

3. Recall that $\mathbb{Z}_p^*$ is cyclic, so by Exercise 1.6(c), there exists an integer m such that $u^m \equiv v \bmod p$. Also, m is coprime to q, since both u and v have order q modulo p. Hence b^m has order q. Also, $b^{-m}ab^m = a^{u^m} = a^v$. Let $b' = b^m$. Then

$$G_1 = \langle a, b' : a^p = b'^q = 1, b'^{-1}ab' = a^v \rangle.$$

Hence $G_1 \cong G_2$.

4. (a) Note that -1 is the only element of order 2 in $\mathbb{Z}_p^*$. Hence

$$u^m \equiv -1 \bmod p \text{ for some } m$$

$$\Leftrightarrow \text{ the element } u \text{ of } \mathbb{Z}_p^* \text{ has even order}$$

$$\Leftrightarrow q \text{ is even}$$

$$\Leftrightarrow p \equiv 1 \bmod 4.$$

(b) By Proposition 25.9, $a^G = \{a^{u^m} : m \in \mathbb{Z}\}$. Therefore

$$a^{-1} \in a^G \Leftrightarrow u^m \equiv -1 \bmod p \text{ for some } m \Leftrightarrow p \equiv 1 \bmod 4.$$

(c) $\chi_i(a) = 1$ for all the q linear characters χ_i of G. Hence

$$0 = \sum_{\chi \text{ irred}} \chi(1)\chi(a) = q + q\phi_1(a) + q\phi_2(a).$$

Therefore $\phi_1(a) + \phi_2(a) = -1$. Also, $|C_G(a)| = p$, so

$$p = \sum_\chi \chi(a)\overline{\chi(a)} = q + \phi_1(a)\overline{\phi_1(a)} + \phi_2(a)\overline{\phi_2(a)}.$$

Hence $\phi_1(a)\overline{\phi_1(a)} + \phi_2(a)\overline{\phi_2(a)} = (p+1)/2$.

If $p \equiv 1 \bmod 4$, then a is conjugate to a^{-1} and so $\chi(a)$ is real for all characters χ, so $(\phi_1(a))^2 + (\phi_2(a))^2 = (p+1)/2$. Hence $\phi_1(a)$ and $\phi_2(a)$ are $(-1 \pm \sqrt{p})/2$.

If $p \equiv -1 \bmod 4$, then a is not conjugate to a^{-1}, and it follows from Corollary 15.6 that $\phi_1(a)$ and $\phi_2(a)$ are not both real. Hence $\phi_2(a) = \overline{\phi_1(a)}$. This time, $2\phi_1(a)\overline{\phi_1(a)} = (p+1)/2$, and we find that $\phi_1(a)$ and $\phi_2(a)$ are $(-1 \pm i\sqrt{p})/2$.

(d) By Theorem 25.10, $\phi_1(a) = \sum_{m=1}^{(p-1)/2} \varepsilon^{u^m}$. Since $\mathbb{Z}_p^*$ is cyclic of order $p - 1$ and u has order $(p - 1)/2$, it follows that $\{u, u^2, \ldots, u^{(p-1)/2}\}$ is precisely the set of quadratic residues modulo p. The result now follows from part (c).

5. Let $H = \langle a, b \rangle$. Then for all $h \in H$, the conjugacy class h^E consists of h and h^{-1}. All the elements outside H form a single conjugacy class of E. Also, $E' = H$, so E has exactly two linear characters, say χ_1 and χ_2. A typical non-trivial linear character of H is

	1	a	a^2	b	b^2	ab	ab^2	a^2b	a^2b^2
χ	1	1	1	ω	ω^2	ω	ω^2	ω	ω^2

where $\omega = e^{2\pi i/3}$. Then $\chi \uparrow E$ is the irreducible character χ_3 given in the table which follows. The characters χ_4, χ_5 and χ_6 are obtained similarly.

Character table of E

g_i	1	a	b	ab	a^2b	c
$|C_G(g_i)|$	18	9	9	9	9	2
χ_1	1	1	1	1	1	1
χ_2	1	1	1	1	1	-1
χ_3	2	2	-1	-1	-1	0
χ_4	2	-1	2	-1	-1	0
χ_5	2	-1	-1	2	-1	0
χ_6	2	-1	-1	-1	2	0

6. $Z(E) = \{1\}$, and for all i with $1 \leqslant i \leqslant 6$, there exist $g_i \in E$ such that $g_i \neq 1$ but $\chi_i(g_i) = \chi_i(1)$ (so $g_i \in \mathrm{Ker}\,\chi_i$).

7. (a) $F_{13,3}$ (see Theorem 25.10).
   (b) $C_2 \times F_{13,3}$ (see Theorem 19.18).
   (c) $D_6 \times F_{13,3}$ (see Theorem 19.18).

8. The conjugacy classes of G are

$$\{1\}, \{a^3, a^6\}, \{a^r: 3 \nmid r\}, \{a^r b^2: 3 \nmid r\}, \{a^r b^4: 3 \nmid r\},$$

$$\{a^r b^2: r = 0, 3, 6\}, \{a^r b^4: r = 0, 3, 6\},$$

$$\{a^r b: 0 \leqslant r \leqslant 8\}, \{a^r b^3: 0 \leqslant r \leqslant 8\}, \{a^r b^5: 0 \leqslant r \leqslant 8\}.$$

Let $H_1 = \langle a \rangle$. Then $H_1 \lhd G$ and $G/H_1 \cong C_6$. Hence we get six linear characters $\chi_1, \ldots, \chi_6$ of G, as shown.

Let $H_2 = \langle a^3, b^2 \rangle$. Then $H_2 \lhd G$ and $G/H_2 = \langle H_2a, H_2b \rangle \cong D_6$. Lift the irreducible character of D_6 of degree 2 to obtain χ_7 in the table below. Then $\chi_8 = \chi_7\chi_2$ and $\chi_9 = \chi_7\chi_3$ are also irreducible. The final irreducible character χ_{10} can be found by using the column orthogonality relations.

Character table of $G = \langle a, b: a^9 = b^6 = 1, b^{-1}ab = a^2 \rangle$

g_i	1	a^3	a	ab^2	ab^4	b	b^2	b^3	b^4	b^5
$\|C_G(g_i)\|$	54	27	9	9	9	6	18	6	18	6
χ_1	1	1	1	1	1	1	1	1	1	1
χ_2	1	1	1	ω^2	ω	$-\omega$	ω^2	-1	ω	$-\omega^2$
χ_3	1	1	1	ω	ω^2	ω^2	ω	1	ω^2	ω
χ_4	1	1	1	1	1	-1	1	-1	1	-1
χ_5	1	1	1	ω^2	ω	ω	ω^2	1	ω	ω^2
χ_6	1	1	1	ω	ω^2	$-\omega^2$	ω	-1	ω^2	$-\omega$
χ_7	2	2	-1	-1	-1	0	2	0	2	0
χ_8	2	2	-1	$-\omega^2$	$-\omega$	0	$2\omega^2$	0	2ω	0
χ_9	2	2	-1	$-\omega$	$-\omega^2$	0	2ω	0	$2\omega^2$	0
χ_{10}	6	-3	0	0	0	0	0	0	0	0

Note: $\omega = e^{2\pi i/3}$

Chapter 26

1. Let χ be an irreducible character of G. Then $\langle \chi \downarrow H, \psi \rangle_H \neq 0$ for some irreducible character ψ of H. Therefore $\langle \chi, \psi \uparrow G \rangle_G \neq 0$ by the Frobenius Reciprocity Theorem. But $\psi(1) = 1$, since H is abelian; and $(\psi \uparrow G)(1) = p$, by Corollary 21.20. Hence $\chi(1) \leqslant p$, and so $\chi(1) = 1$ or p by Theorem 22.11. Assume that G has r linear characters and s irreducible characters of degree p. Then

$$r = p^m \text{ for some } m, \text{ by Theorem 17.11, and}$$

$$r + sp^2 = p^n, \text{ by Theorem 11.12.}$$

Since $s = p^{n-2} - p^{m-2}$ and s is an integer, m is at least 2.

2. $\{1\}$, $\{z\}$ and $\{z^2\}$ are conjugacy classes of H. For all other elements h of H, the conjugacy class $h^H = \{h, hz, hz^2\}$.

Character table of H (a non-abelian group of order 27)

g_i	1	z	z^2	a	a^2	b	ab	a^2b	b^2	ab^2	a^2b^2
$\|C_H(g_i)\|$	27	27	27	9	9	9	9	9	9	9	9
χ_{00}	1	1	1	1	1	1	1	1	1	1	1
χ_{01}	1	1	1	1	1	ω	ω	ω	ω^2	ω^2	ω^2
χ_{02}	1	1	1	1	1	ω^2	ω^2	ω^2	ω	ω	ω
χ_{10}	1	1	1	ω	ω^2	1	ω	ω^2	1	ω	ω^2
χ_{11}	1	1	1	ω	ω^2	ω	ω^2	1	ω^2	1	ω
χ_{12}	1	1	1	ω	ω^2	ω^2	1	ω	ω	ω^2	1
χ_{20}	1	1	1	ω^2	ω	1	ω^2	ω	1	ω^2	ω
χ_{21}	1	1	1	ω^2	ω	ω	1	ω^2	ω^2	ω	1
χ_{22}	1	1	1	ω^2	ω	ω^2	ω	1	ω	1	ω^2
ϕ_1	3	3ω	$3\omega^2$	0	0	0	0	0	0	0	0
ϕ_2	3	$3\omega^2$	3ω	0	0	0	0	0	0	0	0

Note: $\omega = e^{2\pi i/3}$

Character table of $G = \langle a, b: a^{16} = 1, b^2 = a^8, b^{-1}ab = a^{-1} \rangle$

g_i	1	a^8	a	a^2	a^3	a^4	a^5	a^6	a^7	b	ab
$\|C_G(g_i)\|$	32	32	16	16	16	16	16	16	16	4	4
χ_1	1	1	1	1	1	1	1	1	1	1	1
χ_2	1	1	1	1	1	1	1	1	1	-1	-1
χ_3	1	1	-1	1	-1	1	-1	1	-1	1	-1
χ_4	1	1	-1	1	-1	1	-1	1	-1	-1	1
χ_5	2	2	0	-2	0	2	0	-2	0	0	0
χ_6	2	2	$\sqrt{2}$	0	$-\sqrt{2}$	-2	$-\sqrt{2}$	0	$\sqrt{2}$	0	0
χ_7	2	2	$-\sqrt{2}$	0	$\sqrt{2}$	-2	$\sqrt{2}$	0	$-\sqrt{2}$	0	0
ψ_j	2	-2	c_j	c_{2j}	c_{3j}	c_{4j}	c_{5j}	c_{6j}	c_{7j}	0	0
$(j = 1, 3, 5, 7)$											

Note: $c_m = e^{2\pi i m/16} + e^{-2\pi i m/16} = 2\cos(m\pi/8)$

3. G has 11 conjugacy classes: $\{1\}$, $\{a^8\}$, $\{a^r, a^{-r}\}$ $(1 \leqslant r \leqslant 7)$, $\{a^r b: r \text{ even}\}$, $\{a^r b: r \text{ odd}\}$. Here, the group K which appears in Theorem 26.4 is $\{1, a^8\}$, and $G/K \cong D_{16}$. The character table of D_{16} is given in Section 26.8 $(D_{16} = G_1)$ and in Section 18.3. By lifting the irreducible characters of D_{16}, we obtain the characters $\chi_1, \ldots, \chi_7$ of G as shown in the table at the top of this page. Then the four characters ψ_j $(j = 1, 3, 5, 7)$ come from inducing to G those linear characters χ of $\langle a \rangle$ for which $\chi(a^8) = -1$.

4. (a) Check that $AB = -BA$, $AC = -CA$, $AD = DA$, $BC = CB$, $BD = -DB$, $CD = -DC$. Hence $Z \in G$, and $G/\langle Z \rangle$ is abelian while G is non-abelian. Therefore $G' = \langle Z \rangle$ (see Proposition 17.10).

   (b) $A^2 = -B^2 = -C^2 = D^2 = I$. Since $G/\langle Z \rangle$ is abelian, it follows that $g^2 \in \langle Z \rangle$ for all $g \in G$. Hence every element of G has the form $A^i B^j C^k D^l Z^m$ for some $i, j, k, l, m \in \{0, 1\}$, so $|G| \leqslant 32$; also G is a 2-group, since $g^4 = 1$ for all $g \in G$.

   (c) A routine calculation shows that every matrix which commutes with each of A, B, C and D has the form λI for some $\lambda \in \mathbb{C}$. Hence by Corollary 9.3, the given representation is irreducible.

   (d) Since G has irreducible representations of degrees 1 and 4, $|G| \geqslant 1^2 + 4^2 = 17$. Combined with part (b), this shows that $|G| = 32$. Since $G' = \langle Z \rangle$, G has precisely 16 representations of degree 1. These are as follows: for each (r, s, t, u) with $r, s, t, u \in \{0, 1\}$, we get a representation

$$A^i B^j C^k D^l Z^m \to (-1)^{ir+js+kt+lu}.$$

   Together with the irreducible representation of degree 4, these are all the irreducible representations of G, by Theorem 11.12.

5. (a) Let $\varepsilon = e^{2\pi i/8}$. We obtain representations as follows:

$$G_1: a \to \begin{pmatrix} \varepsilon & 0 \\ 0 & \varepsilon^{-1} \end{pmatrix}, b \to \begin{pmatrix} 0 & 1 \\ 1 & 0 \end{pmatrix};$$

$$G_2: a \to \begin{pmatrix} \varepsilon & 0 \\ 0 & \varepsilon^{-1} \end{pmatrix}, b \to \begin{pmatrix} 0 & 1 \\ -1 & 0 \end{pmatrix};$$

$$G_3: a \to \begin{pmatrix} \varepsilon & 0 \\ 0 & \varepsilon^3 \end{pmatrix}, b \to \begin{pmatrix} 0 & 1 \\ 1 & 0 \end{pmatrix};$$

$$G_4: a \to \begin{pmatrix} \varepsilon & 0 \\ 0 & \varepsilon^5 \end{pmatrix}, b \to \begin{pmatrix} 0 & 1 \\ 1 & 0 \end{pmatrix};$$

$$G_5: a \to \begin{pmatrix} 1 & 0 \\ 0 & -1 \end{pmatrix}, b \to \begin{pmatrix} 0 & 1 \\ 1 & 0 \end{pmatrix}, z \to \begin{pmatrix} i & 0 \\ 0 & i \end{pmatrix}.$$

(b) If $j = 5, 6, 7$ or 8 then $Z(G_j) = \{1, a^2, z, a^2z\} \cong C_2 \times C_2$. Since $Z(G_j)$ is not cyclic, G_j has no faithful irreducible representation, by Proposition 9.16.

(c) Check that the matrices satisfy the required relations, so give representations. It is easy to see that the matrices generate groups with more than eight elements, so the representations are faithful.

(d) The following give faithful representations:

$$G_7: a \to \begin{pmatrix} i & 0 & 0 \\ 0 & -i & 0 \\ 0 & 0 & 1 \end{pmatrix}, b \to \begin{pmatrix} 0 & 1 & 0 \\ 1 & 0 & 0 \\ 0 & 0 & 1 \end{pmatrix}, z \to \begin{pmatrix} 1 & 0 & 0 \\ 0 & 1 & 0 \\ 0 & 0 & -1 \end{pmatrix};$$

$$G_8: a \to \begin{pmatrix} i & 0 & 0 \\ 0 & -i & 0 \\ 0 & 0 & 1 \end{pmatrix}, b \to \begin{pmatrix} 0 & 1 & 0 \\ -1 & 0 & 0 \\ 0 & 0 & 1 \end{pmatrix}, z \to \begin{pmatrix} 1 & 0 & 0 \\ 0 & 1 & 0 \\ 0 & 0 & -1 \end{pmatrix}.$$

6. The following table records the numbers of elements of orders 1, 2, 4 and 8 in $G_1, \ldots, G_9$:

	G_1	G_2	G_3	G_4	G_5	G_6	G_7	G_8	G_9
Order 1	1	1	1	1	1	1	1	1	1
Order 2	9	1	5	3	3	7	11	3	5
Order 4	2	10	6	4	12	8	4	12	10
Order 8	4	4	4	8	0	0	0	0	0

Therefore no two of $G_1, \ldots, G_9$ are isomorphic, except possibly G_5 and G_8. But $G_5/G_5' \cong C_2 \times C_4$, while $G_8/G_8' \cong C_2 \times C_2 \times C_2$, so $G_5 \ncong G_8$.

7. (a) By Lemma 26.1(1) we have $\{1\} \neq Z(G) \neq G$. Also $|G/Z(G)| \neq p$ by Lemma 26.1(2). Therefore $|Z(G)| = p$ or p^2. Assume that $|Z(G)| = p^2$. If $g \notin Z(G)$ then $Z(G) \leqslant C_G(g) \neq G$, and $g \in C_G(g)$. Hence

$|C_G(g)| = p^3$ and $|g^G| = p$. Therefore, G has $p^2 + (p^4 - p^2)/p$ conjugacy classes.

(b) Assume that G has r irreducible characters of degree 1 and s irreducible characters of degree p. Since $\sum \chi(1)^2 = p^4$ (Theorem 11.12), there are no irreducible characters of degree greater than p, so $r + sp^2 = p^4$. Therefore $|G/G'| = r = p^2$ or p^3, and if $r = p^2$ then $r + s = 2p^2 - 1$. Part (b) follows, as $r + s$ is equal to the number of conjugacy classes of G.

(c) Note that $G' \cap Z(G) \neq \{1\}$ by Lemma 26.1(1). By parts (a) and (b), if $|Z(G)| = p^2$ then $|G'| = p$; and if $|G'| = p^2$ then $|Z(G)| = p$. Hence $|G' \cap Z(G)| = p$.

8. (a) Let $Z = Z(G)$, and assume that

$$G/Z = \langle aZ, bZ \rangle, \text{ with } a^4 \in Z, \ a^2 Z = b^2 Z, \ b^{-1} a b Z = a^{-1} Z.$$

Then $a^2 = b^2 z$ for some $z \in Z$, and hence $ba^2 = b^3 z = b^2 zb = a^2 b$. Since a^2 commutes with a, b and all elements in Z, we have $a^2 \in Z$. Therefore $G/Z \not\cong Q_8$.

(b) If G is a non-abelian group of order 16, then by Exercise 7, either $G' \cap Z(G) = G'$, in which case $G/(G' \cap Z(G))$ is abelian, or $G' \cap Z(G) = Z(G)$, in which case $G/(G' \cap Z(G)) \not\cong Q_8$ by part (a).

Chapter 27

1. Assume that

$$z = \begin{pmatrix} a & b \\ c & d \end{pmatrix} \in Z(\text{SL}(2, p)).$$

Then

$$z \begin{pmatrix} 1 & 1 \\ 0 & 1 \end{pmatrix} = \begin{pmatrix} 1 & 1 \\ 0 & 1 \end{pmatrix} z \Rightarrow c = 0, \text{ and}$$

$$z \begin{pmatrix} 0 & 1 \\ -1 & 0 \end{pmatrix} = \begin{pmatrix} 0 & 1 \\ -1 & 0 \end{pmatrix} z \Rightarrow c = -b, \ a = d.$$

Therefore $z = aI$; and since $z \in \text{SL}(2, p)$, we have $a^2 = 1$, so $a = \pm 1$.

2. Check that

$$\begin{pmatrix} 1 & 1 \\ 0 & 1 \end{pmatrix} \quad \text{and} \quad \begin{pmatrix} 1 & -1 \\ 0 & 1 \end{pmatrix}$$

are elements of $G = \text{SL}(2, 3)$ of order 3 which are not conjugate to each other. The element

$$\begin{pmatrix} -1 & 0 \\ 0 & -1 \end{pmatrix}$$

lies in $Z(G)$, and

$$\begin{pmatrix} 0 & 1 \\ -1 & 0 \end{pmatrix}$$

has order 4. Hence the following are conjugacy class representatives:

	g_1	g_2	g_3	g_4
	$\begin{pmatrix} 1 & 0 \\ 0 & 1 \end{pmatrix}$	$\begin{pmatrix} -1 & 0 \\ 0 & -1 \end{pmatrix}$	$\begin{pmatrix} 0 & 1 \\ -1 & 0 \end{pmatrix}$	$\begin{pmatrix} 1 & 1 \\ 0 & 1 \end{pmatrix}$
Order of g_i	1	2	4	3
$|C_G(g_i)|$	24	24	4	6

	g_5	g_6	g_7
	$\begin{pmatrix} 1 & -1 \\ 0 & 1 \end{pmatrix}$	$\begin{pmatrix} -1 & 1 \\ 0 & -1 \end{pmatrix}$	$\begin{pmatrix} -1 & -1 \\ 0 & -1 \end{pmatrix}$
Order of g_i	3	6	6
$|C_G(g_i)|$	6	6	6

We now describe how to construct the character table of G, which is given below.

First observe that the vector space $(\mathbb{Z}_3)^2$ has exactly four 1-dimensional subspaces, namely the spans of the vectors $(0, 1)$, $(1, 1)$, $(2, 1)$ and $(1, 0)$. The group G permutes these subspaces among themselves, so we obtain a homomorphism $\phi: G \to S_4$. Check that $\operatorname{Ker} \phi = \{\pm I\}$. Hence $G/\{\pm I\} \cong \operatorname{Im} \phi$, a subgroup of S_4 of order 12; therefore $G/\{\pm I\} \cong A_4$. The characters χ_1, χ_2, χ_3, χ_4 of G are obtained by lifting to G the irreducible characters of A_4 (which are given in Section 18.2).

The values of χ_5, χ_6, χ_7 on the elements g_1, g_2, g_3 can be deduced from the column orthogonality relations. Note that G has three real conjugacy classes, so by Theorem 23.1, one of χ_5, χ_6 and χ_7 must be real. Assume, without loss of generality, that χ_5 is real. Then $\chi_5(g_4) = \alpha$, where α is real. Also $\alpha \neq 0$, by Corollary 22.27. Since $\chi_5\chi_2$ and $\chi_5\chi_3$ are irreducible characters of G of degree 2, whose values on g_4 are $\alpha\omega$ and $\alpha\omega^2$, they must be χ_6 and χ_7 in some order, say $\chi_5\chi_2 = \chi_6$ and $\chi_5\chi_3 = \chi_7$. The equation $\sum_j \chi_j(g_4)\overline{\chi_j(g_4)} = 6$ gives $\alpha\bar{\alpha} = 1$. Since α is real, $\alpha = \pm 1$. Then $\alpha = -1$ since $\chi_5(g_4) \equiv \chi_5(1) \bmod 3$. Now note that for $j = 5, 6, 7$, Exercise 13.5 implies that $\chi_j(g_7) = -\chi_j(g_4)$. Finally, $\chi(g_5) = \overline{\chi(g_4)}$ and $\chi(g_6) = \overline{\chi(g_7)}$ for all χ.

Character table of SL(2, 3)

g_i $\|C_G(g_i)\|$	g_1 24	g_2 24	g_3 4	g_4 6	g_5 6	g_6 6	g_7 6
χ_1	1	1	1	1	1	1	1
χ_2	1	1	1	ω	ω^2	ω^2	ω
χ_3	1	1	1	ω^2	ω	ω	ω^2
χ_4	3	3	-1	0	0	0	0
χ_5	2	-2	0	-1	-1	1	1
χ_6	2	-2	0	$-\omega$	$-\omega^2$	ω^2	ω
χ_7	2	-2	0	$-\omega^2$	$-\omega$	ω	ω^2

Note: $\omega = e^{2\pi i/3}$

3. Apply Proposition 17.6.

4. (a) For the character table of T, notice that T is isomorphic to the group of order 21 whose character table is found in Exercise 17.2 and Example 21.25. Representatives of the conjugacy classes of T are $h_1, \ldots, h_5$, where

$$h_1 = \begin{pmatrix} 1 & 0 \\ 0 & 1 \end{pmatrix} Z, \ h_2 = \begin{pmatrix} 2 & 0 \\ 0 & 4 \end{pmatrix} Z, \ h_3 = \begin{pmatrix} 4 & 0 \\ 0 & 2 \end{pmatrix} Z,$$

$$h_4 = \begin{pmatrix} 1 & 1 \\ 0 & 1 \end{pmatrix} Z, \ h_5 = \begin{pmatrix} 1 & -1 \\ 0 & 1 \end{pmatrix} Z.$$

Two of the linear characters of T are

h_i $\|C_T(h_i)\|$	h_1 21	h_2 3	h_3 3	h_4 7	h_5 7
1_T	1	1	1	1	1
λ	1	ω	ω^2	1	1

where $\omega = e^{2\pi i/3}$. The values of $1_T \uparrow G$ and $\lambda \uparrow G$ are as follows (see Proposition 21.23):

g_i $\|C_G(g_i)\|$	g_1 168	g_2 8	g_3 4	g_4 3	g_5 7	g_6 7
$1_T \uparrow G$	8	0	0	2	1	1
$\lambda \uparrow G$	8	0	0	-1	1	1

We find that $\langle 1_T \uparrow G, \ 1_T \uparrow G \rangle = 2$ and $\langle 1_T \uparrow G, \ 1_G \rangle = 1$. Hence $1_T \uparrow G = 1_G + \chi$, where χ is an irreducible character of G. Also, $\langle \lambda \uparrow G, \lambda \uparrow G \rangle = 1$, so $\lambda \uparrow G$ is irreducible; write $\phi = \lambda \uparrow G$.

(c) The values of χ and χ_S are as shown below (see Proposition 19.14):

g_i $|C_G(g_i)|$	g_1 168	g_2 8	g_3 4	g_4 3	g_5 7	g_6 7
χ	7	-1	-1	1	0	0
χ_S	28	4	0	1	0	0
ζ	12	4	0	0	-2	-2
ψ	6	2	0	0	-1	-1

We find that $\langle \chi_S, 1_G \rangle = \langle \chi_S, \phi \rangle = \langle \chi_S, \chi \rangle = 1$. Hence there is a character ζ of G such that

$$\chi_S = 1_G + \phi + \chi + \zeta.$$

The values of ζ are as shown above. We calculate that $\langle \zeta, \zeta \rangle = 4$, so either $\zeta = 2\psi$ for some irreducible character ψ, or ζ is the sum of four distinct irreducible characters (cf. Exercise 14.7).

Now 1_G, ϕ and χ are three of the six irreducible characters of G, and none is a constituent of ζ. Since there are only six irreducible characters in all, ζ cannot therefore be the sum of four irreducible characters, and so $\zeta = 2\psi$ with ψ irreducible. The values of ψ are as shown above.

(d) The characters 1_G, ϕ, χ and ψ are the characters χ_1, χ_3, χ_2 and χ_6, respectively, in the character table of G given at the end of Chapter 27. The remaining irreducible characters χ_4, χ_5 can readily be calculated using the column orthogonality relations (noting that g_i is real if and only if $1 \leqslant i \leqslant 4$).

5. (a) Compare the proof of Lemma 27.1. Note that because g_2 lies in $Z(G)$, g_i and $g_i g_2$ have the same centralizer for all i.

(b) By lifting, we obtain the characters $\chi_1, \ldots, \chi_6$ in the character table shown below.

(c) Use Exercise 13.5 (noting that $\chi_j(-I) \neq \chi_j(I)$ for $7 \leqslant j \leqslant 11$, since $-I$ is not in the kernel of these characters).

(d) Since $\sum_{j=1}^{11} \chi_j(g_3) \overline{\chi_j(g_3)} = 8$, we deduce that $\chi_j(g_3) = 0$ for $7 \leqslant j \leqslant 11$. (Alternatively, apply part (c).) Also, by Corollary 22.27, $\chi_j(1)$ is even.

(e) By Theorem 22.16, $\chi(g_6) \in \mathbb{Z}$ for all characters χ. Since $\sum_{j=1}^{11} (\chi_j(g_6))^2 = 6$, the values of $\chi_j(g_6)$ for $7 \leqslant j \leqslant 11$ must be ± 1, ± 1, ± 1, 0, 0, in some order. By Corollary 22.27 again, two of $\chi_7, \ldots,$ χ_{11}, say χ_9 and χ_{10}, have degrees divisible by 6. Further, $\sum_{j=1}^{11} (\chi_j(1))^2 = 168$, and $12^2 + 6^2 > 168$, so $\chi_9(1) = \chi_{10}(1) = 6$.

Next, $\chi_7(1)^2 + \chi_8(1)^2 + \chi_{11}(1)^2 = 96$. The only possibility is that two of $\chi_7(1)$, $\chi_8(1)$, $\chi_{11}(1)$, say $\chi_7(1)$ and $\chi_8(1)$, are equal to 4, and $\chi_{11}(1) = 8$. The congruences $\chi(1) \equiv \chi(g_6) \bmod 3$ now give the remaining values on g_6. Use part (c) to fill in the values on g_2 and g_7.

(f) By Proposition 19.14, ψ_A has the following values on g_1, g_2, g_3 and g_6:

	g_1	g_2	g_3	g_6
ψ_A	6	6	2	0

Character table of SL $(2, 7)$

g_i	g_1	g_2	g_3	g_4	g_5	g_6	g_7	g_8	g_9	g_{10}	g_{11}
Order of g_i	1	2	4	8	8	3	6	7	14	7	14
$\lvert C_G(g_i)\rvert$	336	336	8	8	8	6	6	14	14	14	14
χ_1	1	1	1	1	1	1	1	1	1	1	1
χ_2	7	7	-1	-1	-1	1	1	0	0	0	0
χ_3	8	8	0	0	0	-1	-1	1	1	1	1
χ_4	3	3	-1	1	1	0	0	α	α	$\bar{\alpha}$	$\bar{\alpha}$
χ_5	3	3	-1	1	1	0	0	$\bar{\alpha}$	$\bar{\alpha}$	α	α
χ_6	6	6	2	0	0	0	0	-1	-1	-1	-1
χ_7	4	-4	0	0	0	1	-1	$-\alpha$	α	$-\bar{\alpha}$	$\bar{\alpha}$
χ_8	4	-4	0	0	0	1	-1	$-\bar{\alpha}$	$\bar{\alpha}$	$-\alpha$	α
χ_9	6	-6	0	$\sqrt{2}$	$-\sqrt{2}$	0	0	-1	1	-1	1
χ_{10}	6	-6	0	$-\sqrt{2}$	$\sqrt{2}$	0	0	-1	1	-1	1
χ_{11}	8	-8	0	0	0	-1	1	1	-1	1	-1

Note: $\alpha = (-1 + i\sqrt{7})/2$

The values of ψ_A on g_1 and g_2 show that ψ_A is a linear combination of χ_1, χ_4, χ_5 and χ_6. Then the value of ψ_A on g_6 shows that χ_1 is not a constituent of ψ_A; finally, the value on g_3 forces $\psi_A = \chi_6$.

Now g_4^2 is conjugate to g_3. Hence $(\psi(g_4)^2 - \psi(g_3))/2 = \psi_A(g_4) = \chi_6(g_4) = 0$, and therefore, $\psi(g_4) = 0$. Similarly, $\psi(g_5) = 0$.

Let $x = \psi(g_8)$. Since g_8^2 is conjugate to g_8, we get $(x^2 - x)/2 = \psi_A(g_8) = -1$. Therefore $x = (1 \pm i\sqrt{7})/2$. Say $\chi_7(g_8) = (1 - i\sqrt{7})/2$. Then $\chi_8 = \bar{\chi}_7$. For all χ, $\chi(g_{10}) = \chi(g_8)$; using this fact and part (c), we fill in the values of χ_7 and χ_8.

(g) For $i \neq 6$, we have $\sum_{j=1}^{11} \chi_j(g_i)\overline{\chi_j(g_6)} = 0$. This allows us to fill in the values of χ_{11}. Since g_4 is conjugate to g_4^{-1}, $\chi(g_4)$ is real for all χ. Then $\sum_{j=1}^{11} \chi_j(g_1)\overline{\chi_j(g_4)} = 0$ and $\sum_{j=1}^{11} \chi_j(g_4)^2 = 8$ imply that $\chi_9(g_4) = -\chi_{10}(g_4) = \pm\sqrt{2}$. Say $\chi_9(g_4) = \sqrt{2}$. The column orthogonality relations now let us find the remaining values of χ_9 and χ_{10}, thereby completing the character table of G.

6. Let $Z = \{\pm I\}$ and define the subgroup T of G, of order 55, by

$$T = \left\{ \begin{pmatrix} a & b \\ 0 & a^{-1} \end{pmatrix} Z : a \in \mathbb{Z}_{11}^*, b \in \mathbb{Z}_{11} \right\}.$$

Then T is generated by

$$x = \begin{pmatrix} 1 & 1 \\ 0 & 1 \end{pmatrix} Z \text{ and } y = \begin{pmatrix} 2 & 0 \\ 0 & 6 \end{pmatrix} Z,$$

and T has five linear characters ζ_j $(0 \leq j \leq 4)$, where

$$\zeta_j \colon x^u y^v \to e^{2\pi i jv/5}.$$

The characters $\zeta_1 \uparrow G$ and $\zeta_2 \uparrow G$ are irreducible; they are χ_3 and χ_4 in the table. (In calculating $\chi_3(g_5)$, note that $e^{2\pi i/5} + e^{-2\pi i/5} = (-1 + \sqrt{5})/2$.) Let $\chi_1 = 1_G$. We have $\langle \zeta_0 \uparrow G, \chi_1 \rangle = 1$ and $\langle \zeta_0 \uparrow G, \zeta_0 \uparrow G \rangle = 2$. Hence $\zeta_0 \uparrow G = \chi_1 + \chi_2$ for an irreducible character χ_2 of G.

We have now found four of the eight irreducible characters of G, namely $\chi_1, \chi_2, \chi_3, \chi_4$.

Character table of PSL $(2, 11)$

g_i	g_1	g_2	g_3	g_4	g_5	g_6	g_7	g_8
Order of g_i	1	2	3	6	5	5	11	11
$\lvert C_G(g_i) \rvert$	660	12	6	6	5	5	11	11
χ_1	1	1	1	1	1	1	1	1
χ_2	11	-1	-1	-1	1	1	0	0
χ_3	12	0	0	0	α	β	1	1
χ_4	12	0	0	0	β	α	1	1
χ_5	10	2	1	-1	0	0	-1	-1
χ_6	10	-2	1	1	0	0	-1	-1
χ_7	5	1	-1	1	0	0	γ	$\bar{\gamma}$
χ_8	5	1	-1	1	0	0	$\bar{\gamma}$	γ

Note: $\alpha = (-1 + \sqrt{5})/2$, $\beta = (-1 - \sqrt{5})/2$ and $\gamma = (-1 + i\sqrt{11})/2$

Since $\sum_{j=1}^{8} \chi_j(g_5)\overline{\chi_j(g_5)} = 5$, we deduce that the remaining irreducible characters $\chi_5, \chi_6, \chi_7, \chi_8$ take the value 0 on g_5. By Corollary 22.27, $\chi_j(1) \equiv 0 \bmod 5$ for $5 \leqslant j \leqslant 8$. But $\sum_{j=5}^{8} (\chi_j(1))^2 = 250$; hence, without loss of generality, $\chi_5(1), \chi_6(1), \chi_7(1), \chi_8(1)$ are 10, 10, 5, 5, respectively.

Note that $\chi(g_j)$ is an integer for $1 \leqslant j \leqslant 4$ and all characters χ, by Theorem 22.16.

Since $\chi(1) \equiv \chi(g_3) \bmod 3$ for all characters χ, and $\sum_{j=1}^{8} \chi_j(g_3)^2 = 6$, the values of $\chi_j(g_3)$ ($5 \leqslant j \leqslant 8$) are as shown.

Now $\chi(g_4) \equiv \chi(g_3) \bmod 2$ for all χ, and $\sum_{j=1}^{8} \chi_j(g_4)^2 = 6$, so $\chi_j(g_4) = \pm 1$ for $5 \leqslant j \leqslant 8$. Next, $\chi(g_2) \equiv \chi(g_4) \bmod 3$ for all χ, and $\sum_{j=1}^{8} \chi_j(g_2)^2 = 12$; hence $\lvert \chi(g_2) \rvert < 3$ for all irreducible χ. We may now conclude from the facts that $\chi(g_2) \equiv \chi(g_1) \bmod 2$ and $\sum_{j=1}^{8} \chi_j(g_2)^2 = 12$, that $\chi_j(g_2) = \pm 2$ for $j = 5$, 6 and $\chi_j(g_2) = \pm 1$ for $j = 7, 8$. By considering $\sum_{j=1}^{8} \chi_j(1)\chi_j(g_2) = 0$, we see that $\chi_7(g_2) = \chi_8(g_2) = 1$, and $\chi_5(g_2)$, $\chi_6(g_2)$ have opposite signs; without loss of generality, $\chi_5(g_2) = 2 = -\chi_6(g_2)$.

We have now completed columns 1, 2, 3 and 5 of the character table.

Since $\chi(g_4) \equiv \chi(g_2) \bmod 3$ for all characters χ, and $\sum_{j=1}^{8} \chi_j(g_4)^2 = 6$, we can complete column 4.

The column orthogonality relations now enable us to finish the character table.

Chapter 28

1. We take $g_1, \ldots, g_8$ as representatives of the conjugacy classes, where

$$g_1 = \begin{pmatrix} 1 & 0 \\ 0 & 1 \end{pmatrix} \quad g_2 = \begin{pmatrix} 2 & 0 \\ 0 & 2 \end{pmatrix} \quad g_3 = \begin{pmatrix} 1 & 1 \\ 0 & 1 \end{pmatrix} \quad g_4 = \begin{pmatrix} 2 & 1 \\ 0 & 2 \end{pmatrix}$$

$$g_5 = \begin{pmatrix} 1 & 0 \\ 0 & 2 \end{pmatrix} \quad g_6 = \begin{pmatrix} 0 & 1 \\ 2 & 0 \end{pmatrix} \quad g_7 = \begin{pmatrix} 0 & 1 \\ 1 & 2 \end{pmatrix} \quad g_8 = \begin{pmatrix} 0 & 1 \\ 1 & 1 \end{pmatrix}.$$

The character table of GL(2, 3) is then as follows.

g_i	g_1	g_2	g_3	g_4	g_5	g_6	g_7	g_8
$|C_G(g_i)|$	48	48	6	6	4	8	8	8
λ_0	1	1	1	1	1	1	1	1
λ_1	1	1	1	1	−1	1	−1	−1
ψ_0	3	3	0	0	1	−1	−1	−1
ψ_1	3	3	0	0	−1	−1	1	1
$\psi_{0,1}$	4	−4	1	−1	0	0	0	0
χ_1	2	−2	−1	1	0	0	$i\sqrt{2}$	$-i\sqrt{2}$
χ_2	2	2	−1	−1	0	2	0	0
χ_4	2	−2	−1	1	0	0	$-i\sqrt{2}$	$i\sqrt{2}$

2. Every element r of $\mathbb{F}_q$ can be expressed as a square, since $r = r^q$ and q is even.

Suppose that $\begin{pmatrix} a & b \\ c & d \end{pmatrix} \in GL(2, q)$. We may write $ad - bc$ as s^2 for some s in $\mathbb{F}_q^*$. Then

$$\begin{pmatrix} a & b \\ c & d \end{pmatrix} = \begin{pmatrix} s & 0 \\ 0 & s \end{pmatrix} \begin{pmatrix} a/s & b/s \\ c/s & d/s \end{pmatrix}.$$

The first matrix in the product is sI and the second belongs to SL(2, q). It now follows easily that $GL(2, q) \cong Z \times SL(2, q)$ where $Z = \{sI : s \in \mathbb{F}_q^*\}$.

You should have no difficulty in proving that the conjugacy classes of SL(2, q) have representatives as follows.

(a) The identity I has centralizer of order $q^3 - q$.

(b) The matrix $u_1 = \begin{pmatrix} 1 & 1 \\ 0 & 1 \end{pmatrix}$ has centralizer of order q.

(c) There are $(q - 2)/2$ conjugacy classes with representatives

$$d_{s,s^{-1}} = \begin{pmatrix} s & 0 \\ 0 & s^{-1} \end{pmatrix}, \text{ indexed by unordered pairs } \{s, s^{-1}\} \text{ of elements}$$

from $\mathbb{F}_q \backslash \mathbb{F}_2$. Each such element has centralizer of order $q - 1$.

(d) There are $q/2$ conjugacy classes with representatives

$$v_r = \begin{pmatrix} 0 & 1 \\ 1 & r + r^{-1} \end{pmatrix}, \text{ indexed by unordered pairs } \{r, r^{-1}\} \text{ of elements}$$

from $\mathbb{F}_{q^2} \backslash \mathbb{F}_q$ such that $r^{1+q} = 1$. Each such element has centralizer of order $q + 1$.

By restricting characters from $GL(2, q)$ to $SL(2, q)$ you will quickly be able to prove that the character table of $SL(2, q)$ is as follows.

	I	u_1	$d_{s,s^{-1}}$	v_r
λ_0	1	1	1	1
ψ_0	q	0	1	-1
$\psi_{0,i}$	$q + 1$	1	$\bar{s}^i + \bar{s}^{-i}$	0
χ_i	$q - 1$	-1	0	$-(\bar{r}^i + \bar{r}^{-i})$

Here, we have used the function $r \to \bar{r}$ defined in (28.3). The subscripts for $\psi_{0,i}$ satisfy $1 \leqslant i \leqslant (q - 2)/2$, and the subscripts for χ_i satisfy $1 \leqslant i \leqslant q/2$.

If $q \neq 2$ then the kernel of every non-trivial character is the identity subgroup, and therefore $SL(2, q)$ is simple.

3. Note first that $PSL(2, 8) \cong SL(2, 8)$.

The polynomial $x^3 + x + 1$ is irreducible over $\mathbb{F}_2$. Hence we may write

$$\mathbb{F}_8 = \{a + b\eta + c\eta^2 : a, b, c \in \mathbb{F}_2 \text{ and } \eta^3 = 1 + \eta\}.$$

The pairs $\{s, s^{-1}\}$ of elements from $\mathbb{F}_8 \backslash \mathbb{F}_2$ are

$$\{\eta, 1 + \eta^2\}, \{\eta^2, 1 + \eta + \eta^2\}, \{1 + \eta, \eta + \eta^2\}.$$

These give us the conjugacy class representatives g_3, g_4, g_5 below.

The irreducible monic quadratics over $\mathbb{F}_8$ with constant term 1 are

$$x^2 + x + 1, \, x^2 + \eta x + 1, \, x^2 + \eta^2 x + 1, \, x^2 + (\eta + \eta^2)x + 1.$$

The companion matrices for these quadratics give use the conjugacy class representatives g_6, g_7, g_8, g_9 below

We can now list representatives $g_1, \ldots, g_9$ of the conjugacy classes of $SL(2, 8)$, as follows.

$$g_1 = \begin{pmatrix} 1 & 0 \\ 0 & 1 \end{pmatrix} \quad g_2 = \begin{pmatrix} 1 & 1 \\ 0 & 1 \end{pmatrix}$$

$$g_3 = \begin{pmatrix} \eta & 0 \\ 0 & 1 + \eta^2 \end{pmatrix} \quad g_4 = \begin{pmatrix} \eta^2 & 0 \\ 0 & 1 + \eta + \eta^2 \end{pmatrix} \quad g_5 = \begin{pmatrix} 1 + \eta & 0 \\ 0 & \eta + \eta^2 \end{pmatrix}$$

$$g_6 = \begin{pmatrix} 0 & 1 \\ 1 & 1 \end{pmatrix} \quad g_7 = \begin{pmatrix} 0 & 1 \\ 1 & \eta \end{pmatrix} \quad g_8 = \begin{pmatrix} 0 & 1 \\ 1 & \eta^2 \end{pmatrix} \quad g_9 = \begin{pmatrix} 0 & 1 \\ 1 & \eta + \eta^2 \end{pmatrix}.$$

We may choose a generator ε of $\mathbb{F}_{64}^*$ so that $\varepsilon^7 + \varepsilon^{-7} = \eta$. Then

$\varepsilon^{14} + \varepsilon^{-14} = \eta^2$, $\varepsilon^{21} + \varepsilon^{-21} = 1$ and $\varepsilon^{28} + \varepsilon^{-28} = \eta^4 = \eta + \eta^2$. The character table of SL(2, 8) is then as follows.

g_i $|C_G(g_i)|$	g_1 504	g_2 8	g_3 7	g_4 7	g_5 7	g_6 9	g_7 9	g_8 9	g_9 9
λ_0	1	1	1	1	1	1	1	1	1
ψ_0	8	0	1	1	1	-1	-1	-1	-1
$\psi_{0,1}$	9	1				0	0	0	0
$\psi_{0,2}$	9	1		A		0	0	0	0
$\psi_{0,3}$	9	1				0	0	0	0
χ_3	7	-1	0	0	0	-2	1	1	1
χ_1	7	-1	0	0	0	1			
χ_2	7	-1	0	0	0	1		B	
χ_4	7	-1	0	0	0	1			

Here, the 3×3 submatrices A and B are given by

$$A = \begin{pmatrix} 2\cos(2\pi/7) & 2\cos(4\pi/7) & 2\cos(6\pi/7) \\ 2\cos(4\pi/7) & 2\cos(6\pi/7) & 2\cos(2\pi/7) \\ 2\cos(6\pi/7) & 2\cos(2\pi/7) & 2\cos(4\pi/7) \end{pmatrix}$$

$$B = \begin{pmatrix} -2\cos(2\pi/9) & -2\cos(4\pi/9) & -2\cos(8\pi/9) \\ -2\cos(4\pi/9) & -2\cos(8\pi/9) & -2\cos(2\pi/9) \\ -2\cos(8\pi/9) & -2\cos(2\pi/9) & -2\cos(4\pi/9) \end{pmatrix}.$$

Chapter 29

1. (a) It is straightforward to check that ϕ is a homomorphism. For $x, y \in G$, the element $(x, y) \in G \times G$ sends x to y, so that action is transitive.

   (b) $(G \times G)_1 = \{(g, g): g \in G\}$, and $\ker \phi = \{(z, z): z \in Z(G)\}$.

   (c) Every orbit of $G \times G$ on $G \times G$ contains an ordered pair of the form $(1, x)$, and if (g, h) sends $(1, x)$ to $(1, y)$ then $g = h$ and $y = g^{-1}xg$. Hence if $C_1, \ldots, C_k$ are the conjugacy classes of G, and $x_i \in C_i$, then $(1, x_i)$ $(1 \le i \le k)$ are orbit representatives for the action of $G \times G$ on $G \times G$, and so the rank is equal to k.

   Since $x((g, h)\phi) = x$ if and only if $xhx^{-1} = g$, we see that $\pi(g, h) = |\text{fix}_G(g, h)|$ is equal to 0 if g is not conjugate to h, and is equal to $|C_G(x)|$ if g is conjugate to h (since in the latter case, if $xhx^{-1} = g$ then an arbitrary element $y \in G$ such that $yhy^{-1} = g$ is of the form $y = xc$ with $c \in C_G(x)$). Hence using the column orthogonality relations we see that $\pi = \sum \chi \times \bar{\chi}$.

2. There are $q^2 - 1$ non-zero vectors in V, and each 1-dimensional subspace

contains $q - 1$ of them; also two 1-dimensional subspaces have no non-zero vectors in common. Hence $|\Omega| = (q^2 - 1)/(q - 1) = q + 1$.

3. Use the notation for the conjugacy classes and irreducible characters of GL(2, q) given in Proposition 28.4 and Theorem 28.5. It is easy to check that π takes the values $q^2 - 1$, $q - 1$, $q - 1$ on the classes with representatives I, u_1, $d_{1,t}$ respectively, and takes the value 0 on all other classes. Taking inner products we find

$$\langle \pi, 1_G \rangle = \langle \pi, \psi_0 \rangle = \langle \pi, \psi_{0,j} \rangle = 1 \ (1 \leqslant j \leqslant q - 2).$$

As $1_G + \psi_0 + \sum_1^{q-2}\psi_{0,j}$ has degree $q^2 - 1 = \pi(1)$, we conclude that $\pi = 1_G + \psi_0 + \sum_1^{q-2}\psi_{0,j}$.

4. Observe that the coset H_1x is fixed by g if and only if $xgx^{-1} \in H_1$. If G is abelian this amounts to $g \in H_1$, and hence we see that $\pi_1(g) = 0$ if $g \notin H_1$ and $\pi_1(g) = |G : H_1|$ if $g \in H_1$. Thus $H_1 = \{g \in G : \pi_1(g) \neq 0\}$. Likewise for H_2; since $\pi_1 = \pi_2$ we deduce that $H_1 = H_2$.

   As a counterexample for G non-abelian, take $G = D_8 = \langle a, b : a^4 = b^2 = 1, \ b^{-1}ab = a^{-1} \rangle$ with $H_1 = \langle b \rangle$, $H_2 = \langle a^2b \rangle$. Then $\pi_1 = \pi_2$ but $H_1 \neq H_2$.

5. By Proposition 29.4 we have $1 = \frac{1}{|G|}\sum_{g \in G}|\text{fix}_\Omega(g)|$, hence $\sum|\text{fix}_\Omega(g)| = |G|$. Since $|\text{fix}_\Omega(g)|$ is a non-negative integer for each g, and $|\text{fix}_\Omega(1)| = |\Omega| > 1$, we must have $|\text{fix}_\Omega(g)| = 0$ for some g.

6. Write $\pi = \pi^{(n-2,1,1)}$. Calculating inner products using Proposition 29.6, as in the proof of Theorem 29.13, we find

$$\langle \pi, \pi \rangle = 7, \ \langle \pi, 1 \rangle = 1, \ \langle \pi, \pi^{(n-1,1)} \rangle = 3, \ \langle \pi, \pi^{(n-2,2)} \rangle = 4.$$

Using Theorem 29.13 it follows that $\pi^{(n-2,1,1)} = 1 + 2\chi^{(n-1,1)} + \chi^{(n-2,2)} + \chi$ with χ irreducible. Hence $\chi = 1 + \pi - \pi^{(n-1,1)} - \pi^{(n-2,2)}$, from which it is easy to calculate that $\chi(1) = \frac{1}{2}(n - 1)(n - 2)$, $\chi(12) = \frac{1}{2}(n - 2)(n - 5)$, $\chi(123) = \frac{1}{2}(n - 4)(n - 5)$. For $n = 6$, $\chi^{(4,1,1)}$ is the character χ_5 in Example 19.17.

Chapter 30

1. By Theorem 30.4, $a_{245} = 168/(8 \cdot 3) = 7$. Hence, by (30.3), PSL$(2, 7)$ contains elements a and b such that a has order 2, b has order 3 and ab has order 7.

2. No: $a_{225} = (1 + (-1 + i\sqrt{7})/6 + (-1 - i\sqrt{7})/6 - 4/6)168/(8 \cdot 8) = 0$, and similarly $a_{226} = 0$. Hence PSL$(2,7)$ does not contain two involutions whose product has order 7.

3. Yes. Number the conjugacy classes of PSL$(2, 11)$ as in the solution to Exercise 27.6. Then

$$a_{235} = \frac{660}{12 \cdot 6}\left(1 + \frac{1}{11}\right) = 10.$$

Therefore PSL$(2, 11)$ contains elements x and y such that x, y and xy have orders 2, 3 and 5, respectively. Let H be the subgroup $\langle x, y \rangle$ of PSL$(2, 11)$.

There is a homomorphism ϑ from A_5 onto H (ϑ sends $a \to x$, $b \to y$). Since Ker $\vartheta \times A_5$ and A_5 is simple, we deduce that $H \cong A_5$.

4. Suppose that G is a group whose character table is where $\alpha = (1 + \sqrt{5})/2$, $\beta = (1 - \sqrt{5})/2$.

	g_1	g_2	g_3	g_4	g_5
χ_1	1	1	1	1	1
χ_2	4	1	0	-1	-1
χ_3	5	-1	1	0	0
χ_4	3	0	-1	α	β
χ_5	3	0	-1	β	α

For $1 \le i \le 5$ we have $|C_G(g_i)| = \sum_{j=1}^{5} \chi_j(g_i)\overline{\chi_j(g_i)}$. Therefore the centralizers of g_1, g_2, g_3, g_4, g_5 have orders 60, 3, 4, 5, 5, respectively. Hence the orders of g_2, g_4 and g_5 are 3, 5 and 5; also the order of g_3 must be 2, since for no other i (except $i = 1$) is $|C_G(g_i)|$ even.

Now $a_{324} = 60/(4 \cdot 3)$. Therefore G contains elements x and y such that x has order 2, y has order 3 and xy has order 5. As in the solution to Exercise 3, G has a subgroup H with $H \cong A_5$. Since $|G| = 60$, we have $G \cong A_5$.

5. (a) Using the fact that $\sum_{j=1}^{7} \chi_j(g_i)\overline{\chi_j(g_i)} = |C_G(g_i)|$, we find that the centralizer orders and class sizes are as follows:

	g_1	g_2	g_3	g_4	g_5	g_6	g_7
$|C_G(g_i)|$	360	8	4	9	9	5	5
$|g_i^G|$	1	45	90	40	40	72	72

Hence $|G| = 360$. Also G is simple, by Proposition 17.6.

(b) By the Frobenius–Schur Count of Involutions (Corollary 23.17), the number of involutions t in G is bounded by

$$1 + t \le \sum_{j=1}^{7} \chi_j(1) = 46.$$

By considering $|C_G(g_i)|$, we see that g_i has even order only for $i = 2$ and 3. Since $t \le 45$, only g_2 can be an involution. Hence g_3 has order 4. (This information about the orders of g_2 and g_3 can also be deduced using Sylow's Theorem.)

(c) Clearly g_6 and g_7 have order 5, and at least one of g_4 and g_5 has order 3. If $j = 4$ or 5 and $k = 6$ or 7 then

$$a_{2jk} = \frac{|G|}{|C_G(g_2)|\,|C_G(g_j)|} \sum_{\chi} \frac{\chi(g_2)\chi(g_j)\overline{\chi(g_k)}}{\chi(1)}$$

$$= \frac{360}{8 \cdot 9} = 5.$$

Therefore G contains elements x and y such that x has order 2, y has order 3 and xy has order 5. As in the solution to Exercise 3, the subgroup $H = \langle x, y \rangle$ of G is isomorphic to A_5.

(d) If $g, h \in G$ then

$$(gh)\rho: Hx \rightarrow Hxgh, \text{ and}$$

$$(g\rho)(h\rho): Hx \rightarrow Hxg \rightarrow Hxgh.$$

Hence ρ is a homomorphism.

(e) Since G is simple, $\text{Ker}\,\rho = \{1\}$. Hence G is isomorphic to a subgroup K of S_6. Since $|S_6: K| = 6!/360 = 2$, K must be A_6.

6. Consider the figure in Example 30.6(3). We shall explain how to label the vertices by elements of G.

Choose a vertex and label it 1. Label the vertices according to the following inductive rule. Assume that v is a vertex and an adjacent vertex u is labelled by g. Then label v by

ga	if the edge uv has no arrow,
gb	if the edge uv has an arrow from u to v,
gb^{-1}	if the edge uv has an arrow from v to u.

For example, if you decided to label the bottom left-hand vertex by 1, then part of the labelling would be

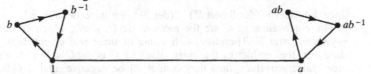

The relation $a^2 = 1$ ensures that the labelling is consistent along unmarked edges; since $b^3 = 1$, the labelling is consistent around triangles; and the relation $abababab = 1$ deals with the octagons.

Every element in G has the form given by the label of one of the 24 vertices, so $|G| \leqslant 24$.

7. The conjugacy classes of PSL(2, 7) are given in Lemma 27.1 The element g_2 is an involution with centralizer of order 8 given in the proof of

Lemma 27.1. Letting $a = \begin{pmatrix} 2 & 2 \\ -2 & 2 \end{pmatrix}$, $b = \begin{pmatrix} 2 & 4 \\ 4 & -2 \end{pmatrix}$, we see that the

centralizer is generated by a and b, and $a^4 = b^2 = 1$, $b^{-1}ab = a^{-1}-$, hence the centralizer is isomorphic to D_8.

As in Exercises 12.3 and 12.4, we see that $C_{A_6}((12)(34))$ has order 8 and is generated by (1324) and (13)(24), hence as above is isomorphic to D_8.

8. Let G be the simple group PSL(2, 17). In the field $\mathbb{Z}_{17}$ the element 4 is a fourth root of unity, so $t = \begin{pmatrix} 4 & 0 \\ 0 & -4 \end{pmatrix} Z$ is an involution. Calculate that $C_G(t)$ is generated by the group of diagonal matrices together with $b = \begin{pmatrix} 0 & 1 \\ -1 & 0 \end{pmatrix} Z$, hence is generated by b and $a = \begin{pmatrix} 3 & 0 \\ 0 & 6 \end{pmatrix} Z$. As $a^8 = b^2 = 1$ and $b^{-1}ab = a^{-1}$, we have $C_G(t) \cong D_{16}$.

Chapter 31

1. Assume that G has an abelian subgroup H of index p^r (p prime), and that $|G| > p$. If $H = \{1\}$ then $|G| = p^r$ and so G is not simple (see Lemma 26.1(1)). So assume that $H \neq \{1\}$; pick $1 \neq h \in H$. Then $H \leqslant C_G(h)$ as H is abelian, so $|G:C_G(h)|$ is a power of p. If $|G:C_G(h)| = 1$ then $\langle h \rangle \lhd G$ and G is not simple. And if $|G:C_G(h)| > 1$, then G is not simple by Theorem 31.3.

2. By Burnside's Theorem, $|G|$ is divisible by at least three distinct primes. Since $3 \cdot 5 \cdot 7 > 80$, $|G|$ is even. Then by Exercise 13.8, $|G|$ is divisible by 4. Since $4 \cdot 3 \cdot 7 > 80$, the only possibility is that $|G| = 4 \cdot 3 \cdot 5 = 60$.

Chapter 32

1. (a) The fact that $BB^t = I$ follows from the observation that for all i, j,
$$d(e_i b, e_j b) = d(e_i, e_j) = \delta_{ij}.$$
Since $1 = \det I = (\det B)(\det B^t) = (\det B)^2$, we have $\det B = \pm 1$.
   (b) (i) The eigenvalues of C are the roots of $\det(C - xI)$, which is a cubic polynomial over $\mathbb{R}$. Therefore, C has one or three real eigenvalues. (ii) Moreover, the product of the eigenvalues of C is $\det C = 1$. If C has three real eigenvalues, then they cannot all be negative; and if C has one real eigenvalue λ and a pair of conjugate non-real eigenvalues $\mu, \bar{\mu}$, then $\lambda \mu \bar{\mu} = 1$, and hence $\lambda > 0$. Therefore C has a real positive eigenvalue, say λ. (iii) Let v be an eigenvector for λ. Then
$$d(v, v) = d(vC, vC) = d(\lambda v, \lambda v) = \lambda^2 d(v, v),$$
and so $\lambda = 1$.
   (c) Let c be the isometry $v \to vC$. By (b), c fixes a vector v; it is now easy to convince yourself that c must be a rotation about the axis through v. The required result for b follows from the definition of c.
   (d) Take three orthogonal axes, one of which is the axis of the rotation b. With respect to these axes, the matrix of b is
$$\begin{pmatrix} 1 & 0 & 0 \\ 0 & \cos\phi & \sin\phi \\ 0 & -\sin\phi & \cos\phi \end{pmatrix}.$$
Hence $\operatorname{tr} B = 1 + 2\cos\phi$.

2. We regard G as a subgroup of $O(\mathbb{R}^3)$. It is easy to see that the translation submodule T (which consists of all the translation modes) is isomorphic to the $\mathbb{R}G$-module given by the natural action of G on $\mathbb{R}^3$. Hence by part (d) of Exercise 1,

$$\chi_T(g) = \begin{cases} 1 + 2\cos\phi, & \text{if } g \text{ is a rotation through } \phi, \text{ about some} \\ & \text{axis,} \\ -(1 + 2\cos\phi), & \text{if the element } -g \text{ of } O(\mathbb{R}^3) \text{ is a rotation} \\ & \text{through } \phi. \end{cases}$$

Now consider the rotation submodule R, which consists of all the rotation modes. A rotation mode is specified by a 3-dimensional vector ϕv, where v is a unit vector along the axis of the rotation and ϕ denotes the angle of rotation, taken positive in the right-hand screw sense. Let $g \in G$, and consider g acting on ϕv. It sends v to vg, and if g is a rotation, it preserves the sense of the rotation; however, if g is a reflection then it transforms a right-hand screw to a left-hand screw, and hence sends ϕv to $(-\phi)(vg)$. Therefore

$$\chi_R(g) = \begin{cases} \chi_T(g), & \text{if } g \text{ is a rotation,} \\ -\chi_T(g), & \text{if } g \text{ is not a rotation,} \end{cases}$$

and so $\chi_T + \chi_R$ has the required form.

3. The matrix A is

$$\frac{k}{m}\begin{pmatrix} -3/2 & 0 & 3/4 & -\sqrt{3}/4 & 3/4 & \sqrt{3}/4 \\ 0 & -1/2 & -\sqrt{3}/4 & 1/4 & \sqrt{3}/4 & 1/4 \\ 3/4 & -\sqrt{3}/4 & -3/4 & \sqrt{3}/4 & 0 & 0 \\ -\sqrt{3}/4 & 1/4 & \sqrt{3}/4 & -5/4 & 0 & 1 \\ 3/4 & \sqrt{3}/4 & 0 & 0 & -3/4 & -\sqrt{3}/4 \\ \sqrt{3}/4 & 1/4 & 0 & 1 & -\sqrt{3}/4 & -5/4 \end{pmatrix}.$$

4. A simpler basis is given by

$$\tfrac{1}{2}(r_1 + r_2) = (v_{12} + v_{21}) - (v_{34} + v_{43}),$$

$$\tfrac{1}{2}(r_1 + r_3) = (v_{13} + v_{31}) - (v_{24} + v_{42}),$$

$$\tfrac{1}{2}(r_1 + r_4) = (v_{14} + v_{41}) - (v_{23} + v_{32}).$$

We chose r_1, r_2, r_3, r_4 to be the images of w_1, w_2, w_3, w_4 under an $\mathbb{R}G$-isomorphism. (Compare the construction of the matrix B, towards the end of Example 32.20.)

5. (a) This is a routine geometrical exercise.
   (b) Let the displaced positions of the atoms be $0', 1', 2', 3', 4'$. The distance of $1'$ from the plane through 1 perpendicular to 12 is $x_{12} + \frac{1}{2}(x_{13} + x_{14})$. Similarly, the distance of $2'$ from the plane through 2 perpendicular to 12 is $x_{21} + \frac{1}{2}(x_{23} + x_{24})$. Therefore 12 has decreased by $x_{12} + x_{21} + \frac{1}{2}(x_{13} + x_{14} + x_{23} + x_{24})$. The other calculations can be done in the same way.
   (c) We express the force at each atom as a vector, and then write this vector

as a linear combination of our three chosen unit vectors at the initial position of the atom. Let d_{ij} denote the decrease in the length ij, as calculated in part (b). Then, for example, at vertex 1, the contributions to the component of the force vector in the direction 12 are as follows: $-k_1 d_{12}$ from the force between atoms 1 and 2; zero from the force between atoms 1 and 3 and from the force between atoms 1 and 4; and $-k_2 d_{10} \sec(\angle 012)$ from the force between atoms 1 and 0. Hence

$$m_1 \ddot{x}_{12} = -k_1 d_{12} - \tfrac{1}{3}\sqrt{(3/2)}k_2 d_{10}.$$

Upon substituting for d_{12} and d_{10} from part (b), we obtain the given expression for $\ddot{x}_{12}$.

The other accelerations are calculated in the same way.

(d) The entries in the 15×15 matrix A are the coefficients which appear in the equations of motion $\ddot{x} = xA$. If you write down the matrix A, then you will easily verify that the given vectors are eigenvectors of A (with eigenvalues

$$-(4k_1 + k_2)/m_1,\ 0,\ 0,\ 0,\ -k_1/m_1,\ -k_1/m_1,$$

respectively).

(e) The matrix B is

$$\begin{pmatrix} -(6k_1 + k_2)/3m_1 & -2k_2/3m_1 & -4k_2\sqrt{2}/(m_2\sqrt{3}) \\ -(3k_1 + k_2)/3m_1 & -2k_2/3m_1 & -4k_2\sqrt{2}/(m_2\sqrt{3}) \\ -k_2\sqrt{3}/(9m_1\sqrt{2}) & -2k_2\sqrt{3}/(9m_1\sqrt{2}) & -4k_2/3m_2 \end{pmatrix}.$$

(f) You will find that $(1, -2, \sqrt{6})$ is an eigenvector of B, with eigenvalue 0. This agrees with the statement in Example 32.20 that the translation vector

$$r_1 - 2s_1 + 3\cos\vartheta w_1$$

is an eigenvector of A.

6. (a) Assign coordinate axes along the edges of the square, as shown below.

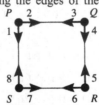

The symmetry group is $G = D_8$. Let a denote the rotation sending $P \to Q \to R \to S \to P$, and let b denote the reflection in the axis PR. The character χ of the $\mathbb{R}G$-module $\mathbb{R}^8$ is

	1	a^2	a	b	ab
χ	8	0	0	0	0

We refer to the character table of D_8 which is given in Example 16.3(3), and see that

$$\chi = \chi_1 + \chi_2 + \chi_3 + \chi_4 + 2\chi_5.$$

The rotation mode is $(t + \beta)v$, where

$$v = (1, -1, 1, -1, 1, -1, 1, -1) \in V_{\chi_2}.$$

The translation modes are $(t + \beta)v$, where v is in the span of v_1, v_2 and

$$v_1 = (1, 0, 0, 1, -1, 0, 0, -1), \quad v_2 = (0, 1, -1, 0, 0, -1, 1, 0).$$

The homogeneous components V_{χ_1}, V_{χ_3} and V_{χ_4} are spanned by the vectors $(1, 1, 1, 1, 1, 1, 1, 1)$, $(1, 1, -1, -1, 1, 1, -1, -1)$ and $(1, -1, -1, 1, 1, -1, -1, 1)$, respectively. The final set of eigen-vectors is given by $V_{\chi_5} \cap \mathbb{R}^8_{\text{vib}}$ which is spanned by

$$(1, 0, 0, -1, -1, 0, 0, 1) \text{ and } (0, 1, 1, 0, 0, -1, -1, 0).$$

(b) The matrix A is

$$-\frac{k}{m} \begin{pmatrix} 1 & 0 & 0 & 0 & 0 & 0 & 0 & 1 \\ 0 & 1 & 1 & 0 & 0 & 0 & 0 & 0 \\ 0 & 1 & 1 & 0 & 0 & 0 & 0 & 0 \\ 0 & 0 & 0 & 1 & 1 & 0 & 0 & 0 \\ 0 & 0 & 0 & 1 & 1 & 0 & 0 & 0 \\ 0 & 0 & 0 & 0 & 0 & 1 & 1 & 0 \\ 0 & 0 & 0 & 0 & 0 & 1 & 1 & 0 \\ 1 & 0 & 0 & 0 & 0 & 0 & 0 & 1 \end{pmatrix}.$$

7. (a, b) Let ε_i $(1 \leqslant i \leqslant m)$ be the projection which is given by

$$\varepsilon_i : u_1 + \ldots + u_m \to u_i$$

(where $u_k \in U_k$ for all k). Then

$$w \to w A \varepsilon_j \vartheta_j^{-1} \vartheta_i \quad (w \in U_i)$$

gives an $\mathbb{R}G$-homomorphism from U_i to U_i. By Exercise 23.8, there exist $\lambda_{ij} \in \mathbb{R}$ such that for all $w \in U_i$,

$$w A \varepsilon_j = \lambda_{ij} w \vartheta_i^{-1} \vartheta_j.$$

Since $\sum_{j=1}^{m} \varepsilon_j$ is the identity endomorphism of $U_1 \oplus \ldots \oplus U_m$, we have

$$w A = \sum_{j=1}^{m} \lambda_{ij} w \vartheta_i^{-1} \vartheta_j \quad \text{for all } w \in U_i.$$

Now take in turn $w = u\vartheta_i$ and $w = v\vartheta_i$ to obtain the results of parts (a) and (b) of the question.

(c) Take a basis $u_1, \ldots, u_n$ of U_1. Assume that the eigenvectors of A_u are known. For all k with $1 \leqslant k \leqslant n$, the eigenvectors of A in $\text{sp}(u_k\vartheta_1, \ldots, u_k\vartheta_m)$ are given by the eigenvectors of A_u (see part (b)). Hence we know all the eigenvectors of A in $U_1 \oplus \ldots \oplus U_m$.

Bibliography

Books mentioned in the text

H. S. M. Coxeter and W. J. O. Moser, *Generators and Relations for Discrete Groups* (Fourth Edition), Springer-Verlag, 1980.

J. B. Fraleigh, *A First Course in Abstract Algebra* (Third Edition), Addison-Wesley, 1982.

D. Gorenstein, *Finite Simple Groups: An Introduction to their Classification*, Plenum Press, New York, 1982.

G. D. James, *The Representation Theory of the Symmetric Groups*, Lecture Notes in Mathematics No. 682, Spring-Verlag, 1978.

D. S. Passman, *Permutation Groups*, Benjamin, 1968.

H. Pollard and H. G. Diamond, *The Theory of Algebraic Numbers* (Second Edition), Carus Mathematical Monographs No. 9, Mathematical Association of America, 1975.

J. J. Rotman, *An Introduction to the Theory of Groups* (Third Edition), Allyn and Bacon, 1984.

D. S. Schonland, *Molecular Symmetry – an Introduction to Group Theory and its uses in Chemistry*, Van Nostrand, 1965.

Suggestions for further reading

M. J. Collins, *Representations and Characters of Finite Groups*, Cambridge University Press, 1990.

C. W. Curtis and I. Reiner, *Methods of Representation Theory with Applications to Finite Groups and Orders*, Volume I, Wiley-Interscience, 1981.

W. Feit, *Characters of Finite Groups*, Benjamin, 1967.

I. M. Isaacs, *Character Theory of Finite Groups*, Academic Press, 1976.

W. Ledermann, *Introduction to Group Characters* (Second Edition), Cambridge University Press, 1987.

J. P. Serre, *Linear Representations of Finite Groups*, Springer-Verlag, 1977.

Index

Printed in the United States
By Bookmasters